Geographical and Fingerprinting Data for Positioning and Navigation Systems

Series Editor

Fatos Xhafa

Geographical and Fingerprinting Data for Positioning and Navigation Systems

Challenges, Experiences, and Technology Roadmap

Edited by

Jordi Conesa
Faculty of Computer Sciences, Multimedia and Telecommunication, eHealth Center at Universitat Oberta de Catalunya (UOC), Barcelona, Spain

Antoni Pérez-Navarro
Faculty of Computer Sciences, Multimedia and Telecommunication at Universitat Oberta de Catalunya (UOC), Barcelona, Spain; Internet Interdisciplinary Institute (IN3) at UOC, Castelldefels, Spain

Joaquín Torres-Sospedra
Institute of New Imaging Technologies, Jaume I University, Castellón, Spain

Raul Montoliu
Institute of New Imaging Technologies, Jaume I University, Castellón, Spain

ELSEVIER

ACADEMIC PRESS
An imprint of Elsevier

Academic Press is an imprint of Elsevier
125 London Wall, London EC2Y 5AS, United Kingdom
525 B Street, Suite 1650, San Diego, CA 92101, United States
50 Hampshire Street, 5th Floor, Cambridge, MA 02139, United States
The Boulevard, Langford Lane, Kidlington, Oxford OX5 1GB, United Kingdom

Library of Congress Cataloging-in-Publication Data
A catalog record for this book is available from the Library of Congress

British Library Cataloguing-in-Publication Data
A catalogue record for this book is available from the British Library

ISBN 978-0-12-813189-3

For information on all Academic Press publications
visit our website at https://www.elsevier.com/books-and-journals

Publisher: Mara Conner
Acquisition Editor: Sonnini R Yura
Editorial Project Manager: Gabriela D Capille
Production Project Manager: R. Vijay Bharath
Designer: Victoria Pearson

Typeset by SPi Global, India

Dedicatories

Gràcies Eduard i Pilar per haver alimentat aquesta curiositat que em permet aprendre i gaudir dia rere dia.

Gràcies Neus i Aniol per ser la meva font d'inspiració diària, per tots els somriures que m'heu robat i pels que encara m'heu de robar.

Jordi

Gràcies Eli, Guillem i Miquel. Aquí trobareu molts moments en què estàveu amb mi, però no em vèieu. Gracias a mis padres y a mi familia, por ver más de lo que veo.

Toni

Gràcies Ana per regalar-me tants moments en aquest viatge.

Gracias a mis padres y familia por haberme educado para ser la persona que ahora soy.

Ximo

Gracias a mi familia por vuestras sonrisas de todos los días.

Raúl

Contents

Contributors

Amanda Antunes Geodetic Sciences Graduate Program, Federal University of Paraná, Curitiba, Brazil

Panagiotis Bamidis Medical School of the Aristotle University of Thessaloniki, Thessaloniki, Greece

Óscar Belmonte-Fernández Institute of New Imaging Technologies, Jaume I University, Castellón, Spain

Maria-Gabriella Di Benedetto Department of Information Engineering, Electronics and Telecommunications (DIET), Sapienza University of Rome, Rome, Italy

Mauri Benedito-Bordonau Estudios GIS, C/Albert Einstein, Vitoria-Gasteiz, Spain

Rafael Berkvens Faculty of Applied Engineering, University of Antwerp—IMEC IDLab, Antwerp, Belgium

Dina Bousdar Ahmed Institute of Communications and Navigation, German Aerospace Center (DLR), Oberpfaffenhofen, Germany

Ignacio Bravo-Muñoz Department of Electronics, University of Alcalá, Madrid, Spain

Thomas Burgess indoo.rs GmbH, Wien, Austria

Andrea Calia Department BE-OP, European Organization for Nuclear Research, CERN Cedex, France

Oscar Canovas Faculty of Computer Science, University of Murcia, Murcia, Spain

Giuseppe Caso Department of Information Engineering, Electronics and Telecommunications (DIET), Sapienza University of Rome, Rome, Italy

Jordi Conesa Faculty of Computer Sciences, Multimedia and Telecommunication at Universitat Oberta de Catalunya (UOC), Barcelona, Spain

Giuseppe Conti Nively, Nice, France

António Costa Algoritmi Research Center, University of Minho, Guimarães, Portugal

Constantinos Costa Department of Computer Science, University of Cyprus, Nicosia, Cyprus

Antonino Crivello Institute of Information Science and Technologies "A. Faedo" National Research Council, ISTI-CNR, Pisa, Italy

Álvaro De-La-Llana-Calvo Department of Electronics, University of Alcalá, Madrid, Spain

Luciene S. Delazari Geodetic Sciences Graduate Program, Federal University of Paraná, Curitiba, Brazil

Estefania Munoz Diaz Institute of Communications and Navigation, German Aerospace Center (DLR), Oberpfaffenhofen, Germany

Nicola Dorigatti Trilogis srl, Rovereto, Italy

Pedro Paulo Farias Cartographic Engineering Undergraduate Course, Federal University of Paraná, Curitiba, Brazil

Leonardo Ercolin Filho Department of Geomatics, Federal University of Paraná, Curitiba, Brazil

Fernando J. Álvarez Franco Sensory Systems Research Group, University of Extremadura, Badajoz, Spain

Felix J. Garcia Faculty of Computer Science, University of Murcia, Murcia, Spain

Alfredo Gardel-Vicente Department of Electronics, University of Alcalá, Madrid, Spain

Noelia Hernández Intelligent Vehicles and Traffic Technologies Group, University of Alcalá, Madrid, Spain

A.K.M. Mahtab Hossain Department of Computing and Information Systems, University of Greenwich, London, United Kingdom

Joaquín Huerta Institute of New Imaging Technologies, Jaume I University, Castellón, Spain

Susanna Kaiser Institute of Communications and Navigation, German Aerospace Center (DLR), Oberpfaffenhofen, Germany

Stefan Knauth HFT Stuttgart, University of Applied Sciences, Stuttgart, Germany

Evdokimos Konstantinidis Nively, Nice, France

José Luis Lázaro-Galilea Department of Electronics, University of Alcalá, Madrid, Spain

Elina Laitinen Laboratory of Electronics and Communications Engineering, Tampere University of Technology, Tampere, Finland

Ki-Joune Li Pusan National University, Busan, South Korea

Elena Simona Lohan Laboratory of Electronics and Communications Engineering, Tampere University of Technology, Tampere, Finland

Jose María Cabero Lopez Tecnalia Research & Innovation, Derio, Spain

Pedro E. Lopez-de-Teruel Faculty of Computer Science, University of Murcia, Murcia, Spain

Juraj Machaj Department of Multimedia and Information-Communication Technologies, University of Zilina, Žilina, Slovakia

Germán M. Mendoza-Silva Institute of New Imaging Technologies, Jaume I University, Castellón, Spain

Filipe Meneses Algoritmi Research Center, University of Minho, Guimarães, Portugal

Raul Montoliu Institute of New Imaging Technologies, Jaume I University, Castellón, Spain

Adriano Moreira Algoritmi Research Center, University of Minho, Guimarães, Portugal

Luca De Nardis Department of Information Engineering, Electronics and Telecommunications (DIET), Sapienza University of Rome, Rome, Italy

Maria João Nicolau Algoritmi Research Center, University of Minho, Guimarães, Portugal

Filippo Palumbo Institute of Information Science and Technologies "A. Faedo" National Research Council, ISTI-CNR, Pisa, Italy

Antoni Pérez-Navarro Faculty of Computer Sciences, Multimedia and Telecommunication at Universitat Oberta de Catalunya (UOC), Barcelona; Internet Interdisciplinary Institute (IN3), Castelldefels, Spain

Francesco Potortì Institute of Information Science and Technologies "A. Faedo" National Research Council, ISTI-CNR, Pisa, Italy

Adrián Puertas-Cabedo Soluciones Cuatroochenta S.L., ESPAITEC2, Castellón, Spain

Philipp Richter Laboratory of Electronics and Communications Engineering, Tampere University of Technology, Tampere, Finland

Luis E. Rodríguez-Pupo Institute of New Imaging Technologies, Jaume I University, Castellón, Spain

David Rodríguez-Navarro Department of Electronics, University of Alcalá, Madrid, Spain

Emilio Sansano-Sansano Institute of New Imaging Technologies, Jaume I University, Castellón, Spain

Scarlet Barbosa dos Santos Cartographic Engineering Undergraduate Course,Federal University of Paraná, Curitiba, Brazil

Rhaissa Viana Sarot Geodetic Sciences Graduate Program, Federal University of Paraná, Curitiba, Brazil

Lorenz Schauer Mobile and Distributed Systems Group, LMU Munich, Munich, Germany

Pedro Figueiredo e Silva Laboratory of Electronics and Communications Engineering, Tampere University of Technology, Tampere, Finland

Jukka Talvitie Laboratory of Electronics and Communications Engineering, Tampere University of Technology, Tampere, Finland

Joaquín Torres-Sospedra Institute of New Imaging Technologies, Jaume I University, Castellón, Spain

Sergio Trilles Institute of New Imaging Technologies, Jaume I University, Castellón, Spain

Pawel Wilk Samsung R&D Poland, Warszawa, Poland

Sisi Zlatanova University of New South Wales, Sydney, Australia

Preface

Geolocation systems have been present during decades thank to Global Navigation Satellite System (GNSS), of which the best known is the GPS. However, the eruption of mobile technologies and the availability of geographical data have driven to an eclosion of geolocation services and its popularization. Nowadays, most of people daily use services that deal with geographical information. These services, known as Location-Based Services (LBS), include navigation systems that guide users by car, foot, or bicycle; help to evacuation systems; social network services, among others.

Nevertheless, despite its popularization, these services can be considered useless in most of the situations, since they are based mostly on GNSS and people spend 80%-90% of their time in places where GNSS do not work: indoor environments. These environments include offices, undergrounds, shopping malls, airports, etc. The main reasons for GNSS not working in indoor environments are two: satellite signal cannot reach with enough intensity in indoor spaces; and in the case signal would be strong enough, maps of the buildings are not available in the same way that nowadays are outdoors maps, that is, the maps should have to be available and they should have to be in a digital format that LBS could understand and interpret. Thus, indoor positioning has two challenges: location, to know the coordinates of an object within a building; and position, for which development and creation of indoor maps is needed. Therefore, new solutions should be taken into account.

Another relevant issue when dealing with indoor positioning is the creation and access of indoor maps. Although several solutions have been proposed, like *IndoorGML, IndoorOSM, Indoor Here Maps,* there is not a unique way to access and to create maps. If indoor positioning and navigation should be used as democratically as outdoors, new specifications to define maps in a unified semantics should be provided. In addition, the publication of indoor maps arise new serious issues that should be taken into account carefully: (1) privacy, that is, which maps should be available and for whom?; and (2) security, that is, which techniques are need to enforce the detected privacy constraints?

It is important to note that several solutions and techniques are already available to get positioning in indoor environments with acceptable accuracy, such as Bluetooth, radio frequency, infrared, ultra-wide-band, etc. But all these solutions need the deployment of dedicated sensors within the building. Therefore, although they are efficient, they may not be economically scalable and should need maintenance. Furthermore, most of them do not provide continuity outdoor-indoor, some of them do not even work in smartphones, and also they only work properly in the single environment for which they have been designed.

There are other techniques that are able to work in current indoor environments without adding new infrastructure: they just use the signals already present in the environment, named *opportunity signals*, like WLAN, magnetic field, or even GPS. To get location with these signals, one of the most popular techniques nowadays is fingerprinting. Fingerprinting is a data-centric approach based on two phases: a training phase, when fingerprints are taken for further reference; and a location phase, when the location of an object (or user) is estimated by using the map of fingerprints generated in the previous phase. The training phase is very tedious and consists in gathering a great amount of signal lectures of different locations within buildings. To simplify the process of gathering data, some collaboration mechanisms have been proposed that take advantage of crowds in order to gather the required information with low cost. But once the data are collected, typical problems of big data arises, due to the amount of data gathered and its heterogeneity. The location phase uses machine-learning techniques to forecast the location of objects according to the data gathered in the previous phase. Although this is an apparent affordable technique, it has some drawbacks: accuracy is usually lower than sensor-based techniques; obtaining the fingerprints is very costly and not scalable; the system is not robust against changes in the infrastructure (like changes in the WiFi access points), that can be done without notice; environment conditions, like the number of people in the room or the mobile used, affect the results. Another challenge in fingerprinting is to find out what signals should be taken into account for locating objects. Nowadays, there is multitude of signals available in, and for each signal there is a huge amount of data that can be gathered, cleaned, integrated, and analyzed.

Thus, fingerprinting is a very promising technique, whose applications provide location services in indoor environments without extra infrastructure. However, some challenges should be taken into account in order to facilitate the creation of fingerprints; to infer more accurate positioning; to define sustainable and scalable environments; to deal efficiently with the big deal of data required for fingerprinting; and to define specifications in order to take into account indoor maps in positioning and navigation.

This book addresses the challenge of developing positioning and navigation systems within indoor environments by using fingerprinting techniques, but also of working in indoor and outdoor locations seamlessly. It covers scientific, technical and practical perspectives that will contribute to advance of the state of the art, to provide better understanding of the different problems and challenges when dealing with indoor environments and to facilitate the design and implementation of indoor positioning and navigation systems to practitioners. It is important to note that, although the main technique with which the book deals is fingerprinting, it offers also introductions to other techniques.

The book is composed by 18 chapters organized in 8 sections:

1. *Challenges of fingerprinting in indoor positioning and navigation.* This section introduces the main topics and concepts related to indoor positioning and navigation, summarizes the current state of the art in indoor positioning and navigation, and presents the current challenges faced.

2. *Privacy and security aspects of indoor positioning and navigation.* This section reviews some of the issues that indoor positioning faces regarding security and privacy, since, for example, making indoor maps available in the same way that outdoor maps are and what privacy issues should be taken into account, since revealing the map of a floor in a building can reveal the fundamentals of the other floors in the same building.
3. *Creating radiomaps for fingerprinting.* This section shows the main issues found when creating a radiomap. It is a very challenging work and it is important to be very effective in order to maximize the outputs of the process. Thus, this section gives instructions and recommendations on how to create radiomaps.
4. *Fingerprinting positioning.* This section explains in detail the fingerprinting positioning process. It focuses in the use of WiFi and magnetic field signals, since they allow to position users either when they are in movement or when they are static. Nevertheless, the fundamentals of fingerprinting are the same whatever the signals are.
5. *Mapping indoor environments.* This section deals with the base map of indoor spaces. It explains how to create vector maps useful for the systems in order to facilitate the positioning and navigation. The map can also be a piece that can help to improve positioning.
6. *Infrastructures to support fingerprinting services and data.* This section reviews the technologies related with indoor positioning using fingerprinting. It presents also the platform: indoorlocplatform.uji.es, which allows to compare the quality of different solutions of indoor positioning; and makes also a review of the competitions that have taken place last years in Indoor Positioning and Indoor Navigation (IPIN) international conference.
7. *Navigating necessities, solutions, and success cases in indoor positioning.* This section shows some issues related to navigation in real environments and provides successful examples of indoor positioning deployments pointing out the problems faced, the approaches followed and the lessons learnt.
8. *Other technologies to position and navigate in indoor environments.* This section introduces other technologies to indoor positioning, different for fingerprinting. This can help to know the range of validity of fingerprinting, as well as to choose the best of option for every indoor positioning problem. In particular, the techniques explained allow to deal with indoor positioning using ultrasound, optical and inertial signals.

Acknowledgments

The book has its origins in the first international symposium of *Challenges of Fingerprinting in Indoor Positioning and Navigation* (http://symposium.uoc.edu/4323.html), which took place at Barcelona (Spain) on 2016, and sponsored by Universitat Oberta de Catalunya and the program Internationalization at Home from CaixaBank. In the event, the state of the art of fingerprinting indoor positioning and navigation and practical cases were discussed. Results of this discussion are summarized in the first chapter of the book, that constitutes a brief introduction to the chapters that appear in the entire book.

We would like to thank the authors of the chapters for their invaluable collaboration, generosity, and prompt responses to our enquiries. We also would like to acknowledge and thank the feedback, assistance, reminds, and encouragement received from the editorial staff of Elsevier, Gabriela Capille and Sonnini Yura as well as the book series editor Dr. Fatos Xhafa. We would like to thank Vijay Bharath for managing the production process of the book.

Finally, we would like to thank the REPNIN Spanish excellence network in indoor positioning and navigation, TEC2015-71426-REDT, and the Spanish Ministry of Economy and Competitiveness for the project TIN2015-70202-P.

1

Challenges of Fingerprinting in Indoor Positioning and Navigation

Antoni Pérez-Navarro[*,†], Joaquín Torres-Sospedra[‡], Raul Montoliu[‡], Jordi Conesa[*], Rafael Berkvens[§], Giuseppe Caso[¶], Constantinos Costa[‖], Nicola Dorigatti[**], Noelia Hernández[††], Stefan Knauth[‡‡], Elena Simona Lohan[§§], Juraj Machaj[¶¶], Adriano Moreira[‖‖], Pawel Wilk[***]

*FACULTY OF COMPUTER SCIENCES, MULTIMEDIA AND TELECOMMUNICATION AT UNIVERSITAT OBERTA DE CATALUNYA (UOC), BARCELONA, SPAIN †INTERNET INTERDISCIPLINARY INSTITUTE (IN3), CASTELLDEFELS, SPAIN ‡INSTITUTE OF NEW IMAGING TECHNOLOGIES, JAUME I UNIVERSITY, CASTELLÓN, SPAIN §FACULTY OF APPLIED ENGINEERING, UNIVERSITY OF ANTWERP—IMEC IDLAB, ANTWERP, BELGIUM ¶DEPARTMENT OF INFORMATION ENGINEERING, ELECTRONICS AND TELECOMMUNICATIONS (DIET), SAPIENZA UNIVERSITY OF ROME, ROME, ITALY ‖DEPARTMENT OF COMPUTER SCIENCE, UNIVERSITY OF CYPRUS, NICOSIA, CYPRUS **TRILOGIS SRL, ROVERETO, ITALY ††INTELLIGENT VEHICLES AND TRAFFIC TECHNOLOGIES GROUP, UNIVERSITY OF ALCALÁ, MADRID, SPAIN ‡‡HFT STUTTGART, UNIVERSITY OF APPLIED SCIENCES, STUTTGART, GERMANY §§LABORATORY OF ELECTRONICS AND COMMUNICATIONS ENGINEERING, TAMPERE UNIVERSITY OF TECHNOLOGY, TAMPERE, FINLAND ¶¶DEPARTMENT OF MULTIMEDIA AND INFORMATION-COMMUNICATION TECHNOLOGIES, UNIVERSITY OF ZILINA, ŽILINA, SLOVAKIA ‖‖ALGORITMI RESEARCH CENTER, UNIVERSITY OF MINHO, GUIMARÃES, PORTUGAL ***SAMSUNG R&D POLAND, WARSZAWA, POLAND

1 Motivation

Since the beginning of humanity, localization and positioning have been a worry of human beings. This has driven to the creation and search of many different mechanisms to localize: the Sun, the Moon, the Stars, the magnetic field, radio beacons, etc. And this information has been represented in several kind of maps.

Everything changed with the apparition of Global Navigation Satellite Systems (GNSS) in the 1960s that drove to the apparition of the American Global Positioning System (GPS) in 1995, and later to the Russian GLONASS, in 1996 or, more recently, to the European Galileo and the Chinese Beidou. GNSS systems have two important roles: (1) allow to get position with centimeter precision and, in some cases, even milliliter precision[1]; and (2)

[1]See https://www.gps.gov/systems/gps/performance/accuracy/ (Accessed July 2018).

Geographical and Fingerprinting Data for Positioning and Navigation Systems. https://doi.org/10.1016/B978-0-12-813189-3.00001-0

allow to get that precision to anyone who owns the appropriate device, whatever his or her knowledge in positioning would be.

Although GNSS were born with military objectives, by the end of the 1990s receivers of GNSS became commercially available and very popular. By then, military restrictions did not allow high accuracy (more than 10 m), but this restriction disappeared in the 2000s and thus, people became used to have high accuracy localization and navigation.

The next step came with the inclusion of GNSS receivers in smartphones and its popularization which has driven most of people (78% in Europe) to wear a GNSS receiver. This situation has driven to an explosion of location-based systems (LBS). According to a Mobile Life study[2] 19% users use LBSs and 62% think of use LBSs. Navigation is the most popular.

One special kind of applications, Context Aware Recommender Systems have become many popular in marketing and, for example, *FourSquare* allows users to obtain different offers and discounts. In the social area, applications such as *Google Now* send users recommendations of events. There are also projects of tracking of patients such as *Ekahau*[3]; projects to include systems that automatically call emergencies in case of an accident (*eCall*[4]); projects that provide virtual information over reality (*Ingress*[5]); or the project *Fieldtripglass* to add augmented reality to what the user is seeing.[6]

All these LBS need for a reliable and real-time localization, which in most of these systems is obtained via GNSS, although a first localization is performed by using other signals such as Wi-Fi, Bluetooth, or GSM and, once GNSS is available, the high precision position is obtained (Laitinen, 2017).

Despite the successes achieved, the problem of GNSS is that they are affected by Non-Line-Of-Sight problem, multipath propagation issues, signal blockage, intentional and unintentional interferences, etc. (Bhuiyan, 2011). This drives to a signal attenuation that fails to get position in urban canons or indoor environments. Thus, all those LBS fail when going indoors.

Then, the question that arises is: Is indoor positioning and navigation important? This is important, since people spend 80% of their time indoors (Wadden and Scheff, 1983), and, according to Gartner, in 2020 indoor revenues will be as high as 10 billion dollars. Thus, it can be seen that indoor environment has an important social and economic relevance.

The chapter is structured as follows: first, a brief review of indoor positioning systems is performed; then we focus on fingerprinting techniques and show some examples; in the following sections, the problems of indoor maps and privacy and security are presented; and finally, the chapter ends with the conclusions and future challenges.

[2]See http://www.tnsglobal.com/press-release/two-thirds-world%E2%80%99s-mobile-users-signal-they-want-be-found.

[3]See https://www.airistaflow.com/wp-content/uploads/2016/07/AiRISTAFlow_-Ekahau_RTLS_BR.pdf.

[4]See http://europa.eu/rapid/press-release_IP-13-534_en.htm.

[5]See https://www.ingress.com/.

[6]See http://www.fieldtripper.com/glass/.

2 Indoor Positioning Systems

Nowadays, there are still no universal standards for indoor positioning, similarly with what we can find outdoors with GNSS (Lymberopoulos et al., 2015). However, there are several techniques and methodologies that can solve the problem in some specific situations. In this section we will show some generic aspects of indoor positioning systems and which systems exist in the literature or on the market. However, first of all, some important concepts are defined in order to make the chapter (and the book) more understandable.

2.1 Position, Location, and Navigation

The first point to take into account is the difference between three key concepts: position, location, and navigation.

The *position* corresponds to the coordinates of a specific point in a coordinate system, such as the GPS latitude-longitude-altitude coordinates.

The *location* gives the position, but in the context of the specific point, for example: "you are situated in front of H&M shop at third floor of the Mega mall." Location is what gives position its meaning to the user. Location can be given for static elements or for dynamic elements. Although location of dynamic elements might appear like locate the same element several times, the reality is that techniques applied for both types of location (static and dynamic) can be very different.

Finally, *navigation* refers to how one goes from point A to point B. There are two main constraints: (1) there are some rules for going to one point to the other (such as speed limit, or staying in a track); and (2) navigation deals with localized elements (position is not enough). It is important to note that navigation is always associated with elements that are moving (dynamic location).

This chapter is mainly focused on position and location (static and dynamic).

2.2 Classification

All positioning systems can have two parts:

- *Interaction device*: is the device where the user can receive position, localization, navigation instructions, etc., and interact with the information. It can be a specific dedicated device, a computer, a tablet, or, more commonly, a smartphone. It is something that the user takes with himself or herself.
- *Infrastructure*: corresponds to all the devices to be located in the environment in order to get position together with the devices or access nodes used to help the position estimate. Usually, indoor positioning systems can be divided in two main categories, regarding the infrastructure they need:
  - *Infrastructure-based systems*: require equipment (e.g., proprietary transmitters, beacons, antennas, cabling) to provide location signals. They can also be divided in two categories:

  - *Systems with dedicated infrastructure*: use devices added specifically to provide location, such as Bluetooth, Ultra Wide Band (UWB), or RFID. Some examples of systems are:
    - Bluetooth Low Energy (BLE) beacons: iBeacons (Apple) (Newman, 2014)
    - Ultrasound: ALPS (CMU) (Koehler et al., 2014)
    - Visible light: EPSILON (Microsoft Research) (Li et al., 2014)
    - UWB: Decawave (Ye et al., 2012)
  - *Systems that use another infrastructure*: use devices of the environment that have another purpose, like using Wi-Fi (Laitinen and Lohan, 2016) or GSM.
- *Infrastructure-less systems*: do not require dedicated equipment for the provisioning of location signals. These systems are free of any infrastructure in the environment, such as using magnetic field, or inertial navigation systems (INS). In this last case as inertial device can be used the smartphone or an Inertial Measurement Unit (IMU). In the latter case, a smartphone or an IMU can be used as an inertial device.

2.3 Localization Mechanisms

The kind of system determines the methodology to get location. The more common position techniques are (Zekavat and Buehrer, 2011; Liu et al., 2007; Hakan Koyuncu, 2010):

- *Proximity*: This method can be used with GSM, and assigns the smartphone to the cell to which is connected. Accuracy will depend on the distance between cells.
- *Distance based*: This method uses the path loss model of the received signal strength (RSS) to get position. It needs to know the position of the emitters and the mean accuracy got is about 4 m (Bose and Foh, 2007).
- *Time of arrival (ToA)*: This method obtains the position from the time of arrival from the emitters. Its main drawback is that receivers need resolutions lower than $\simeq 1\ \mu s$ (Bocquet et al., 2005).
- *Angle of arrival (AoA)*: Position is obtained by triangulation. The position of the emitters has to be known. It can get accuracies of about 4 m. For GSM, accuracy can be of 150 m, in case of 4 km spacing of BTSs (Wong et al., 2008).
- *Inertial*: This method applies kinematics to obtain position from the sensors carried by the user, which can give information about orientation or speed. This method has the problem that error grows with time and it is important to recalibrate the system from time to time (Jimenez et al., 2010).
- *Fingerprinting*: This method is based on comparing RSS values with a reference map of RSS (radiomap) that associates values with positions (Honkavirta et al., 2009; Kaemarungsi and Krishnamurthy, 2004).

It is important to note that when talking about users, we refer not only to persons, but also to robots or any other element that we are interested in localizing indoors.

These techniques can be applied with different technologies that can be dedicated to indoor positioning, such as UWB (Gigl et al., 2007), RFID (Li and Becerik-Gerber, 2011), Bluetooth (Faragher and Harle, 2014); or not dedicated, such as Wi-Fi, as will be shown in Section 3.1. As it can be seen, there are many different technologies to solve the indoor positioning problem. It is important to note that many times several technologies are fusioned in a system and they can use information from several sensors (what is known as sensor fusion). There are systems that use GNSS outdoors, only if they have good signals. But when this is not the case, one can use GSM or Wi-Fi as backup, since Wi-Fi is generally present in most of urban environments. When indoors, Wi-Fi signals and GSM can be used as backup solution, both combined with inertial information.

Fig. 1 shows several technologies and the range of precision of everyone of them.

Despite their possibilities, using so many different technologies affect interoperability between systems and standardization. On the other hand, which method or methods to choose? To answer this question there are several items to take into account:

- *Application*: Guiding a robot within a building is a different application than guiding a person indoors, since a robot requires much more precision (a robot can crash a wall because a lack of precision, but a person will see the wall). Knowing the application to build will give also information about the accuracy needed.

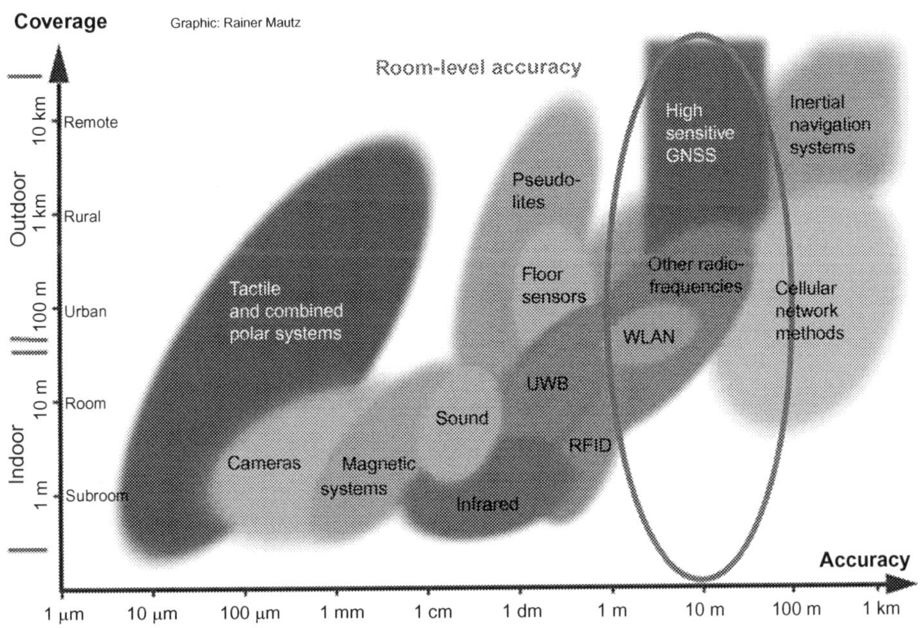

FIG. 1 Methods of positioning regarding their applicability (indoor-outdoor) and their accuracy. *(Adapted from Mautz, R., 2012.Indoor Positioning Technologies (PhD thesis). ETH Zurich, Department of Civil, Environmental and Geomatic Engineering, Institute of Geodesy and Photogrammetry Subject. https://doi.org/10.3929/ethz-a-007313554. Available from: https://www.research-collection.ethz.ch/handle/20.500.11850/54888.)*

- *Cost*: An application that needs to deploy beacons can increase the cost. It is important to take also into account the maintenance cost regarding hardware as well as data.
- *Scalability*: If the application has possibilities to grow to larger scales, it is important to take into account if the system chosen is easily scalable.
- *Environment*: A mall is different than a research facility. Therefore, it is important to observe the place where the system will be deployed in order to detect the drawbacks regarding every technology.

Among all these systems, the symposium "Challenges of Fingerprinting in Indoor Positioning and Navigation" was focused on fingerprinting, that will be presented more deeply in the next section.

3 Fingerprinting Indoor Positioning Techniques

Fingerprinting is becoming one of the most popular indoor positioning systems available. It has the advantage that needs no special infrastructure to be deployed, since uses a signal already present in the environment, such as Wi-Fi or magnetic field. Although Wi-Fi is the most popular nowadays, some solutions deploy their own beacons to make fingerprinting with their own signal.

The technique is based in two steps:

- In a first step, measures of RSS are taken with a device with specialized software. Every measure is associated with a position or zone and the signal to measure can be any signal available (Wi-Fi, magnetic field, etc.), that will be the signal to get position. The set of RSS values and its positions is known as *radiomap*;
- The second step refers to positioning itself and this position is obtained by comparing the value measured by the device to position with the reference radiomap.

To get position from the comparison, a positioning algorithm is needed. There are many different algorithms and, although many of them can be applied to different signals, some of them depend on the kind of signal (Wi-Fi, GSM, magnetic field, and even GPS, that can have enough intensity to make fingerprinting, although not enough to give a GPS position). Here we will focus mainly on Wi-Fi fingerprinting and give some notes about magnetic field fingerprinting.

3.1 Wi-Fi Fingerprinting

Wi-Fi fingerprinting is an indoor positioning system that is able to offer 2–3 m of accuracy in stand-alone mode, although the most common is about 6–7 m. There is a relation between transmitter and receiver, but in indoor propagation we cannot count on this, because Wi-Fi radiation is reflected by many different materials and is absorbed by life bodies, like human beings. The position of access points (AP) is usually not known (although it would be possible to find them by a signal study (Mendoza-Silva et al., 2016)).

Thus, multilateration is not a good option for indoor positioning via Wi-Fi, and using RSS and fingerprinting become the most common alternatives.

In Wi-Fi fingerprinting, in the offline phase, Wi-Fi RSS in specific reference points is collected. Usually more than one measurement at every point is taken, in order to minimize problems of signal propagation in indoor environments, and optionally take the average of all measurements. Since values fluctuate, usually several values during some time are taken and the mean is taken as a representative. Then a vector is built with the strength and the name of the AP. However, since the absolute value of RRS has a strong dependence on the receiver, some times relative values between APs are taken for the vector. The set of all these vectors associated with every position is known as the *radiomap*.

In the online phase, measures are taken by the device to position. Then, these measures are compared with the radiomap and position is obtained with an algorithm. It is important to keep the algorithm as simple as possible and, at the same time, able to give enough accuracy.

There are several algorithms to get positioning but, besides the algorithm, a distance or similarity metric has to be decided. Usually matching algorithms, such as *k*-nearest neighbors (*k*NN), have to be applied in the online phase, which require the aforementioned distance or similarity metric. Several metrics can be used: Euclidean, Minkowsky, Sorensen, Manhattan, inner product between vectors, etc.

Usually, the space where vectors of fingerprints *live* is known as *fingerprints space, RSSI space, features space,* or *phases space* and the distance between two points in this space is defined by the similarity metrics decided. It is important to note that this is not the geometrical indoor space where users and objects live, but Wi-Fi fingerprinting assumes that two vectors near in the fingerprints space, will be near in the geometrical space.

The algorithms that calculate position, work in the fingerprints space, and obtain a *location* in that space, that then is transformed to the geometrical space. There are two main kind of algorithms:

- Probabilistic, such as Bayesian algorithms (Madigan et al., 2005).
- Deterministic, such as the Nearest Neighbor (NN) algorithms, which is the most common (Ma et al., 2008). It has three variants:
    - *NN algorithm:* It takes as position the position of the vector in the radiomap closer to the online vector.
    - k*NN algorithm:* It takes as position the mean of the positions of the *k* vectors closer to the online vector. *k* can be obtained by test-error, or calculated from a formula; and can be the same for all the position calculations, or a dynamic number can be used. Usually it takes a value between 1 and 10.
    - *W*k*WNN algorithm:* It is like the *k*NN, but it gives different weights (W) to vectors. Usually closer vectors have higher W.

Besides all these metrics and algorithms, which can be considered as the base line, many scientists developed their own solutions, such as the *P*-value matrix (Caso and De Nardis, 2015, 2017; Caso et al., 2015a,b; Lemic et al., 2016), or applied some refinements, such as saying that strong values are closer to the vector in the features space to position than weak values (Torres-Sospedra et al., 2017a,b).

In case the position of APs is known, the centroid method can also be used. Then we can have a fingerprint calibrated centroid radiomap and calculate the positions of the access points from this radiomap with a weighted centroid, which will have fewer points (Knauth et al., 2015). Accuracy will not be as good as a detailed radiomap, but this method allows to create a simpler radiomap in situations when high accuracy is not so important.

3.2 Problems of Wi-Fi Fingerprinting

Although Wi-Fi fingerprinting has become very popular and many applications use this method for indoor positioning, these types of systems have several drawbacks that is important to take into account (Torres-Sospedra et al., 2014):

- *Creation and maintenance of the radiomap:* Creating the radiomap is a very laborious, heavy, and time-consuming task. In large buildings, many fingerprints need to be taken and at every point several measures have to be taken. On the other hand, once the radiomap has been build, any change in the Wi-Fi infrastructure (movements or changes of APs), can make the map useless and has to be created again.
  This is the main drawback that Wi-Fi fingerprint systems face and in the next section we will show several approximations that try to minimize it.
- *Lack of uniformity of the signal:* Wi-Fi signal might be irregularly available inside buildings due to poor WLAN planning or due to budget constraints.
- *Energy consumption:* Power efficiency is a crucial element when dealing with smartphone applications. Two tricks that use nowadays algorithms are: reducing scanning intervals and not scanning all the channels. If possible, the positioning algorithm can be moved to the server, although this has the drawback that connectivity is mandatory for getting location. Many algorithms also use mechanisms to reduce complexity, like clustering (Feng et al., 2012).
- *Initial heading:* For magnetic field or INS-based positioning, since the compass does not work as expected in the building and RSS measures depend on the orientation of the smartphone, initial heading affects accuracy and reliability of the system.
- *Outband area:* It is not only the problem of signal strength propagation and triangulation, but also the *k*NN algorithms and similar. We want to detect that we are out of the band and have specific algorithms to detect it.
- *Absorption by living tissues:* Since Wi-Fi is absorbed by water and people are mainly water, fluctuations in the number of people affect the RSS and can affect accuracy (Garcia-Villalonga and Perez-Navarro, 2015).
- *Device dependence:* Differences in sensors of different devices make that measured values are different.

As it has been said, the main drawback among all of these is the radiomap creation. The next section expands on this problem and gives some clues on how to overcome it.

3.2.1 Solutions to Radiomap Creation

Wi-Fi fingerprinting has the advantage that no special infrastructure is needed nor any special device. As has been stated before, the creation and maintenance of the radiomap is the main drawback of Wi-Fi fingerprinting.

As an example, the first experience in the creation of the UJIIndoorLoc database (Torres-Sospedra et al., 2014, 2015), 20,000 fingerprints were collected by 20 people. Then 1000 measures were taken for validations and 5000 for test. The second experience was to collect 20,000 points for a conference. Two mappers mapped a building during 3 days, but 10 people performed route-based mapping.

To overcome the problem of database maintenance, the main solutions are:

- reducing the number of fingerprints to take;
- reducing the time needed to take a fingerprint; and
- simplifying the maintenance of the radiomap.

These solutions are not mutually exclusive, and can be implemented together.

Reduction of the Number of Fingerprints to Take

To decrease the number of measurements to take in the offline phase, the most common approximation is to take few measurements and then create virtual fingerprints. There are several methods to create virtual fingerprints:

- *To estimate the position of APs to apply a path loss model.* Thus, some measurements are needed at the beginning to estimate parameters. The important thing is that initial set of measurements will be uniformly distributed (Caso and De Nardis, 2015, 2017; Caso et al., 2015a,b; Lemic et al., 2016).
- *To increase resolution without increasing site-survey time, by using continuous space estimator (CSE) (Hernández et al., 2015, 2017).* This process is made in two stages:
  - *Training stage:* An RSS map is created using discrete information for each AP. From this information, continuous surfaces are estimated using the vector machine algorithm. Then we have the real information and the virtual information obtained with the vector machine algorithm.
  - *Localization stage:* When the RSS is measured at an unknown position, the algorithm searches those RSS values in the corresponding continuous surface. Thus, possible positions are obtained for the device in the surface corresponding to every AP. After that, all the surfaces are added together and that gives zones with more likely positions. Then, a smooth filter is applied and finally, an environment mask is applied to match the solution with the accessible zones.

Table 1 shows the accuracy obtained with different algorithms in static positioning, and in dynamic positioning in two different trajectories. As can be seen, CSE accuracy is maintained in discrete positions and is better than SVM and *k*NN when measuring trajectories.

Table 1 Accuracy Obtained With Several Algorithms

	CSE (m)	SVM (m)	*k*NN (m)
Discrete positions	1.56	1.53	1.56
Trajectory 1	2.72	4.07	4.13
Trajectory 2	3.60	5.34	5.56

CSE, continuous space estimator; *kNN*, *k* nearest neighbors; *SVM*, support vector machine.

Reduction of the Time Needed to Take a Fingerprint

As has been stated before, taking one fingerprint per point is very time consuming, since several measures have to be taken, and it is important to space them. Thus, every fingerprint takes between 30 s and 1 min, and user has to stand.

However, some applications allow to take fingerprints dynamically while the user is walking, is what is known as *route-based mapping*. This mechanism takes less measures at every single point, but it compensates with the great amount of measures taken. This kind of fingerprints is very useful for applications that try to match trajectories instead of points, that is, for dynamic positioning.

Simplification of the Maintenance of the Radiomap

To simplify the maintenance of the map, crowdsourcing seems the most accepted solution (Laoudias et al., 2013). In this solution, users not only get location from the applications, but also transmit information that helps to update information, as will be shown in the case of Samsung or Anyplace application in Section 3.4. Crowdsourcing collection of data can be passive, and then data are automatically collected by the system, transparent for users; or active, and then users involve themselves in data collection. In order to improve users' commitment, gamification techniques can be useful.

3.3 Magnetic Field Fingerprint

Previous sections have been devoted to Wi-Fi fingerprinting. However, as has been stated before, it is possible to apply fingerprinting techniques to any kind of signal. It has been seen in the last years, an increasing interest in the solutions based on magnetic field.

Magnetic field is generated by Earth and, therefore, it is present everywhere, and structures such as metal objects and columns disturb it locally; as well as magnetic fields created for elements in the environment. Therefore, these disturbances can be used to characterize points with magnetic field measures. On the other hand, many of actual smartphones are able to measure it; thus, the magnetic field is a signal with local features that can be measured with a smartphone.

The main advantages of using magnetic field for positioning are: (1) since the characterization comes from structural elements of the building, it is a system more robust than Wi-Fi, because it is difficult that these elements change with time; and (2) it is possible to capture magnetic field continuously (even more than 10 samples per second), instead of the time needed by Wi-Fi to take every single fingerprint, therefore, it is very suitable to make route-based methods.

Thus, instead of obtaining a vector associated with a single point, such as in the case of Wi-Fi fingerprinting, we obtain a curve associated with a route. The fingerprinting process is performed, then, taking fragments of the curve and associating every fragment to a location. It is important to normalize data and to make the appropriate corrections in order to overcome differences between curves due to different speed.

The process of positioning using magnetic field as fingerprints is the same as that used for Wi-Fi fingerprinting (see Section 3.1). The metrics and algorithms used can be also the same, such as the NN algorithm and its derivatives (Torres-Sospedra et al., 2014, 2015).

However, using magnetic field has some drawbacks:

1. It is a vector and different orientations give different values, therefore, usually the module is used as measure.
2. Although magnetic field allows to capture continuously and using route-based mapping, user speed plays a key role since the same distance can be done in different times. Thus, curves corresponding to the same length are different. To deal with this problem, users try to maintain constant velocity and the device always in the same orientation when creating the radiomap.
3. It is possible that there is not enough local variability of the magnetic field to perform localization.
4. Mobile devices such as smartphones, tablets, or laptops emit magnetic fields that can affect stability and repeatability of the measured values.

To conclude, results of using the magnetic field as a signal for indoor positioning are promising, but this is still a very immature and challenging technology.

3.4 Examples of Solutions

In this section some examples of solutions are presented. These solutions were presented in the symposium "Challenges of Fingerprinting in Indoor Positioning and Navigation."

Samsung Solution

Samsung presented in 2012 a solution based on fingerprinting, with few centimeter accuracy in testbed and the company got the second place in the EVAAL Indoor Competition held in Banff on IPIN 2015 Conference (Wilk et al., 2015).

The focus of Samsung is solutions for smartphones, with simple set-up and easy maintenance. Thus, they look for automatically update the databases to avoid remanufacturing it.

The application proposed has several parts:

- *Positioning engine,* which is within the smartphone. It uses a probabilistic approach with particle filter.

- *Datastore server,* which is responsible for the storage and maintenance of radiomaps. It works always and serves radiomaps to the clients.
- *Set-up tool,* which is an application on the smartphone that allows to walk in predefined paths where the application collects fingerprints and from time to time they collect real positions.
- *Venue creation wizard,* which is the web application, used to prepare venue-related information that will be later used by client applications. It includes few editors to edit the floor plan, which are used to map matching. Locations and access points can also be included.

To maintain radiomaps and avoid multiple surveys, a crowdsourcing application is used. Users do not need to report exact locations and the application, installed on the smartphone, collects: Wi-Fi fingerprints, pedestrians speed, and heading. Then the information is uploaded to the server and the algorithm decides if it wants to apply this maintenance to the map. If yes, it is merged with the radiomap. The process to make the merging is performed in two steps: the first step determines the walk of the pedestrian in the indoor environment and the trajectory goes to the server; in the second step crowdsourcing locations are assigned.

The position algorithm has four inputs: these crowdsourcing locations, fingerprints, the trajectories calculated, and fingerprints similarities. The system assumes that fingerprints near in the fingerprints space are closer in the geometrical space.

Anyplace

Anyplace (born as Airplace) is an indoor information service created by University of Cyprus. It is open source and work in Android, Windows, and iOS (Konstantinidis et al., 2012).[7] In 2012 it received the Best Demo Award at IEEE MDM'12 (Laoudias et al., 2012; Li et al., 2013). In 2014, it was awarded with the second place in the IPSN'14 Indoor Localization Competition (Microsoft Research) in Berlin, Germany, and the first position at EVARILOS Open Challenge, European Union, also in Berlin. Nowadays, it is a Outdoor-to-Indoor Navigation through URL, with 60 Buildings mapped and includes thousands of POIs (stairways, WC, elevators, equipment, etc.)

To easy the creation of the map, Anyplace uses a crowdsourcing mechanism to create and maintain the map (Laoudias et al., 2013). And to reduce computation load, the algorithm uses a clustering algorithm, Bradley-Fayyad-Reina (BFR), which is a variant of k-means designated for large datasets. It has room-level accuracy and is able to detect pretty well the change in floor. Anyplace allows to obtain position by using several algorithms: kNN, WkNN, and Bayesian.

The system has a connectivity threshold to RSS intensity of −30 to −90 dBm. Thus, if the user loses signal (can happen in a mall when a user gets into a shop), loses navigation. This situation is corrected by using historical data to get the trajectories.

[7] See http://anyplace.cs.ucy.ac.cy/.

To overcome device diversity and how it affects RSS values, there is a linear relation between RSS values of devices. The question is if it is possible to exploit this to align reported RSS values.

Modular Localization System

The modular localization system has been proposed by the University of Zilina, in Slovakia. It is based on using the appropriate localization module (Brida et al., 2014). The process followed by the system is:

- Make measurements and analyze the signals. If GPS is available with good quality, it is used.
- If GPS is not available, then Wi-Fi is used for positioning. Positioning with Wi-Fi outdoors is possible when there are signals available, but error is higher than 20 m.
- If GPS is not available, and there are not enough Wi-Fi access points available (three), GSM is used.

Wi-Fi positioning via fingerprinting offers 5 m on average accuracy. And 95% of positions gave less than 6 m accuracy. The best results were obtained with the *k*NN algorithm.

In order to reduce computational load, a two-phase map reduction algorithm is proposed. Thus, the algorithm distinguishes between areas with at least 1 transmitter and areas with highest similarity. However, time required for positioning by Wi-Fi is higher than time required for positioning with GSM.

iLocate

iLocate[8] is an indoor/outdoor location and asset management through open geodata funded by European Union.

The project faces with the lack of standards for localization-based services: there are many technologies for indoor localization, but they do not communicate between them. iLocate mix technologies and support standards. The project uses many different technologies: GPS, triangulation, UWB (that gives under 10 cm precision), etc. It gives real-time information with 100 measurements per seconds.

Since it promotes open technologies, the project has been the first implementation of indoorGML (see Section 4).

Nowadays, iLocate has 13 pilots working around Europe. The 13 pilots are running different localization systems. If a new technology comes into place and comes in iLocate, it is integrated to the system and the technology will function within it.

iLocate incorporates, also, an immersive 3D system that uses virtual reality (glasses) and augmented reality. The user can keep the device pointing to a place and receive indications.

3.4.1 How to Compare Solutions

We have seen in this section several solutions offering indoor positioning based on Wi-Fi fingerprinting. The question is: Which application is better?

[8]See http://www.i-locate.eu/.

The answer is not easy. As has been shown in Section 3.2, there are several drawbacks related with Wi-Fi fingerprinting that make these systems very environment dependent and affect its reproducibility. Thus, although every system gives an accuracy, it is not possible to know if in other environment the same system will have the same accuracy.

Then, how to compare different systems of indoor positioning?

A possible solution is having open public databases with fingerprint radiomaps and validation and test data. The first one of this kind of databases is the UJIIndoorLOC (Torres-Sospedra et al., 2014, 2015).

UJIIndoorLoc was created by University Jaume I and has thousands of fingerprints with their location, and thousands of test data obtained in the campus of the university. It also offers the maps of the zones. Lately, the platform has grown and allows users to register and upload their own databases and their own algorithms.

With this database, it will be possible to compare different systems with the same corpus of data and then knowing which algorithm is faster, which offers the best accuracy, which is more reliable in the application environment, etc.

Nevertheless, it is important to take into account that UJIIndoorLoc will allow to compare algorithms, but still, a deeper study has to be performed in order to know which is the more appropriate system for every situation.

Another comprehensive database containing also references to other open-source Wi-Fi databases and supporting Matlab and Python software has been published recently (Lohan et al., 2017).

4 Indoor Maps

To pass from position to location it is important to have the map of the zone. Therefore, indoor location systems need the map of the building. However, nowadays it is not possible to get maps for every single building, not even the public ones. In addition, many available indoor maps have errors. This is quite different from outdoor situations, where maps of around the world are digitally available since the mid-2000s and in general are very reliable.

On the other hand, maps can also be helpful to improve location by applying map matching techniques. These techniques allow to avoid impossible locations by taking into account the map. Thus, for example, if we are using an indoor system and obtain a position, obtaining the location with the map, we can know if it is out of the building and move it inside. As has been seen in the section with examples of solutions based on Wi-Fi fingerprinting (Section 3.4), many of the solutions make a comparison with the map in order to improve results.

But, which has to be the format of the map? Many solutions give the map only as an image under the position points, that helps user to locate them. However, in order to use the map to improve location, a richer map is needed. As has been shown in Section 3.4, the project iLocate uses indoorGML to create such rich maps.[9]

[9]See http://www.opengeospatial.org/standards/indoorgml.

indoorGML is an XML-based standard that is used to describe indoor spaces and indoor navigation paths. It is based on what is known as the Poincaré duality. Thus, it is possible to map everything to a node or an edge and can create a connectivity graph, much simpler to understand for a computer.

Nevertheless, although indoorGML is defined as an standard, it has not still been adopted by the community.

5 Privacy and Security Issues

Another aspect to take into account regarding indoor positioning is privacy and security issues (Konstantinidis et al., 2015; Lohan et al., 2017).

An indoor service can continuously "know" (survey, track, or monitor) the location of a user while the system is connected. Location tracking can be considered, in some specific cases, unethical and can even be illegal if it is carried out without the explicit user consent.

This is an imminent privacy threat, with greater impact than other privacy concerns, as it can occur at a very fine granularity. It reveals the stores/products of interest in a mall, the book shelves of interest in a library, the artifacts observed in a museum, etc.

On the other hand, when publishing an indoor map, information about the building is being given. Therefore, if a company in one floor publishes the map of its offices, they are giving also some information about the other floors of the building (like the position of bathrooms).

6 Conclusions and Future Challenges of Indoor Positioning

How many people are using indoor positioning technology frequently? If we have Google Android or Apple iOS, maybe we are using it, even without knowing. Who is using indoor navigation? Just a small fraction of us. We have been working in indoor navigation for about 20 years. Maybe we are looking for technology of indoor navigation accuracy, but maybe we have enough accuracy for some applications.

There is a market for indoor positioning and indoor navigation. It is one of the top technologies from Gartner Group: 10 billions dollars for 2020. Google already provides indoor navigation. There is a huge market for factory plans, for locating people, but also for groups or robots.

However, as can be deduced from the chapter, there is still no a standard solution for indoor positioning, although there are many solutions that were shown to work with acceptable precision. Since there is no standard, it is difficult to arrive a massive number of users, since they should have to download a different application for every single environment.

Why is there no standard? In 1996 Wi-Fi standard was published and available technology satisfying that standard offered: 1 Mb/s and 2 Mb/s optional. IBM and other companies had better solutions, but they were proprietary solutions and were not interoperable. Thus, in 1996 it was given more importance to a standard than to the available technology.

Maybe it is time to look for a standard, sacrificing precision and accuracy, but giving more importance to accuracy.

On the other hand, to spread indoor applications we need not only a standard, but available indoor maps of the buildings. Nevertheless, it is not clear how to create an indoor map. Indoor spaces exhibit complex topologies. They are composed of entities that are unique to indoor settings, like rooms and hallways that are connected by doors. In addition, there is not even an standard to keep indoor maps useful for positioning and navigation, since indoorGML is not being used by the community.

Standardization of systems and maps, and spreading of maps are maybe the main challenges that indoor positioning is facing and they will condition the evolution of many other items and even the final standardized technologies.

Besides these two main goals, there are some other challenges to face:

- Seamless transfer between outdoor and indoor: The technology for outdoor is clear (GNSS), and several systems are being tested indoors, but how to make a fluid evolution from outdoors to indoors and the opposite?
- Quality: We find unreliable crowdsourcers, multidevice issues, hardware outliers, temporal decay, etc.
- Improve switching between modules and estimate reliability of positioning system.
- When to use hybrid positioning systems based on combination of heterogeneous networks.
- Regarding Wi-Fi fingerprinting, we need to build a quality radiomap, and we need to build the radiomap everyday. How many buildings, how many hours do we need to spend to collect all data? We have an scalability problem. Probably the solution to this problem is crowdsourcing. Wikipedia or OpenStreetMaps are good examples, even in navigation, such as Waze. It brings new challenges if we rely on crowdsourcing. But then we have to take care of privacy, take care of malicious users, or if the system goes down. How can we do it in large scales?
- The environment changes, even for a magnetic field. In some cases, such as in a factory, the plan can change. We need to take care of these changes and also, with repeated MAC addresses in different places. This can happen when several datasets are combined, which include thousands of APs created by thousands of users.
- Accuracy: It runs between millimeter with some technologies until room accuracy with some others. It is important to know what accuracy is needed for every specific application. We need more accuracy for some applications, like robot navigation; but maybe we already have enough accuracy for some other applications, like airport navigation.
- Standards for protocols to access the floor maps. Service discovery protocols: go inside a building and find the floor maps and indoor positioning engine available for that building.
- Exploitation of the capacities of chips: Wi-Fi chips support ToF measurements, and we should have to learn how to exploit it.

- High-dimensionality vectors. RSS vectors are not the only solution, we can look channel state information and combine it with magnetic field and atmospheric pressure.
- Normalization of quality measurements to do benchmarking.

As future work, maybe the next step in communication, 5G, will change the scenario of positioning and will facilitate time of arrival estimate. In the future we will have ultra-dense 5G networks with devices and can see/hear multiple access nodes. There will be multiple antenna arrays and high bandwidths with extremely high positioning accuracy that will be able to have, potentially, centimeter-scale.

Finally, it is important to take into account that solutions to indoor positioning can also help to solve the problem to outdoor (urban canyons).

Acknowledgments

A. Perez-Navarro, J. Torres-Sospedra, Raul Montoliu, and Jordi Conesa thank the network of excellence REPNIN (ref. number TEC2015-71426-REDT) from the Spanish Ministry of Economy and Competitivity; and also want to thank Obra Social La Caixa and Universitat Oberta de Catalunya as sponsors of the Symposium "Challenges of Fingerprinting in Indoor Positioning and Navigation."

E.S. Lohan thanks the Academy of Finland (project insure-project.org).

References

Bhuiyan, M.Z.H., 2011. Analysis of Multipath Mitigation Techniques for Satellite-based Positioning Applications. Tampere University of Technology, Tampere. ISBN 978-952-15-2624-4.

Bocquet, M., Loyez, C., Benlarbi-Delai, A., 2005. Using enhanced-TDOA measurement for indoor positioning. IEEE Microwave Wireless Compon. Lett. 15 (10), 612–614. ISSN 1531-1309. https://doi.org/10.1109/LMWC.2005.855392.

Bose, A., Foh, C.H., 2007. A practical path loss model for indoor WiFi positioning enhancement. In: 2007 6th International Conference on Information, Communications & Signal Processing. IEEE, pp. 1–5. Available from: http://ieeexplore.ieee.org/document/4449717/.

Brida, P., Machaj, J., Benikovsky, J., 2014. A modular localization system as a positioning service for road transport. Sensors 14 (11), 20274–20296. ISSN 1424-8220. https://doi.org/10.3390/s141120274. Available from: http://www.mdpi.com/1424-8220/14/11/20274/.

Caso, G., De Nardis, L., 2015. On the applicability of multi-wall multi-floor propagation models to WiFi fingerprinting indoor positioning. In: Atanasovski, V., Leon-Garcia, A. (Eds.), Future Access Enablers for Ubiquitous and Intelligent Infrastructures. Springer International Publishing, Cham, pp. 166–172.

Caso, G., De Nardis, L., 2017. Virtual and oriented WiFi fingerprinting indoor positioning based on multi-wall multi-floor propagation models. Mobile Netw. Appl. 22 (5), 825–833. ISSN 1383-469X. https://doi.org/10.1007/s11036-016-0749-x.

Caso, G., de Nardis, L., di Benedetto, M.G., 2015a. A mixed approach to similarity metric selection in affinity propagation-based WiFi fingerprinting indoor positioning. Sensors 15 (11), 27692–27720. ISSN 1424-8220. https://doi.org/10.3390/s151127692. Available from: http://www.mdpi.com/1424-8220/15/11/27692/.

Caso, G., De Nardis, L., Di Benedetto, M.G., 2015b. Frequentist inference for WiFi fingerprinting 3D indoor positioning. In: 2015 IEEE International Conference on Communication Workshop (ICCW). IEEE, pp. 809–814. Available from: http://ieeexplore.ieee.org/document/7247278/.

Faragher, R., Harle, R., 2014. An analysis of the accuracy of Bluetooth Low Energy for indoor positioning applications. In: Proceedings of the 27th International Technical Meeting of the Satellite Division of the Institute of Navigation (ION GNSS+ 2014), pp. 201–210. Available from: https://www.cl.cam.ac.uk/rmf25/papers/BLE.pdf.

Feng, C., Au, W.S.A., Valaee, S., Tan, Z., 2012. Received-signal-strength-based indoor positioning using compressive sensing. IEEE Trans. Mobile Comput. 11 (12), 1983–1993. ISSN 1536-1233. https://doi.org/10.1109/TMC.2011.216. Available from: http://ieeexplore.ieee.org/document/6042868/.

Garcia-Villalonga, S., Perez-Navarro, A., 2015. Influence of human absorption of Wi-Fi signal in indoor positioning with Wi-Fi fingerprinting. In: 2015 International Conference on Indoor Positioning and Indoor Navigation (IPIN). IEEE, pp. 1–10. Available from: http://ieeexplore.ieee.org/document/7346778/.

Gigl, T., Janssen, G.J.M., Dizdarevic, V., Witrisal, K., Irahhauten, Z., 2007. Analysis of a UWB indoor positioning system based on received signal strength. In: 2007 4th Workshop on Positioning, Navigation and Communication. IEEE, pp. 97–101. Available from: http://ieeexplore.ieee.org/document/4167824/.

Hakan Koyuncu, S.H.Y., 2010. A survey of indoor positioning and object locating systems. Int. J. Comput. Sci. Netw. Secur. 10 (5), 121–128. Available from: http://paper.ijcsns.org/07_book/201005/20100518.pdf.

Hernández, N., Ocaña, M., Alonso, J.M., Kim, E., 2015. WiFi-based indoor localization using a continuous space estimator from topological information. In: Proceedings of the 2015 International Conference on Indoor Positioning and Indoor Navigation, pp. 148899-1–148899-4.

Hernández, N., Ocaña, M., Alonso, J., Kim, E., 2017. Continuous space estimation: increasing WiFi-based indoor localization resolution without increasing the site-survey effort. Sensors 17 (1), 147. ISSN 1424-8220. https://doi.org/10.3390/s17010147. Available from: http://www.mdpi.com/1424-8220/17/1/147.

Honkavirta, V., Perala, T., Ali-Loytty, S., Piche, R., 2009. A comparative survey of WLAN location fingerprinting methods. In: 2009 6th Workshop on Positioning, Navigation and Communication. IEEE, pp. 243–251. Available from: http://ieeexplore.ieee.org/document/4907834/.

Jimenez, A.R., Seco, F., Prieto, J.C., Guevara, J., 2010. Indoor pedestrian navigation using an INS/EKF framework for yaw drift reduction and a foot-mounted IMU. In: 2010 7th Workshop on Positioning, Navigation and Communication. IEEE, pp. 135–143. Available from: http://ieeexplore.ieee.org/document/5649300/.

Kaemarungsi, K., Krishnamurthy, P., 2004. Modeling of indoor positioning systems based on location fingerprinting. In: IEEE INFOCOM 2004. IEEE, pp. 1012–1022. Available from: http://ieeexplore.ieee.org/document/1356988/.

Knauth, S., Storz, M., Dastageeri, H., Koukofikis, A., Mahser-Hipp, N.A., 2015. Fingerprint calibrated centroid and scalar product correlation RSSI positioning in large environments. In: 2015 International Conference on Indoor Positioning and Indoor Navigation (IPIN). IEEE, pp. 1–6. Available from: http://ieeexplore.ieee.org/document/7346968/.

Koehler, C., Banovic, N., Oakley, I., Mankoff, J., Dey, A.K., 2014. Indoor-ALPS. In: Proceedings of the 2014 ACM International Joint Conference on Pervasive and Ubiquitous Computing—UbiComp '14 Adjunct. ACM Press, New York, NY, pp. 171–181. Available from: http://dl.acm.org/citation.cfm?doid=2632048.2632069.

Konstantinidis, A., Chatzimiloudis, G., Laoudias, C., Nicolaou, S., Zeinalipour-Yazti, D., 2012. Towards planet-scale localization on smartphones with a partial radiomap. In: Proceedings of the 4th ACM

International Workshop on Hot Topics in Planet-Scale Measurement—HotPlanet '12. ACM Press, New York, NY, p. 9. Available from: http://dl.acm.org/citation.cfm?doid=2307836.2307841.

Konstantinidis, A., Chatzimilioudis, G., Zeinalipour-Yazti, D., Mpeis, P., Pelekis, N., Theodoridis, Y., 2015. Privacy-preserving indoor localization on smartphones. IEEE Trans. Knowl. Data Eng. 27 (11), 3042–3055. ISSN 1041-4347. https://doi.org/10.1109/TKDE.2015.2441724. Available from: http://ieeexplore.ieee.org/document/7118199/.

Laitinen, E., 2017. Physical Layer Challenges and Solutions in Seamless Positioning via GNSS, Cellular and WLAN Systems. Tampere University of Technology, Tampere. ISBN 978-952-15-3937-4.

Laitinen, E., Lohan, E., 2016. On the choice of access point selection criterion and other position estimation characteristics for WLAN-based indoor positioning. Sensors 16 (5), 737. ISSN 1424-8220. https://doi.org/10.3390/s16050737. Available from: http://www.mdpi.com/1424-8220/16/5/737.

Laoudias, C., Constantinou, G., Constantinides, M., Nicolaou, S., Zeinalipour-Yazti, D., Panayiotou, C.G., 2012. The Airplace indoor positioning platform for Android smartphones. In: 2012 IEEE 13th International Conference on Mobile Data Management. IEEE, pp. 312–315. Available from: http://ieeexplore.ieee.org/document/6341408/.

Laoudias, C., Zeinalipour-Yazti, D., Panayiotou, C.G., 2013. Crowdsourced indoor localization for diverse devices through radiomap fusion. In: International Conference on Indoor Positioning and Indoor Navigation. IEEE, pp. 1–7. Available from: http://ieeexplore.ieee.org/document/6817906/.

Lemic, F., Handziski, V., Caso, G., De Nardis, L., Wolisz, A., 2016. Enriched training database for improving the WiFi RSSI-based indoor fingerprinting performance. In: 2016 13th IEEE Annual Consumer Communications & Networking Conference (CCNC). IEEE, pp. 875–881. Available from: http://ieeexplore.ieee.org/document/7444904/.

Li, N., Becerik-Gerber, B., 2011. Performance-based evaluation of RFID-based indoor location sensing solutions for the built environment. Adv. Eng. Inform. 25 (3), 535–546. ISSN 14740346. https://doi.org/10.1016/j.aei.2011.02.004. Available from: http://linkinghub.elsevier.com/retrieve/pii/S1474034611000097.

Li, C.L., Laoudias, C., Larkou, G., Tsai, Y.K., Zeinalipour-Yazti, D., Panayiotou, C.G., 2013. Indoor geolocation on multi-sensor smartphones. In: Proceeding of the 11th Annual International Conference on Mobile Systems, Applications, and Services—MobiSys '13. ACM Press, New York, NY, p. 503. Available from: http://dl.acm.org/citation.cfm?doid=2462456.2465704.

Li, L., Hu, P., Peng, C., Shen, G., Zhao, F., 2014. Epsilon: a visible light based positioning system. In: 11th USENIX Symposium on Networked Systems Design and Implementation (NSDI 14), pp. 331–343.

Liu, H., Darabi, H., Banerjee, P., Liu, J., 2007. Survey of wireless indoor positioning techniques and systems. IEEE Trans. Syst. Man Cybern. C (Appl. Rev.) 37 (6), 1067–1080. ISSN 1094-6977. https://doi.org/10.1109/TSMCC.2007.905750. Available from: http://ieeexplore.ieee.org/document/4343996/.

Lohan, E., Torres-Sospedra, J., Leppäkoski, H., Richter, P., Peng, Z., Huerta, J., 2017. Wi-Fi crowdsourced fingerprinting dataset for indoor positioning. Data 2 (4), 32. ISSN 2306-5729. https://doi.org/10.3390/data2040032. Available from: http://www.mdpi.com/2306-5729/2/4/32.

Lymberopoulos, D., Liu, J., Yang, X., Choudhury, R.R., Handziski, V., Sen, S., 2015. A realistic evaluation and comparison of indoor location technologies. In: Proceedings of the 14th International Conference on Information Processing in Sensor Networks—IPSN '15. ACM Press, New York, NY, pp. 178–189. Available from: http://dl.acm.org/citation.cfm?doid=2737095.2737726.

Ma, J., Li, X., Tao, X., Lu, J., 2008. Cluster filtered kNN: a WLAN-based indoor positioning scheme. In: 2008 International Symposium on a World of Wireless, Mobile and Multimedia Networks. IEEE, pp. 1–8. Available from: http://ieeexplore.ieee.org/document/4594840/.

Madigan, D., Einahrawy, E., Martin, R.P., Ju, W.-H., Krishnan, P., Krishnakumar, A.S., 2005. Bayesian indoor positioning systems. In: Proceedings IEEE 24th Annual Joint Conference of the IEEE Computer and

Communications Societies. IEEE, pp. 1217–1227. Available from: http://ieeexplore.ieee.org/document/1498348/.

Mendoza-Silva, G.M., Torres-Sospedra, J., Huerta, J., Montoliu, R., Benitez, F., Belmonte, O., 2016. Situation goodness method for weighted centroid-based Wi-Fi APs localization. In: Proceedings of the Thirteenth Conference on Location Based Services.

Newman, N., 2014. Apple iBeacon technology briefing. J. Dir. Data Digit. Mark. Pract. 15, 222–225.

Torres-Sospedra, J., Montoliu, R., Martinez-Uso, A., Avariento, J.P., Arnau, T.J., Benedito-Bordonau, M., Huerta, J., 2014. UJIIndoorLoc: a new multi-building and multi-floor database for WLAN fingerprint-based indoor localization problems. In: 2014 International Conference on Indoor Positioning and Indoor Navigation (IPIN). IEEE, pp. 261–270. Available from: http://ieeexplore.ieee.org/document/7275492/.

Torres-Sospedra, J., Rambla, D., Montoliu, R., Belmonte, O., Huerta, J., 2015. UJIIndoorLoc-Mag: a new database for magnetic field-based localization problems. In: 2015 International Conference on Indoor Positioning and Indoor Navigation (IPIN). IEEE, pp. 1–10. Available from: http://ieeexplore.ieee.org/document/7346763/.

Torres-Sospedra, J., Jiménez, A., Knauth, S., Moreira, A., Beer, Y., Fetzer, T., Ta, V.C., Montoliu, R., Seco, F., Mendoza-Silva, G., Belmonte, O., Koukofikis, A., Nicolau, M., Costa, A., Meneses, F., Ebner, F., Deinzer, F., Vaufreydaz, D., Dao, T.K., Castelli, E., 2017a. The smartphone-based offline indoor location competition at IPIN 2016: analysis and future work. Sensors 17 (3), 557. ISSN 1424-8220. https://doi.org/10.3390/s17030557. Available from: http://www.mdpi.com/1424-8220/17/3/557.

Torres-Sospedra, J., Moreira, A., Knauth, S., Berkvens, R., Montoliu, R., Belmonte, O., Trilles, S., João Nicolau, M., Meneses, F., Costa, A., Koukofikis, A., Weyn, M., Peremans, H., 2017b. A realistic evaluation of indoor positioning systems based on Wi-Fi fingerprinting: the 2015 EvAAL-ETRI competition. J. Ambient Intell. Smart Environ. 9 (2), 263–279. ISSN 18761372. https://doi.org/10.3233/AIS-170421.

Wadden, R.A., Scheff, P.A., 1983. Indoor Air Pollution: Characterization, Prediction, and Control. Wiley, New York, NY. ISBN 0471876739, 213 pp.

Wilk, P., Karciarz, J., Swiatek, J., 2015. Indoor radio map maintenance by automatic annotation of crowdsourced Wi-Fi fingerprints. In: 2015 International Conference on Indoor Positioning and Indoor Navigation (IPIN). IEEE, pp. 1–8. Available from: http://ieeexplore.ieee.org/document/7346933/.

Wong, C., Klukas, R., Messier, G.G., 2008. Using WLAN infrastructure for angle-of-arrival indoor user location. In: 2008 IEEE 68th Vehicular Technology Conference. IEEE, pp. 1–5. Available from: http://ieeexplore.ieee.org/document/4656978/.

Ye, T., Walsh, M., Haigh, P., Barton, J., Mathewson, A., O'Flynn, B., 2012. An experimental evaluation of IEEE 802.15.4a ultra wide band technology for precision indoor ranging. Int. J. Ambient Comput. Intell. 4 (2), 48–63. ISSN 1941-6237. https://doi.org/10.4018/jaci.2012040104.

Zekavat, S.A.R., Buehrer, R.M. (Eds.), 2011. Handbook of Position Location. John Wiley & Sons, Inc., Hoboken, NJ. ISBN 9781118104750. https://doi.org/10.1002/9781118104750.

Further Reading

Mautz, R., 2012. Indoor Positioning Technologies (Ph.D. thesis). ETH Zurich, Department of Civil, Environmental and Geomatic Engineering, Institute of Geodesy and Photogrammetry Subject. https://doi.org/10.3929/ethz-a-007313554. Available from: https://www.research-collection.ethz.ch/handle/20.500.11850/54888.

2

Wi-Fi Tracking Threatens Users' Privacy in Fingerprinting Techniques

Lorenz Schauer

MOBILE AND DISTRIBUTED SYSTEMS GROUP, LMU MUNICH, MUNICH, GERMANY

1 Introduction

Due to the immense diffusion of smart mobile devices, the usage of Wi-Fi as state-of-the-art wireless communication standard has increased dramatically. Wi-Fi infrastructures are installed in many public spaces and buildings providing Internet access and local services. They also render Wi-Fi fingerprinting possible, which is one of the most auspicious technique for indoor positioning, first presented by Bahl and Padmanabhan (2000) in the RADAR system. Wi-Fi fingerprinting requires IEEE 802.11 active scans in the online phase, which are performed by users for position estimations. Hence, beside automatic Wi-Fi scans which usually take place every 2 min on average (Bonné et al., 2013), even more signals are sent out by mobile devices when fingerprinting is used. This leads to a serious risk for users' privacy, due to an increased Wi-Fi traffic which can be sniffed by any person in reach and without the users' consent or their awareness (Schauer et al., 2014).

However, only few works can be found in literature, where such privacy issues are investigated. For instance, Li et al. (2014) propose a privacy-preserving Wi-Fi fingerprinting localization scheme protecting both the data privacy of the localization service provider and the user's location. The authors realize an encrypted transmission of online fingerprints from mobile devices to the localization server. Thus, vectors of received signal strengths (RSS) are protected against sniffer attacks. However, the scanning process remains unencrypted and can be easily captured.

This chapter deals with privacy risks when using common Wi-Fi-based indoor positioning techniques, such as fingerprinting. In particular, we focus on the information, which can be extracted out of captured data from IEEE 802.11 active scans. Beside the technical background, we provide an overview of existing Wi-Fi tracking approaches. Furthermore, we describe investigated IEEE 802.11 protocol extensions and the mechanism of MAC address randomization, which was recently established in well-known mobile operating systems. However, these methods do not fulfill the requirements for an overall privacy-preserving positioning approach. Hence, we further investigate our

Geographical and Fingerprinting Data for Positioning and Navigation Systems. https://doi.org/10.1016/B978-0-12-813189-3.00002-2

previous work (Schauer et al., 2016a) proposing a fully passive scanning procedure for Wi-Fi fingerprinting.

In summary, the chapter is structured as follows: Section 2 reveals related work. The technical background for Wi-Fi tracking is described in Section 3 while Section 4 gives an overview of potentials and limitations. In Section 5 existing security mechanisms are presented, whereas Section 6 describes the privacy-preserving fingerprinting approach by Schauer et al. (2016a) and presents further investigations. Finally, Section 7 concludes the chapter giving hints on future work.

2 Related Work

A lot of researches have been done in recent years improving common Wi-Fi indoor positioning systems in terms of accuracy (Bell et al., 2010), precision (Martin et al., 2010), scalability (Ledlie et al., 2012), and efficiency (Sabek et al., 2015). Among these systems, Wi-Fi fingerprinting achieves adequate positioning results in literature (Farshad et al., 2013), but also suffers from high efforts for recording the radio map in the training phase. Therefore, many investigations have focused on reducing these efforts for creating the radio map, for example, Gunawan et al. (2012) or Koweerawong et al. (2013), rather than on solving privacy issues.

By contrast, a vast number of approaches for achieving location and communication privacy in location-based services (LBS) have been developed, such as Dorfmeister et al. (2015) or Yang et al. (2013), most of which are adapting well-established privacy concepts from other fields, such as k-anonymity (Sweeney, 2002), to the particular needs of LBS. However, with the majority of approaches dealing with outdoor LBS, privacy risks concerning a user's location caused by active Wi-Fi scans are not discussed any further. Especially, considering the terminal-based positioning approach of GPS, the mere acquisition of position updates in an outdoor environment usually does not pose any threats to a user's location privacy.

Privacy-preserving approaches are also well studied in the field of indoor positioning and wireless LANs. Jiang et al. (2007) analyze the problem of location privacy in wireless infrastructures and introduce a protocol to protect the user's location by obfuscating privacy compromising information leaking in Wi-Fi communications. They already consider silent attackers capturing Wi-Fi packets within communication range as the strongest attackers for users' privacy. Note that the approach by Schauer et al. (2016a) protects the positioning process against these silent sniffers.

Konstantinidis et al. (2015) propose a privacy-preserving indoor positioning approach for mobile devices protecting users against location tracking by the localization service. Furthermore, they discuss several positioning techniques, including Wi-Fi fingerprinting, and present a framework to protect the user's location against known privacy attacks. However, neither the scanning procedure nor any data sent from the mobile device, such as RSS vectors, are protected in particular.

Only few works can be found in literature, where privacy issues in case of Wi-Fi fingerprinting are investigated, such as Li et al. (2014), Schauer et al. (2016a), or Gschwandtner and Schindhelm (2011). The latter propose a privacy-preserving Wi-Fi fingerprinting approach using enhanced Wi-Fi beacons. The authors add all required positioning information into the vendor-specific elements of IEEE 802.11 beacon frames, such as AP positions, or essential parts of the recorded radio map. Thus, the client is able to calculate its' position locally on the device and no bidirectional communication with the localization service provider is required. The authors also describe the idea to listen on such enhanced beacon frames only, rather than sending probe requests for determining the current RSS vector. However, this is not further investigated, nor evaluated.

Besides our previous investigations (Schauer et al., 2016a), none of the mentioned works concentrate on the commonly used IEEE 802.11 active scanning process itself, which is proven to be a serious privacy risk (Lindqvist et al., 2009). Therefore, we deal with this aspect in the next section, where the technical background of Wi-Fi tracking is explained focusing on related privacy issues for Wi-Fi fingerprinting.

3 Technical Background

The wireless local area network technology, commonly known as Wi-Fi, is standardized as IEEE 802.11 (IEEE, 2007). The standard introduces three different frame types:

1. *Control frames*, in order to support the delivery process of data frames and to manage the medium access
2. *Data frames*, to transport user data for higher layers
3. *Management frames*, such as Beacon, Probe, Authentication, and Association frames, to exchange management information for connection establishment and maintenance

We focus on the latter, due to the fact that only management frames are involved in the 802.11 network discovery procedure, as shown in Fig. 1. From a mobile device's perspective, network discovery can be performed either passively (1) or actively (2). When performing passive scans, a client merely listens on beacon frames, which are periodically transmitted by access points over all channels they are currently operating on. The standard defines a periodic beacon interval of 100 ms. Hence, to receive a beacon frame transmitted on a certain channel, the client's radio must be set to the corresponding channel during transmission. For this purpose, the client iterates over all available channels and listens on each channel for a maximum duration defined in the IEEE 802.11 standard. Note, due to mismatching channels, clients may miss transmitted beacons using this procedure.

In order to bypass this problem and to enable a more efficient network discovery process, most clients—especially mobile devices—prefer active scanning, where clients actively send out probe request frames (2). This is done iteratively for each channel, with the client waiting for a certain period on each channel and listening for corresponding

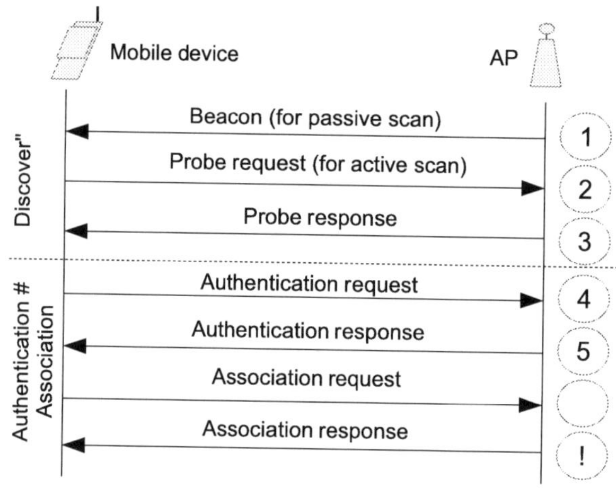

FIG. 1 Frames which are involved in the IEEE 802.11 discovery, authentication, and association process.

probe responses (3) According to Bonné et al. (2013), probe requests are sent out every 2 min on average, regardless of the connection status in order to discover available access points on the fly. Our own experiments confirm these results on average. However, the exact scanning interval depends on various parameters, such as the used chip set or the Wi-Fi driver.

Probe request frames contain unencrypted device-specific information, such as the client's MAC address, supported rates, the destination's network name (SSID), and other management information. With an empty SSID field, the probe request is interpreted as a broadcast message and all access points in reach on the corresponding channel will reply with a probe response. In case a specific network name is contained in the SSID field, only the access point offering the specified network will respond. These so-called directed probe requests are required in case of hidden networks. However, various mobile devices send out directed probe requests for each SSID contained in their preferred network list (PNL) in every active scan. The PNL stores all SSIDs to which the mobile client has tried to connect in the past in order to reconnect to known networks automatically.

All management frames can simply be captured by any Wi-Fi card in reach set into monitor mode. This mode allows to read the content of these frames on application level. Hence, information about stored network names and the device-specific MAC address is accessible for silent attackers within an area of interest. In case of continuous active scans, for example, performed during common Wi-Fi fingerprinting inside buildings, the user's locations as well as complete trajectories can be tracked (Musa and Eriksson, 2012; Schauer et al., 2016b). In summary, it becomes obvious that continuous Wi-Fi active scans in conjunction with indoor positioning systems lead to privacy issues, as it is also stated in the following section.

4 Potentials and Limitations of Wi-Fi Tracking

Due to the increasing percentage of modern mobile devices, Wi-Fi tracking has gathered high interest in both scientific and commercial world. In this section, we discuss potentials and limitations of this technique and present both current research and commercial projects using Wi-Fi tracking for different purposes. A related overview is also given by Schauer (2018).

As stated earlier, IEEE 802.11 probe request frames contain management data, which can be used to gather general information about the crowd in an area of interest. Using the first three bytes of a captured MAC address, one can perform an organizationally unique identifier lookup to determine the manufacturer ID of the sender. Hence, the distribution of present mobile devices according to their Wi-Fi chip set is easily accessible, which is a valuable information about the crowd (Schauer et al., 2014). However, take into account that the results only reflect the composition of detected devices, rather than the real distribution. For instance, a remarkable dominance of Apple devices is often seen in real-world scenarios, for example, in Musa and Eriksson (2012), due to the fact, that these devices show a higher scan frequency.

Due to accessible information about preferred network names, Wi-Fi probes have also been used in literature to determine social links and relationships in the crowd by comparing sets of captured SSIDs from mobile devices (Barbera et al., 2013; Cunche et al., 2012). The authors of the latter come to the conclusion that social relationships can be easily detected and analyzed by just using simple Wi-Fi tracking techniques. Hence, this poses a huge privacy risk for mobile users.

Beside relationships, other types of context information have already been extracted out of Wi-Fi tracking data. In general, such information describes the user's current situation, whereas *location*, *activity*, *time*, and *identity* belong to the four primary types, according to Abowd et al. (1999). Among these, location information plays a key role. It forms the basis of LBS and is often used for activity recognition (Lara and Labrador, 2013). Therefore, many investigations focus on inferring location and trajectory information from captured probe requests (e.g., Musa and Eriksson, 2012 or Schauer et al., 2016b).

However, this is still a challenging task, due to the arbitrary nature of probe request bursts and existing fluctuations in received signal strength indicators (RSSI). Therefore, simplistic positioning approaches are not sufficient for achieving reliable localization results (Schauer et al., 2016b; Bonné et al., 2013). The authors have conducted various empirical experiments in real-world scenarios indicating that pure RSSI values are not suitable to track mobile devices in crowded environments. Hence, many researchers have started to use probabilistic approaches, such as particle filter (Bartoletti et al., 2014) or hidden Markov models (HMM) in combination with the Viterbi algorithm (Musa and Eriksson, 2012; Trogh et al., 2015), in order to track spatiotemporal trajectories using Wi-Fi data.

Activity information, as another primary type of users' context, has also been extracted out of captured probe requests. Among others, Schauer and Linnhoff-Popien (2017) infer the current mobility state of a person by analyzing Wi-Fi traces, which is an important activity information. The authors create a two-state HMM and perform the Viterbi algorithm to detect dwelling and motion periods of mobile users. Muthukrishnan et al. (2009) take deterministic distance functions into account and use the extracted motion information to improve state-of-the-art signal-based localization algorithms. Shen et al. (2016) also determine dwell times and use this information for a feature-based room-level localization approach, which is able to detect the correct room in case of indistinguishable fingerprints.

An extensive work for extracting identity knowledge from Wi-Fi tracking data is presented by Ruiz-Ruiz et al. (2014). The authors perform a real-world study in a hospital and compute several spatio and/or temporal features for classification tasks. They conclude that realistic information reflecting the users' behavior in such a complex environment can be extracted out of passively recorded Wi-Fi data from mobile phones. Beside these works dealing with the extraction of various types of context information, it has proven that Wi-Fi tracking can further be used to determine crowd density and flows (Schauer and Werner, 2015), the users' proximity (Maier et al., 2015), or even for measuring waiting periods in human queues (Wang et al., 2014). It is important to take into account that all of this information can be inferred without the user's consent or even awareness.

Due to low costs and its simplicity, Wi-Fi tracking has also been discovered as an innovative business technology in recent years. Various companies and start-ups (e.g., *42reports*, *sensalytics*, or *walkbase*) have started to use this technology offering different services, such as localization, density and flow analysis, motion inference, return of investment calculations, or marketing optimization. These services are mainly used by retailers or shop owners who try to find similar analytic tools for their business like those which are already well-adopted in online shops. Hence, shoppers in the physical world get tracked for marketing purposes just like in the Internet. The only precondition is that they carry a Wi-Fi enabled mobile device. No complex hardware or software is required, which makes it easy for companies and other persons to use this technique for different purposes.

In summary, it can be seen that Wi-Fi tracking provides great potentials to gather personal information from an unknown crowd and without requiring the users' permission. Hence, this shows high risks concerning the privacy of most smartphone users and their implicit right on personal data. Take into account that in case of common fingerprinting techniques, even more Wi-Fi signals can be captured, due to continuous IEEE 802.11 active scans. But what are the consequences for our modern society where Wi-Fi is the de facto standard for a stable and fast connection to the Internet, which can be seen as a basic need? Which kind of security mechanisms exist or have to be developed in the near future to protect user's privacy completely against Wi-Fi sniffing in real-world scenarios? These questions are discussed in the following section.

5 Security Mechanisms Against Wi-Fi Tracking

Any person who carries a Wi-Fi enabled mobile device can be tracked by passive sniffers without his or her awareness. Hence, the most effective way to bypass this issue is to disable the Wi-Fi interface anytime or at least when the user is on the way. Sure, this is not a suitable solution, due to the loss of Internet connectivity, especially within buildings where mobile data connections may degrade. Beside this, most people are not willing to turn their interfaces on and off all the time. By contrast, the majority of mobile users keep Wi-Fi enabled, as it is shown for an airport environment by Schauer et al. (2014).

Beside this, users could remove network names from the PNL. This would decrease the amount of SSIDs and hence, social profiling like in Barbera et al. (2013) becomes more difficult. However, this does not prevent that users can be generally tracked by Wi-Fi sniffers. In contrast to these simple tricks, more sophisticated security mechanisms are presented in literature. Within these works, two major categories can be identified: the first proposes changes or extensions to the existing IEEE 802.11 protocol in terms of network discovery. The second focuses on MAC address randomization to obfuscate the hardware identifier rendering device recognition challenging.

5.1 Protocol Extensions

One work of this category is presented by Lindqvist et al. (2009). The authors propose a modified 802.11 protocol for the network discovery process which bypass directed probe requests. Furthermore, SSID information is completely obfuscated by using encryption mechanisms. With this concept, a passive attacker should not be able to gather knowledge about saved network names, and thus, social links cannot be determined. However, the software of both client and access point has to be modified requiring small changes at every participating device. Note that a randomization of the client's MAC address is solely discussed, rather than implemented.

By contrast, Greenstein et al. (2008) present a complete and stand-alone protocol, which is called *SlyFi*. The goal is to obfuscate all of the explicit identifiers, such as the MAC address or SSIDs. For this reason, SlyFi applies two mechanisms. Like before, they use encryption techniques for data transmissions. Hence, information about explicit identifiers is no longer accessible by Wi-Fi sniffers and cannot be used for device recognition. Again, this approach requires software modifications for clients and involved access points. In contrast to the former, the complete IEEE 802.11 protocol has to be replaced by the proposed SlyFi protocol.

5.2 MAC Address Randomization

Continuous tracking approaches mainly use the client's MAC address as explicit identifier in order to recognize and trace a particular device over time. Therefore, prior privacy-preserving methods, such as Hu and Wang (2005), or Jiang et al. (2007) proposed randomization of the client's MAC address to obfuscate this identifier rendering unambiguous tracking challenging. However, these solutions have only been investigated theoretically, rather than applying them in real scenarios.

This was changed in 2014, when Apple—as one of the leading companies for mobile platforms—has started to integrate a MAC address randomization mechanism into the mobile operating system *iOS 8* in order to protect users against Wi-Fi sniffing. Hence, for the first time, a simple but also comprehensive privacy-preserving solution was applied to common devices. The mechanism keeps the real hardware address and broadcasts a randomized version during the standard network discovery process. Due to this approach, it is getting difficult to trace an iOS device over time. However, when Apple introduced the first version, MAC address randomization was only activated if the device has changed completely into sleep mode. This occurs only when both mobile data and location services are disabled, which is rarely given in real-world scenarios. Hence, most users were not protected by the proposed randomization mechanism (Beasley, 2014). Therefore, some improvements have been made and released with *iOS 9* where the mechanism was extended to location and auto join scans enabling MAC address randomization also for devices being in active mode (Skinner and Novak, 2015). However, the mechanism is disabled if a device is associated (connected) with an access point, and thus, the real MAC address of the mobile device is sent out by probe requests, like before.

Beside Apple, other well-known providers (e.g., Windows or Google) realized that Wi-Fi sniffing is an existing privacy issue, and hence, they started to develop their own MAC address randomization mechanisms. Since *Windows 10* (Wang, 2015) or *Android 6.0* (Android Developers, 2015) such approaches are integrated in the corresponding operating systems. However, due to the diversity of available mobile devices, the operability depends on the hardware driver. Hence, many users may still not have randomization enabled.

Furthermore, it is clearly stated in literature that randomization of the client's MAC address does not fulfill an overall privacy protection mechanism and it does not prevent users to be tracked by Wi-Fi sniffers. Recent experimental studies (e.g., by Vanhoef et al., 2016 or Martin et al., 2017) show that between 50% and 100% of all devices have been successfully recognized and traced, despite the usage of randomized MAC addresses. Also Pang et al. (2007) have proven that implicit identifiers and certain characteristics of 802.11 data transmissions are sufficient to recognize 64% of mobile devices with obfuscated hardware addresses.

Hence, it can be seen that the mentioned approaches are just a first and necessary step in order to complicate Wi-Fi tracking and to protect users against sniffers. However, and up to now, these attacks cannot be completely avoided by using the existing mechanisms, such as MAC address randomization. An overall protection is only given, if users deactivate their Wi-Fi adapters, which is not an adequate solution, due to the loss of network connectivity. Furthermore, many services and indoor positioning approaches (e.g., Wi-Fi fingerprinting) are not working with disabled Wi-Fi interfaces. Therefore, we present an alternative and privacy-preserving solution for Wi-Fi fingerprinting.

6 Privacy-Preserving Wi-Fi Fingerprinting

In order to preserve the users' privacy, a fully passive positioning process for Wi-Fi fingerprinting is described in this section. The basic idea is that a mobile device just listens for beacon frames in the online phase and determines a valid RSSI fingerprint, rather than performing any IEEE 802.11 active scans. Note that the content of this section is based on our previous work (Schauer et al., 2016a) and presents further investigations.

6.1 Basic Concept

In the online phase of common Wi-Fi fingerprinting systems, IEEE 802.11 active scans are usually performed in order to determine the RSSI vector of all access points in reach. However, this procedure involves probe request frames, and hence, continuous Wi-Fi tracking becomes easily possible. Therefore, a fully passive fingerprint creation is proposed as follows: For a specific time interval Δt, the mobile device listens for incoming beacon frames while iterating over the possible radio channels, switching channels after another interval Δt_{h}. This channel hopping is necessary to capture signals from all access points in reach operating on different channels.

The fingerprint vector v is filled with the mean RSSI values $\overline{v}_i$ of each access point i. Note that different values of Δt impact both the duration of determining v and the amount of information contained in a single fingerprint, which may influence the position accuracy as it is able to planish the impact of an observed RSSI outlier. Hence, Δt can be adapted to different mobility states depending on constraints concerning duration and accuracy of a position fix. For instance, when a persons is moving, Δt should be much smaller than for users who are dwelling.

For the online phase in Wi-Fi fingerprinting approaches, there are two ways of retrieving a location estimation based on a recent online RSSI vector: a deterministic way of comparing the input vector with distinct entries of the radio map, and a probabilistic approach considering the probability of a measurement given the prior knowledge. For evaluation, we investigate both ways which are described in the following section.

6.1.1 Deterministic Approach

According to SMARTPOS by Kessel and Werner (2011), we consider both weighted and nonweighted k-nearest neighbors (kNN) classifiers in signal space for deterministic location estimations during the online phase. Based on the Euclidean distance $d_i = \mathrm{dist}(v, r_i)$ between a passively measured RSSI vector v and a specific record r_i of the fingerprint database, we determine the k nearest candidates of possible user positions. Using the nonweighted kNN classifier, the centroid of these k positions is calculated and returned to the user as his/her current location. In addition, the weighted kNN method multiplies an individual weight w_i to each of the k position candidates, with w_i being calculated as:

$$w_i = \left(d_i \sum_{j=1}^{k} \frac{1}{d_j} \right)^{-1} \tag{1}$$

The user's current location estimate is then calculated as the sum of weighted k position candidates. Note that the authors of SMARTPOS obtained better results using the weighted procedure. For a thorough comparison, we also investigate both types of kNN classifiers for our passive Wi-Fi fingerprinting and compare the results.

Whenever an online RSSI vector is compared to those stored in the radio map, it is possible that any two vectors differ in their lengths, due to RSSI values of a certain access points only being contained in one of them and vice versa. Thus, one has to find a consistent way of dealing with missing values in order to compute the k-nearest neighbors. These values can either simply be ignored or be set to a predefined minimum value (e.g., -100 dBm; Kessel and Werner, 2011). Both ways have their qualification. When ignoring matchless entries, important information for accurate location estimations may be lost, while negative effects caused by changes occurring in the setup of access points may be kept low. On the other hand, a fixed minimum value such as -100 dBm punishes comparisons between strong RSSI and missing values, and favors comparisons between weak RSSI and missing values. However, this is to be expected in real-world scenarios, due to the fact, that strong signals should be measured again at the corresponding position, while weak signals may be missed cause of strong fluctuations of radio signals within buildings. We investigate the impact of treating missing values in our evaluation, described in Section 6.2.2. Take into account that SMARTPOS shows better results when ignoring missing values.

6.1.2 Probabilistic Approach

For probabilistic location estimations, we use a naive Bayes classifier in order to classify RSSI measurements into certain rooms of our building. Unlike before, this approach returns a specific room number to the user rather than a certain position fix as a coordinate-pair. To this end, room information has to be saved together with corresponding RSSI vectors in the radio map during offline phase.

More specifically, our naive Bayes classifier is based on the Bayes theorem and assigns the most probable class to a problem instance represented by a feature vector. In our case, we treat rooms as classes, and vectors of RSSI measurements as problem instances and feed them to Bayes theorem:

$$P(R|v) = \frac{P(v|R) \cdot P(R)}{P(v)} \tag{2}$$

calculating the posteriori probability $P(R|v)$ of being in a certain room R under the condition that fingerprint v is observed. The probability $P(R)$ is the prior probability, which is based on our knowledge of frequencies in the training set, and hence, it can be easily estimated by counting the occurrence of each room. $P(v|R)$ is the likelihood function determining the probability of observing v in case of being in room R, and $P(v)$ is called the evidence which can be calculated assuming a normal distribution with mean μ and the standard deviation σ for each one-dimensional parameter.

The naive Bayes classifier is simple to use, given the fact that it always assumes conditional independent features. Hence, we are allowed to express the probability of being in a certain room R in case of observing fingerprint v consisting of n access points' RSSI values v_i as follows:

$$P(R|v_1,\ldots,v_n) = \frac{1}{Z}P(R)\prod_{i=1}^{n} P(v_i|R) \tag{3}$$

with evidence $Z = P(v)$ treated as a constant in this case, because the values of RSSI measurements are known. Therefore, and due to the fact that we are only interested in the most probable room R_j with $j \in 1,\ldots,|R|$ using the maximum-a-posteriori decision rule, the naive Bayesian classifier can be directly derived from Eq. (3) and is expressed as follows:

$$R = \underset{R_j}{\operatorname{argmax}} \left\{ P(R_j)\prod_{i=1}^{n} P(v_i|R_j) \right\} \tag{4}$$

Hence, applying an online measured RSSI vector of length n to Eq. (4), the most probable room R_j out of all labeled rooms R is returned by the proposed naive Bayes classifier. Note that a similar probabilistic estimator is also used in SMARTPOS, but is not described by the authors how they treat missing values in this case.

Being linked by means of multiplication, however, observed RSSI values that are lacking their counterpart in the radio map for a certain room R_j would rigorously lead to zero-probability of R_j. In our case, this would lead to false classifications, due to the fact that a room will consequently show the probability of zero even if only one RSSI value is missing. In order to solve this problem, a small sample correction is added to all probability estimations guaranteeing that no probability is ever set to zero. These corrections are called pseudocounts or additive smoothing, which is commonly used with naive Bayes classifiers in order to treat missing values. In our case, we use Laplace smoothing for all of our measurements v_i taken in a certain room R_j and smooth $P(v_i|R_j)$ with a pseudocount $\gamma = 1$. This is done according to the following formula:

$$\hat{P}(v_i|R_j) = \frac{\text{count}(v_i)_{R_j} + \gamma}{|v|_{R_j} + \gamma \cdot (v)_{R_j}} \tag{5}$$

where $\text{count}(v_i)_{R_j}$ is the number of occurrences of the measurement v_i in room R_j, $|v|_{R_j}$ is the amount of all measurements made in room R_j, and $(v)_{R_j}$ is the domain of all measurements observed in room R_j. By using this Laplace smoothing technique, we are able to consider all of our rooms for classification even when the measurement data differ from the corresponding entries in the radio map. Thus, Eq. (4) is still correct for the whole set of possible rooms and returns the most probable room as the user's location estimate derived from a certain measured RSSI vector.

6.2 Evaluation

The proposed passive Wi-Fi fingerprinting approach was implemented for a mobile device using the existing wireless infrastructure in our office environment. For evaluation, it is compared to common active Wi-Fi fingerprinting in terms of well-known performance metrics, which are also used by SMARTPOS. Both deterministic and probabilistic location estimations are performed and confronted with the results of active scanning. Furthermore, we enhance our previous investigations and consider both static and dynamic users. We investigate how passive Wi-Fi fingerprinting differs in terms of accuracy and precision when users are moving. In order to compare the results of active and passive Wi-Fi fingerprinting, we perform both types of online scans using identical parameter setups.

6.2.1 Implementation and Setup

We will now give a closer look to our implementation and experimental setup. As first step of Wi-Fi fingerprinting, a database of RSSI measures on certain reference points has to be built up during an offline training phase. We use common active scans for recording the radio map, as this step is not sensitive to users' privacy. The active scans are performed directly by an application on a Samsung Galaxy S2 (I9100), which is used as our test device. At each reference point, we perform 20 active Wi-Fi scans for each of the four main orientations 0, 90, 180, and 270 degrees. A series of scans is always annotated with the position of the corresponding reference point on the map and the user's orientation when the fingerprint was taken. For radio map generation in the deterministic approach, each entry in the fingerprint database represents the vector of the means of 20 consecutively measured RSSI values per reachable access point. For the probabilistic approach, we determine the fingerprint as normal distribution over the measurements of each access point and add the corresponding room label information. In total, we recorded 332 fingerprints on 83 reference points located within one aisle of our office building and its main corridor, as shown in Fig. 2A. The distance between two consecutive reference points is always under 1.5 m.

For a fully privacy preserving fingerprint approach, any active bidirectional communication with a central location server has to be avoided. Instead, the radio map can either be locally stored on the device or might be transmitted in the beacon frames, as successfully shown by Gschwandtner and Schindhelm (2011). For our evaluation the radio map is directly stored on our mobile test device, which is used for location estimations. Due to the fact that beacon frames are only used for management purposes, they are not forwarded to the application layer, and thus usually cannot be read or further processed by user-level programs. In order to make them usable for passive fingerprinting, the device's Wi-Fi card has to be set into monitor mode, which is not possible with all common phones. Hence, we first rooted the phone and installed a patch for the Wi-Fi card using the Android application package of Bcmon[1], in order to be able to use 802.11 monitor mode for recording beacons. It is important to note that when the Wi-Fi card is set into

[1]https://code.google.com/p/bcmon/

monitor mode, the device is only listening and does not send out any packages that could be captured by a malicious party or infrastructure provider. Hence, this can be seen as the highest level of protecting users from Wi-Fi tracking, while still being able to offer indoor positioning and navigation.

In order to create passive fingerprints during the online phase, the mobile device listens on incoming beacon frames for a specific time interval Δt, switching between the most commonly used Wi-Fi channels 1, 6, and 11 after Δt_h. If not explicitly stated otherwise, we set Δt to 3 s, and Δt_h to 1 s in our experiments. All necessary information, such as hardware addresses of access points and corresponding RSSIs, is then extracted from the resulting dump file and a fingerprint is constructed containing the mean values $\overline{v}_i$ of the observed RSSI values for each seen access point i.

For the creation of common active fingerprints the same application as for generating the radio map was used. Take into account that one active scan required about 4.5 s on average. Thus, our passive approach with $\Delta t = 3$ needs less time to collect the data necessary for formulating a position fix query. To allow for direct comparisons of the active and passive fingerprinting approaches, we successively apply both methods during the online phase on the same device at 19 randomly chosen locations across the test environment, as indicated in Fig. 2B. For determining the user's orientation at the moment a fingerprint was taken, we use a digital compass derived from the smartphone's accelerometer and magnetometer sensor readings. Both the orientation information and the observed fingerprint are compared fed to the locally stored radio map using either deterministic or probabilistic location estimation. The following sections indicate the achieved results, separately.

6.2.2 Deterministic Location Estimation

We now investigate the position accuracy and precision using various deterministic methods. More precisely, we calculate the mean, minimum and maximum positioning

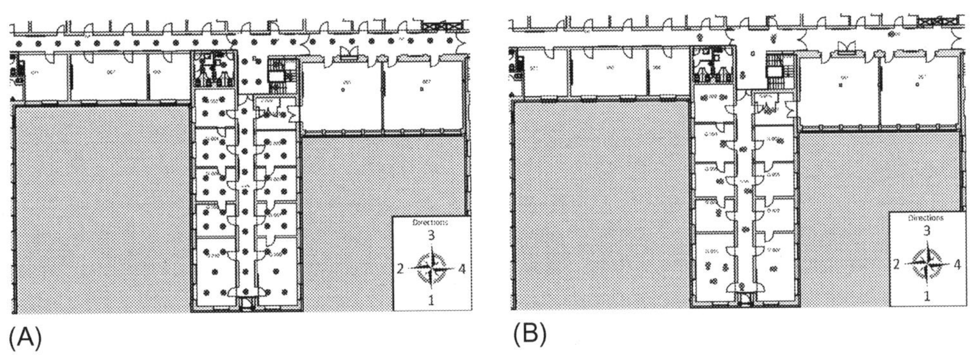

FIG. 2 Schematic overview of our test setup indicating reference points and the locations of position fixes with users' orientation. (A) Reference points for the radio map marked as *dots*. (B) Locations of position fixes marked as *dots* with user's orientation shown as *black pinnacles*.

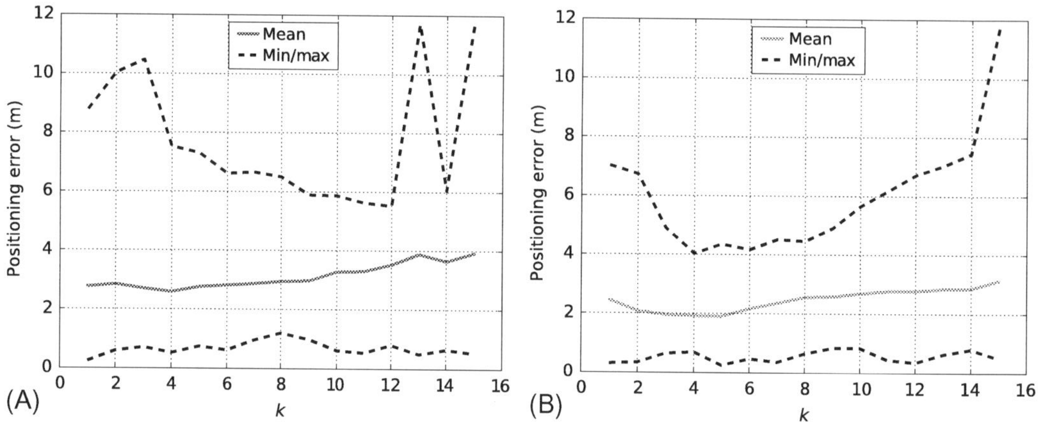

FIG. 3 Comparison of active versus passive Wi-Fi fingerprinting considering missing values, weighted *k*NN, and user orientation. (A) Active scanning. (B) Passive scanning.

error, as well as the standard deviation considering all of the 19 active versus the 19 passive position fixes while iterating over different values of *k*. Fig. 3 indicates the results of this real-world experiment when both the user's orientation and missing RSSI values are considered and a weighted *k*NN approach is used for deterministic location estimation. The best results are obtained with $k = 4$, showing a mean positioning error of 1.92 m and a standard deviation of 0.92 m for passive scans, and 2.59 m with a standard deviation of 1.68 m in case of common active scanning. Hence, these results indicate that the passive approach performs more accurately within our test set. Furthermore, and as expected, it can be observed that for both scan types, the mean positioning error tends to increase for higher values of *k*.

In order to investigate the impact of the used parameters, we now successively compare the results obtained by ignoring missing values, using nonweighted *k*NN, and completely neglecting the user's orientation. Eventually, the optimal parameter setting that results in the lowest average positioning error will be determined and discussed. Fig. 4 indicates the results for active and passive fingerprinting, when missing RSSI values are ignored, but relative weighting and orientation are still considered for location estimation. It is clearly shown that missing values should be treated by applying a minimal value. Otherwise, as shown in Fig. 4, the obtained values show unfeasible positioning results both for the active and the passive approach. In our case, we observe a mean position error of over 10 m for both scan types and $k < 10$. The standard deviation is greater than 8 m for active and greater than 6 m for passive scanning, which is not suitable for most indoor position scenarios. These observations are contrary to SMARTPOS, where the authors decided to ignore missing values in order to achieve slightly better positioning results.

As next step, we use a nonweighted *k*NN while considering the user's orientation and treating missing RSSI values. Again, the obtained results are shown in Fig. 5 for both scan types. It can be seen that the mean positioning error is a bit higher for each value of *k* and

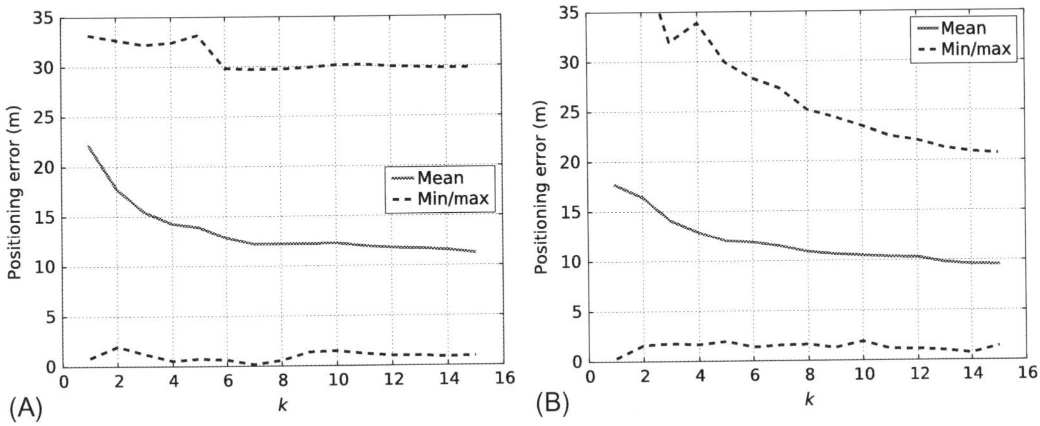

FIG. 4 Comparison of active versus passive Wi-Fi fingerprinting using weighted *k*NN, and users' orientation, but ignoring missing values. (A) Active, ignoring missing values. (B) Passive, ignoring missing values.

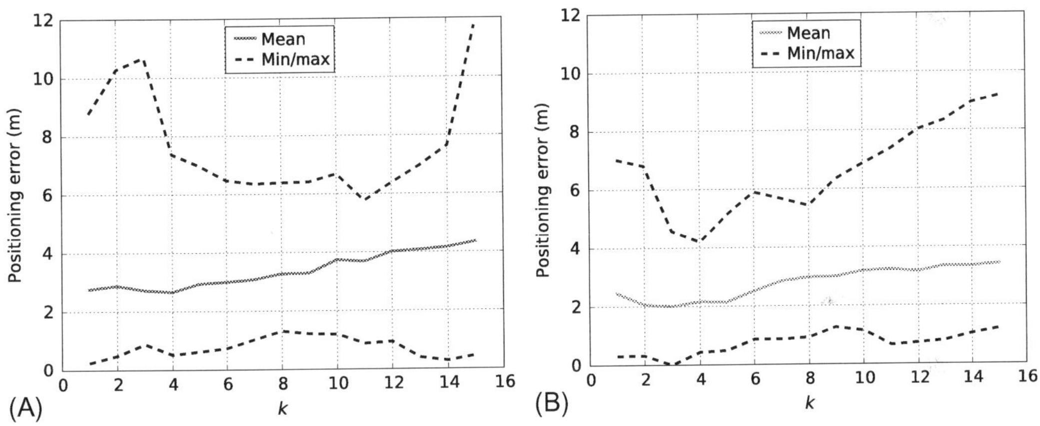

FIG. 5 Comparison of active versus passive Wi-Fi fingerprinting using weighted and nonweighted *k*NN. (A) Active, with nonweighted *k*NN. (B) Passive, with nonweighted *k*NN.

for both approaches when using nonweighted *k*NN instead of weighted *k*NN. With $k = 4$, the mean accuracy lies at 2.66 m for active and at 2.16 m for passive fingerprinting with a standard deviation of 1.2 and 1.7 m, respectively. Hence, our passive approach returns more accurate position fixes even when a nonweighted *k*NN location estimator is used. Overall, weighted *k*NN is to be preferred for both scan types, like in SMARTPOS.

As a last parameter, we investigate the impact of considering or ignoring the user's orientation. Thus, we apply weighted *k*NN, consider missing values, but now ignore the orientation information for our location estimation. Fig. 6 shows the corresponding results for active and passive fingerprinting. It can be observed that the overall positioning error on average is slightly lower for both scan types, especially for higher values of *k*, and again

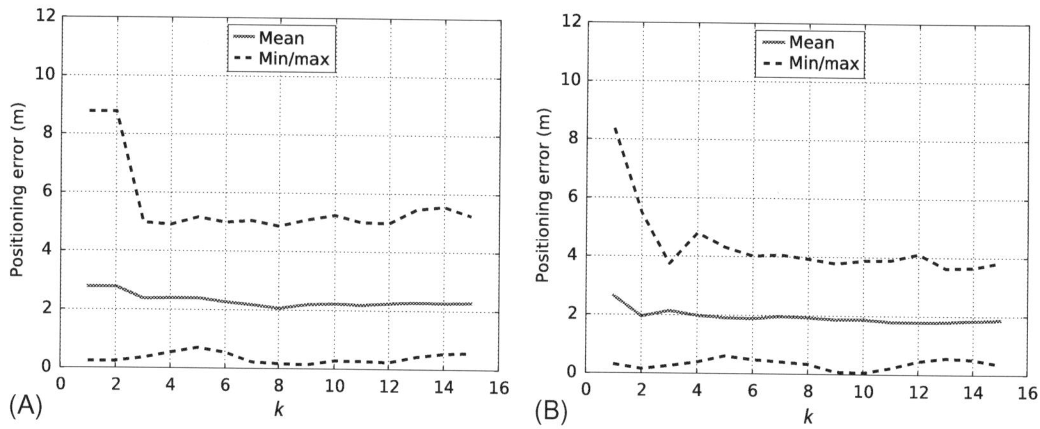

FIG. 6 Comparison of active versus passive Wi-Fi fingerprinting when considering and ignoring the users' orientation. (A) Active, without orientation. (B) Passive, without orientation.

passive scanning results in a lower positioning error than the common approach. For active scanning with $k = 4$, the mean accuracy is at 2.40 m and a standard deviation of 1.52 m is obtained. This is a tiny improvement of 0.19 m in terms of accuracy and 0.16 m in terms of precision for active fingerprinting. In case of our passive approach with $k = 4$, we observe a little degradation of 0.06 m for both accuracy and precision when ignoring the user's orientation instead of considering it. However, for $k = 6$ a mean positioning error of 1.88 m and a standard deviation of 0.99 m is obtained, indicating a slightly improvement of 0.04 m. An interesting observation is that for higher values of k, the mean accuracy remains constant when ignoring orientation while it is increasing when considering the orientation information. The explanation is that in case of ignoring orientation, the RSSI vector of online measurements is compared to the complete database, rather than comparing only entries with corresponding orientation. Thus, more similar fingerprints are available for k-nearest neighbors, and hence, an increasing k is less likely to negatively influence the positioning result. However, when neglecting orientation information, deterministic location estimation requires four times more database comparisons, and thus, a position request takes more time to be served.

In summary, we obtain the best results for both scan types within our experiment when using weighted kNN, considering missing values instead of ignoring them, but ignoring the user's orientation. These findings are contrary to SMARTPOS, where the information of users' orientation helped to increase the positioning accuracy. In our case for passive fingerprinting, the mean positioning error remains constantly below 2 m, the standard deviation below 1.1 m for $k > 3$. In comparison, the best results for active scanning were achieved with $k = 8$, showing an accuracy on average of 2.06 m and a standard deviation of 1.8 m. Hence, with respect to these results based on deterministic location estimation, we conclude that our passive approach performs slightly better than common active Wi-Fi fingerprinting within our test set. In the next section, we describe our evaluation using probabilistic location estimations.

6.2.3 Probabilistic Location Estimation

We now apply the naive Bayes classifier as described in Section 6.1.2 to the same 19 position fixes of our online phase. In order to use the classifier, we first partition our test environment into 19 different rooms and corridor segments as shown in Fig. 7. Each segment contains four to six reference points marked with the corresponding room label. A room segment is classically divided by its walls, except the segments mapped onto the corridors of the building, which are quite long and are hence further divided into several parts to allow for a fine-grained positioning.

Overall, we investigate the correctness of the classification result for each position fix in three categories:

1. *Correct:* The online RSSI vector is classified to the room where the user is actually located.
2. *Nearby:* The online RSSI vector is classified to a direct neighbor of the actual room.
3. *False:* The classification returns a room far away from the actual room.

As before, we evaluate the impact of user's orientation by considering (+o) and ignoring (−o) the orientation information in the database. This is done for both active and passive scans. The classification results for the complete test set are depicted in Table 1, where

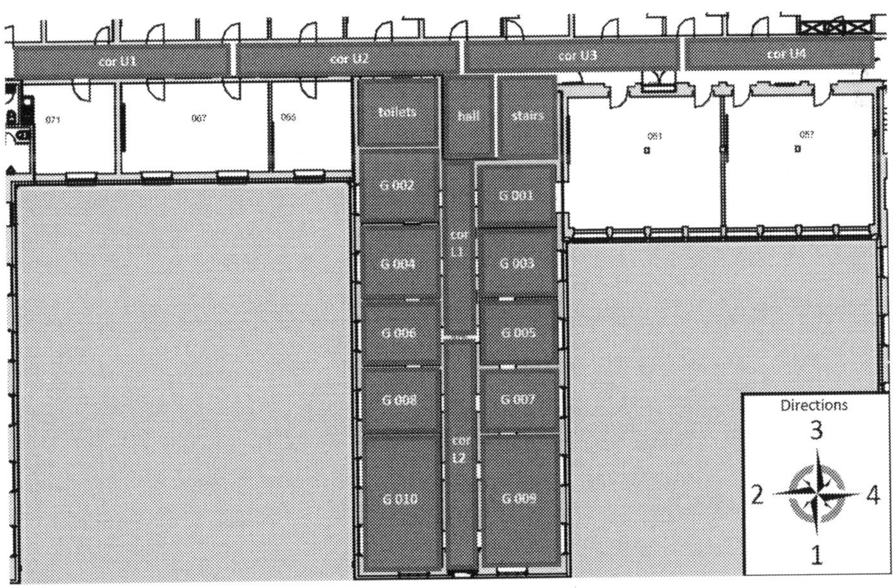

FIG. 7 Schematic overview of test environment divided into room segments.

Table 1 Classification Results for Active and Passive Scanning

	Active Scanning		Passive Scanning	
Real Room	**−o**	**+o**	**−o**	**+o**
hall	**hall**	*corU3*	*g001*	**hall**
corL2	**corL2**	**corL2**	**corL2**	**corL2**
g002	*g004*	*g004*	corU1	*toilets*
corL1	*g006*	*g006*	**corL1**	**corL1**
corU2	stairs	g004	g001	g003
g002	**g002**	*toilets*	g001	g001
g009	**g009**	**g009**	**g009**	**g009**
g008	**g008**	toilets	**g008**	g001
g006	**g006**	*corL2*	**g006**	**g006**
g003	*corL1*	*corL1*	**g003**	**g003**
g001	**g001**	**g001**	*g003*	**g001**
corL2	*g010*	*g010*	*g009*	**corL2**
corL1	*g007*	*g008*	*g007*	*g007*
corU2	*toilets*	*toilets*	**corU2**	g007
g010	g006	g006	*corL2*	*corL2*
g010	g006	g006	**g010**	*corL2*
g010	**g010**	**g010**	**g010**	**g010**
g004	toilets	g001	**g004**	*corL1*
corU3	g001	g001	**corU3**	g001
		Summary		
Correct	8	4	11	9
Nearby	5	8	5	5
False	6	7	3	5

emphases denote the correctness, such as bold values = correct, italic values = nearby, and underlined values = false.

It can easily be seen that the best results are obtained for both scan types when the information about users' orientation is ignored. This observation confirms to SMARTPOS, where the authors conclude that orientation information should not be used as a filter in a naive Bayesian estimator. When orientation is ignored, 42% of all rooms are classified correctly and 31% are false results using common active fingerprint. In comparison, when using our passive approach, 58% of all rooms are classified correctly with only 16% being misclassified. Based on these results, we conclude that passive Wi-Fi fingerprinting performs more accurately for both deterministic and probabilistic location estimations, and furthermore, it is capable to completely preserve mobile users' privacy during the whole positioning process, as the device's transceiver is only passively used for receiving beacons.

An explanation for the improvement in terms of accuracy is that even quick passive scans are able to aggregate more information about received signal strengths for calculating an online fingerprint than common active scans which cycle through all Wi-Fi channels but typically only report one observed RSSI value per access point. We will review this assumption in the following section by performing longer scanning periods.

6.2.4 Considering User Movement

Up to this point, we have only been considering a user moving through our building and hence requiring a short passive scanning process. For this purpose, we set $\Delta t = 3$ and $\Delta t_h = 1$. This means that for the creation of a passive Wi-Fi fingerprint, the mobile device listens for beacon frames for a period of 3 s, while staying on the most commonly used Wi-Fi channels for 1 s each. As a next step, we investigate whether the positioning results can be improved by allowing a longer period of time for both Δt and Δt_h. We thereby increase the information for calculating an RSSI vector, which in return can be expected to further reduce the possible impact of observed RSSI outliers as the number of overheard beacons grows. This directly applies to the scenario of a static user, for example, a person remaining in a certain location for a longer period of time, which of course is very common within buildings. Note that this type of a user's context (i.e., activity) can easily be inferred by modern mobile devices using its integrated sensors (Maier and Dorfmeister, 2013) or even by passive Wi-Fi captures (Schauer and Linnhoff-Popien, 2017). Hence, while the user remains static, for example, sitting in an office or cafeteria, we propose a longer-time period, experimentally set to $\Delta t = 20$ and $\Delta t_h = 2$.

In order to compare the obtained positioning errors of dynamic and static users, we conduct another experiment. Our test device captures Wi-Fi data at a fixed position for a long duration of 30 min, which simulates a person sitting in an office. We distinguish between two types of users computing an online fingerprint as follows:

1. *Dynamic user:* Every 20 s, an RSSI vector is calculated based on a 3-s capture with $\Delta t_h = 1$.
2. *Static user:* Every 20 s, an RSSI vector is calculated based on all information captured during the last 20 s with $\Delta t_h = 2$.

Both types of calculated RSSI vectors are used for deterministic location estimation with weighted *k*NN, considering the user's orientation, and missing RSSI values. Fig. 8 depicts the results of the obtained positioning errors for this experiment.

It can clearly be seen that in case of a static user's device performing a longer scanning period, the achieved positioning error fluctuates considerably less and shows a higher accuracy on average of 2.2 m. By contrast, position fixes of the dynamic user suffer from high fluctuations in terms of their accuracy during the experiment. Hence, we conclude that the precision of our positioning approach is lower when using a shorter scanning period, which was expected. Thus, in order to obtain position fixes with higher precision when the user does not move, it is useful to determine the user's behavior in a first step in order to adjust the scanning period.

Finally, we investigate the accuracy of our naive Bayes classifier for a static user and compare the performed classifications with the results obtained in Table 1. For this

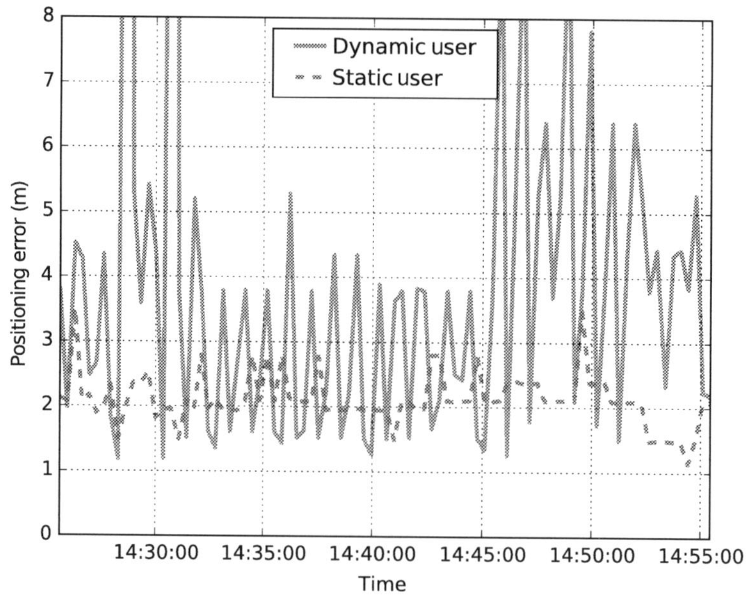

FIG. 8 Comparing positioning errors for dynamic and static users.

purpose, we repeat the 19 online position fixes with $\Delta t = 20$ and $\Delta t_{\mathrm{h}} = 2$. The obtained results indicate that one more room is classified correctly. Overall, 12 (63%) rooms were correct, 4 (21%) rooms nearby, and, still, 3 (16%) rooms falsely classified. This indicates a marginal improvement for static scenarios when probabilistic location estimations are performed. Hence, we conclude that for both deterministic and probabilistic fingerprinting, a longer scanning period can help to improve the accuracy and precision of our passive approach. This confirms the previous assumption that passive scans are able to aggregate more information for localization tasks than common IEEE 802.11 active scans, while fully preserving users' privacy.

In summary, the proposed approach presents one possible solution to overcome the privacy problem in IEEE 802.11 fingerprinting techniques. However, it requires root privileges and small driver manipulations to set the phone's Wi-Fi card into monitor mode. Furthermore, network connectivity is disabled in this mode, and hence, the proposed method is not applicable in real life for most smartphone users.

7 Conclusion and Future Work

In this chapter, we have focused on existing privacy problems for mobile users due to simple and low-cost Wi-Fi tracking techniques. Beside the technical background, we have demonstrated the wide range of possibilities to analyze an unknown crowd without the users' consent or awareness by just capturing probe requests. Overall, it has been shown that every person who carries a Wi-Fi-enabled mobile device risks to be tracked and analyzed involuntarily. This threat even increases when performing Wi-Fi fingerprinting using additional active scans. Therefore, several exiting approaches have been presented for the protection of users' privacy against this attack. Beside protocol extensions and MAC address randomization, a fully passive Wi-Fi fingerprinting approach was discussed in this chapter.

However, it is clear that none of the described solutions is both practical and suitable for real-world scenarios. While protocol extensions need software manipulations, the randomization of the hardware address is not sufficient for completely obfuscating the identity of the device. On the one hand, our passive Wi-Fi fingerprinting approach ensures the highest level of privacy preservation, due to the fact that no signals are sent out by the mobile device. On the other hand, it requires software modifications and disables network connectivity.

Hence, we conclude that an overall and practical privacy-preserving approach for fingerprinting techniques is still missing. Further research has to be performed to prevent Wi-Fi sniffing completely while providing the full functionality. Therefore, future work should enhance MAC address randomization and extend it to other implicit and explicit identifiers in order to fully obfuscate the user's identity.

References

Abowd, G.D., Dey, A.K., Brown, P.J., Davies, N., Smith, M., Steggles, P., 1999. Towards a better understanding of context and context-awareness. In: International Symposium on Handheld and Ubiquitous Computing, pp. 304–307.

Android Developers, 2015. Android 6.0 Changes. Available from: https://developer.android.com/about/versions/marshmallow/android-6.0-changes.html#behavior-notifications (Accessed 12 July 2017).

Bahl, P., Padmanabhan, V.N., 2000. RADAR: an in-building RF-based user location and tracking system. In: INFOCOM 2000. Nineteenth Annual Joint Conference of the IEEE Computer and Communications Societies. Proceedings. IEEE, vol. 2, pp. 775–784.

Barbera, M.V., Epasto, A., Mei, A., Perta, V.C., Stefa, J., 2013. Signals from the crowd: uncovering social relationships through smartphone probes. In: Proceedings of the 2013 Conference on Internet Measurement Conference, pp. 265–276.

Bartoletti, S., Conti, A., Giorgetti, A., Win, M.Z., 2014. Sensor radar networks for indoor tracking. IEEE Wireless Commun. Lett. 3 (2), 157–160.

Beasley, M., 2014. More details on how iOS 8's MAC address randomization feature works (and when it doesn't). Available from: https://9to5mac.com/2014/09/26/more-details-on-how-ios-8s-mac-address-randomization-feature-works-and-when-it-doesnt/ (Accessed 12 July 2017).

Bell, S., Jung, W.R., Krishnakumar, V., 2010. WiFi-based enhanced positioning systems: accuracy through mapping, calibration, and classification. In: Proceedings of the 2nd ACM SIGSPATIAL International Workshop on Indoor Spatial Awareness, pp. 3–9.

Bonné, B., Barzan, A., Quax, P., Lamotte, W., 2013. WiFiPi: involuntary tracking of visitors at mass events. In: 2013 IEEE 14th International Symposium and Workshops on a World of Wireless, Mobile and Multimedia Networks (WoWMoM), pp. 1–6.

Cunche, M., Kaafar, M.A., Boreli, R., 2012. I know who you will meet this evening! Linking wireless devices using Wi-Fi probe requests. In: 2012 IEEE International Symposium on a World of Wireless, Mobile and Multimedia Networks (WoWMoM), pp. 1–9.

Dorfmeister, F., Wiesner, K., Schuster, M., Maier, M., 2015. Preventing restricted space inference in online route planning services. In: MOBIQUITOUS'15 Proceedings of the 12th EAI International Conference on Mobile and Ubiquitous Systems: Computing, Networking and Services. ACM.

Farshad, A., Li, J., Marina, M.K., Garcia, F.J., 2013. A microscopic look at WiFi fingerprinting for indoor mobile phone localization in diverse environments. In: 2013 International Conference on Indoor Positioning and Indoor Navigation (IPIN), pp. 1–10.

Greenstein, B., McCoy, D., Pang, J., Kohno, T., Seshan, S., Wetherall, D., 2008. Improving wireless privacy with an identifier-free link layer protocol. In: Proceedings of the 6th International Conference on Mobile Systems, Applications, and Services, pp. 40–53.

Gschwandtner, F., Schindhelm, C.K., 2011. Spontaneous privacy-friendly indoor positioning using enhanced WLAN beacons. In: 2011 International Conference on Indoor Positioning and Indoor Navigation (IPIN), pp. 1–8.

Gunawan, M., Li, B., Gallagher, T., Dempster, A.G., Retscher, G., 2012. A new method to generate and maintain a WiFi fingerprinting database automatically by using RFID. In: 2012 International Conference on Indoor Positioning and Indoor Navigation (IPIN), pp. 1–6.

Hu, Y.C., Wang, H.J., 2005. A framework for location privacy in wireless networks. In: ACM SIGCOMM Asia Workshop.

IEEE, 2007. IEEE Std 802.11: Wireless LAN Medium Access Control (MAC) and Physical Layer (PHY) Specifications. IEEE Computer Society, New York, NY.

Jiang, T., Wang, H.J., Hu, Y.-C., 2007. Preserving location privacy in wireless LANs. In: Proceedings of the 5th International Conference on Mobile Systems, Applications and Services, pp. 246–257.

Kessel, M., Werner, M., 2011. SMARTPOS: accurate and precise indoor positioning on mobile phones. In: Proceedings of the 1st International Conference on Mobile Services, Resources, and Users, Barcelona, pp. 158–163.

Konstantinidis, A., Chatzimilioudis, G., Zeinalipour-Yazti, D., Mpeis, P., Pelekis, N., Theodoridis, Y., 2015. Privacy-Preserving Indoor Localization on Smartphones. IEEE Transactions on Knowledge and Data Engineering 27 (11), 3042–3055. IEEE.

Koweerawong, C., Wipusitwarakun, K., Kaemarungsi, K., 2013. Indoor localization improvement via adaptive RSS fingerprinting database. In: 2013 International Conference on Information Networking (ICOIN), pp. 412–416.

Lara, O.D., Labrador, M.A., 2013. A survey on human activity recognition using wearable sensors. IEEE Commun. Surv. Tutorials 15 (3), 1192–1209.

Ledlie, J., Park, J.G., Curtis, D., Cavalcante, A., Camara, L., Costa, A., Vieira, R., 2012. Molé: a scalable, user-generated WiFi positioning engine. J. Locat. Based Serv. 6 (2), 55–80.

Li, H., Sun, L., Zhu, H., Lu, X., Cheng, X., 2014. Achieving privacy preservation in WiFi fingerprint-based localization. In: INFOCOM, 2014 Proceedings IEEE, pp. 2337–2345.

Lindqvist, J., Aura, T., Danezis, G., Koponen, T., Myllyniemi, A., Mäki, J., Roe, M., 2009. Privacy-preserving 802.11 access-point discovery. In: Proceedings of the Second ACM Conference on Wireless Network Security, pp. 123–130.

Maier, M., Dorfmeister, F., 2013. Fine-grained activity recognition of pedestrians travelling by subway. In: 5th International Conference on Mobile Computing, Applications and Services (MobiCASE 2013), vol. 130. Springer, Paris, France, pp. 122–139.

Maier, M., Schauer, L., Dorfmeister, F., 2015. ProbeTags: privacy-preserving proximity detection using Wi-Fi management frames. In: 2015 IEEE 11th International Conference on Wireless and Mobile Computing, Networking and Communications (WiMob), pp. 756–763.

Martin, E., Vinyals, O., Friedland, G., Bajcsy, R., 2010. Precise indoor localization using smart phones. In: Proceedings of the International Conference on Multimedia, pp. 787–790.

Martin, J., Mayberry, T., Donahue, C., Foppe, L., Brown, L., Riggins, C., Rye, E.C., Brown, D., 2017. A Study of MAC Address Randomization in Mobile Devices and When It Fails. Proceedings on Privacy Enhancing Technologies 4, 365–383. De Gruyter Open. ArXiv preprint arXiv:1703.02874.

Musa, A.B.M., Eriksson, J., 2012. Tracking unmodified smartphones using Wi-Fi monitors. In: Proceedings of the 10th ACM Conference on Embedded Network Sensor Systems, pp. 281–294.

Muthukrishnan, K., van der Zwaag, B., Havinga, P., 2009. Inferring motion and location using WLAN RSSI. In: Mobile Entity Localization and Tracking in GPS-Less Environments, pp. 163–182.

Pang, J., Greenstein, B., Gummadi, R., Seshan, S., Wetherall, D., 2007. 802.11 user fingerprinting. In: Proceedings of the 13th Annual ACM International Conference on Mobile Computing and Networking, pp. 99–110.

Ruiz-Ruiz, A.J., Blunck, H., Prentow, T.S., Stisen, A., Kjærgaard, M.B., 2014. Analysis methods for extracting knowledge from large-scale WiFi monitoring to inform building facility planning. In: 2014 IEEE International Conference on Pervasive Computing and Communications (PerCom), pp. 130–138.

Sabek, I., Youssef, M., Vasilakos, A.V., 2015. ACE: an accurate and efficient multi-entity device-free WLAN localization system. IEEE Trans. Mobile Comput. 14 (2), 261–273.

Schauer, L., 2018. Analyzing the digital society by tracking mobile customer devices. In: Digital Marketplaces Unleashed. Springer, pp. 467–478.

Schauer, L., Linnhoff-Popien, C., 2017. Extracting context information from Wi-Fi captures. In: Proceedings of the 10th International Conference on PErvasive Technologies Related to Assistive Environments, pp. 123–130.

Schauer, L., Werner, M., 2015. Analyzing pedestrian flows based on Wi-Fi and Bluetooth captures. EAI Endorsed Trans. Ubiquit. Environ. 1, e4.

Schauer, L., Werner, M., Marcus, P., 2014. Estimating crowd densities and pedestrian flows using Wi-Fi and Bluetooth. In: Proceedings of the 11th International Conference on Mobile and Ubiquitous Systems: Computing, Networking and Services, pp. 171–177.

Schauer, L., Dorfmeister, F., Wirth, F., 2016a. Analyzing passive Wi-Fi fingerprinting for privacy-preserving indoor-positioning. In: 2016 International Conference on Localization and GNSS (ICL-GNSS), pp. 1–6.

Schauer, L., Marcus, P., Linnhoff-Popien, C., 2016b. Towards feasible Wi-Fi based indoor tracking systems using probabilistic methods. In: 2016 International Conference on Indoor Positioning and Indoor Navigation (IPIN), pp. 1–8.

Shen, J., Cao, J., Liu, X., Wen, J., Chen, Y., 2016. Feature-based room-level localization of unmodified smartphones. In: Smart City 360°, pp. 125–136.

Skinner, K., Novak, J., 2015. Privacy and your app. In: Apple Worldwide Dev. Conf. (WWDC).

Sweeney, L., 2002. K-anonymity: a model for protecting privacy. Int. J. Uncertain. Fuzziness Knowl. Based Syst. 10 (5), 557–570. ISSN 0218-4885.

Trogh, J., Plets, D., Martens, L., Joseph, W., 2015. Advanced real-time indoor tracking based on the Viterbi algorithm and semantic data. Int. J. Distrib. Sens. Netw. 11 (10), 271818.

Vanhoef, M., Matte, C., Cunche, M., Cardoso, L.S., Piessens, F., 2016. Why MAC address randomization is not enough: an analysis of Wi-Fi network discovery mechanisms. In: Proceedings of the 11th ACM on Asia Conference on Computer and Communications Security, pp. 413–424.

Wang, W., 2015. Wireless networking in Windows 10. In: Windows Hardware Engineering Community conference (WinHEC).

Wang, Y., Yang, J., Chen, Y., Liu, H., Gruteser, M., Martin, R.P., 2014. Tracking human queues using single-point signal monitoring. In: Proceedings of the 12th Annual International Conference on Mobile Systems, Applications, and Services, pp. 42–54.

Yang, D., Fang, X., Xue, G., 2013. Truthful incentive mechanisms for *k*-anonymity location privacy. In: INFOCOM 2013 Proceedings IEEE, pp. 2994–3002.

3

Lessons Learned in Generating Ground Truth for Indoor Positioning Systems Based on Wi-Fi Fingerprinting

Joaquín Torres-Sospedra*, Óscar Belmonte-Fernández*, Germán M. Mendoza-Silva*, Raul Montoliu*, Adrián Puertas-Cabedo†, Luis E. Rodríguez-Pupo*, Sergio Trilles*, Andrea Calia‡, Mauri Benedito-Bordonau§, Joaquín Huerta*

*INSTITUTE OF NEW IMAGING TECHNOLOGIES, JAUME I UNIVERSITY, CASTELLÓN, SPAIN †SOLUCIONES CUATROOCHENTA S.L., ESPAITEC2, CASTELLÓN, SPAIN ‡DEPARTMENT BE-OP, EUROPEAN ORGANIZATION FOR NUCLEAR RESEARCH, CERN CEDEX, FRANCE §ESTUDIOS GIS, C/ALBERT EINSTEIN, VITORIA-GASTEIZ, SPAIN

1 Introduction

The development of applications and systems relying on localization-based services has boosted over the last few years. These service providers are eager to exploit customer information with the widespread adoption of smartphones, the proliferation of affordable mobile devices, and the ubiquity of Internet connections (4G and Wi-Fi). Automatic user localization can be considered a hot research and business topic with an expected $3.96 billion market by 2019 (Markets&Markets, 2014). Indoor positioning and communications will be crucial elements in next-gen location-based services.

Applications based on users location need to know the position or localization for providing customized services, tracking assets and people, among others. Although outdoor localization is already solved due to the inclusion of GNSS support in localization devices, obtaining high precision in indoor positioning is still an unsolved problem for general-purpose applications. Two different approaches have been commonly applied to develop an indoor positioning system (IPS): infrastructure-based and infrastructure-less solutions. The former solutions need the deployment and installation of specific custom hardware in-site, while the later solutions only use a sensory system embedded in the location device to estimate the current location.

Geographical and Fingerprinting Data for Positioning and Navigation Systems. https://doi.org/10.1016/B978-0-12-813189-3.00003-4

For both infrastructure-based and infrastructure-less approaches, the fingerprinting technique has been widely used in indoor positioning. In fingerprinting, measurements (fingerprints) of a physical quantity are collected at known locations, in order to characterize the target environment. The environment characterization is then used in estimating the location where new measurements are taken. The measured quantities most often used in indoor positioning studies include wireless network signals strength (BLE (Čabarkapa et al., 2015), ZigBee (Marti et al., 2012), Wi-Fi (He and Chan, 2016)), which mostly use an infrastructure-based approach, and magnetic field directed strength, which uses a infrastructure-less approach. This chapter focuses on the development of IPS based on Wi-Fi fingerprinting to support heterogeneous mobile applications: indoor localization, pedestrian indoor navigation, and monitoring people at home.

Traditional Wi-Fi fingerprinting requires two steps or phases (Marques et al., 2012): the *training phase*, where the database with the ground truth or reference data are generated, and the *operational phase*, where the location algorithm estimates the location using previous knowledge. The training phase, or calibration phase, usually requires to measure the received signal strength indicator of the surrounding Wi-Fi access points at many (usually predefined) locations. After this phase, a reference database (or radio map) is available to estimate the positions of users at the operational phase. This simple approach is the base of many working IPS.

It is known that the process of site survey to generate the radio map is really time-consuming and labor intensive (Liu et al., 2014; Zhang et al., 2016a; Hossain and Soh, 2015). To overcome this issue, some approaches apply regression techniques to obtain a dense consistent radio map from less measurements (Hernández et al., 2017; Ezpeleta et al., 2015), whereas other authors artificially generate the radio map by using equations based on the radio signal (path loss) propagation (Chiou et al., 2010; Deasy and Scanlon, 2007) or ray tracing (Raspopoulos et al., 2012; Ayadi et al., 2015; El-Kafrawy et al., 2010). Although these advanced approaches to radio map generation provide interesting results, real measurements are still required in most of the works based on Wi-Fi fingerprinting.

Most of the scientific solutions found in the research literature are tested and evaluated on small- or medium-sized controlled environments (Zhang et al., 2016b; Li et al., 2016; Mizmizi and Reggiani, 2016), also known as laboratory environments, where the time required to generate the radio map might be affordable depending on the experimental setup. However, there is currently a raising interest in the development and deployment of realistic IPS in large realistic environments (Marques et al., 2012; Torres-Sospedra et al., 2014, 2015; Mathisen et al., 2016; Berkvens et al., 2016; Liu et al., 2016; Guimarães et al., 2016). In large scenarios, the calibration might be critical step since it requires the collection of many actual measurements. However, as far as we know, there is little information regarding how data are collected, the effort required to generate the radio map, and the problems that arise at this calibration stage in such large scenarios.

This chapter fills this gap and describes the lessons learned regarding collecting the necessary labeled measurements for real indoor positioning applications related to in-home monitoring and pedestrian navigation. Both contexts have been selected since they are uncontrolled environments (no control about AP placement and management) and they are representative enough of the current research and commercial trends. Moreover, the interest on these two main environments can be seen in the competitions held in related top conferences, where researchers have to deploy and evaluate their system on realistic large scenarios, including conference venues, university facilities, and living labs.

2 Lessons Learned at In-Home Scenarios

This section presents the application of indoor positioning to two study cases related to in-home monitoring. For both of them, details about each solution and experiences gathered with it are presented.

2.1 Calibration and Experimental Setup

The demand for higher comfort levels at home is increasing since some segments of the population, such as patients with mental disorders and senior citizens, spend most of their time at home. Remote monitoring, including in-home monitoring, and remote health care are valid alternatives for disease management in order to reduce frequent hospitalizations, including emergency visits, and to improve the patients quality of life. An IPS for remote monitoring operates over a very large environment, which is the aggregation of many "small" contexts (particular flats or houses) characterized by a specific user or device.

2.1.1 Smartphone-Based Patient Monitoring

An Android application was developed in 2014 for monitoring patients at their homes, with a preliminary feasibility study being conducted to determine whether nonobtrusive Wi-Fi fingerprinting was suitable for in-home monitoring in Spanish flats. The indoor localization service developed for this application was a infrastructure-less solution since we relied on the already existing Wi-Fi network topology, that is, we did not deploy any extra device to support positioning, and we did not know the layout of each home. The IPS only processed the data gathered by smartphones to estimate the users' position with room-level precision.

Eight different volunteers accepted to participate in the evaluation of target in-home monitoring system. We met them to provide the monitoring application and gave them some basic instructions about the four stages of our evaluation: installation, configuration, mapping, and operation. The volunteers were untrained people who had not used any Wi-Fi mapping application before. The only requirement to participate in the evaluation was to own an Android smartphone.

The feasibility study had eight very distinct in-home scenarios scattered in three cities (province of Castellón, Spain). The exact location and distribution of five of them were not provided to us due to privacy concerns. However, the detected APs indicated that they were not close to each other. Each AP was seen only at one of the eight scenarios. All volunteers installed the application when we met them and then they were distributed in four groups depending on the number of weeks (one or two) spent on the mapping and operation stages. Using such approach, we could get some additional feedback from the interactive stages.

In the configuration stage, the user introduced some important data about the flat: number of different rooms, room labels, room pictures, and rooms distribution, among others. To avoid mapping issues and improve the mapping application usability, supplying the room labels and taking pictures were compulsory. The configuration data and pictures remained in the volunteer phone and they were not provided to us. This stage was used to facilitate the mapping stage for the user and it took, an average, less than 6 min.

In the mapping stage, the user had to collect Wi-Fi fingerprints at each room within a period of 1 or 2 weeks (see Fig. 1). The application interface was as simple as possible, and they only had to select the room where they were using a combo box and click on a button in order to collect up to 10 consecutive Wi-Fi fingerprints (see Fig. 1). To avoid human errors while selecting the room, the picture attached to the room was shown before letting the user click on the *Collect* button. This approach considerably reduced the chances of selecting a wrong room label during the mapping procedure, thus avoiding entering errors into the IPS reference database. After this stage, a basic model based on 1-NN was created in each volunteer's phone and the reference database (Wi-Fi fingerprints and location labels) was sent to a centralized server. Only one volunteer quit the evaluation at this stage due to external factors.

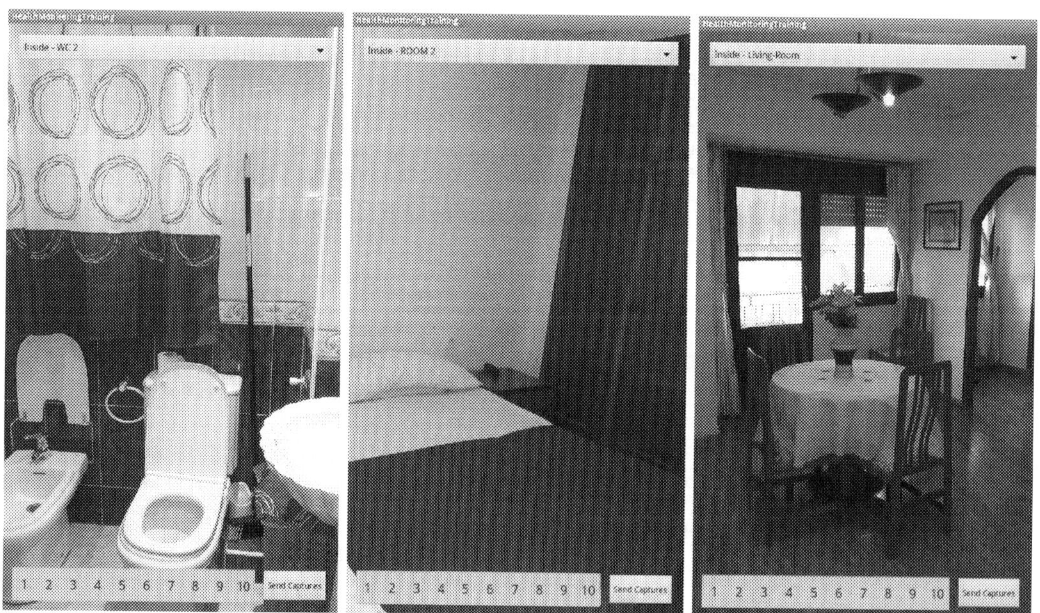

(A) (B) (C)

FIG. 1 Example of the Health Monitoring application at the mapping stage (smartphone) in the guest bath room (A), guest bed room (B) and living room (C).

Once the mapping stage finished and the in-home monitoring mode was activated on the phone, the system collected fingerprints every 10 min. At the operation stage, the fingerprints were processed on the phone to estimate the user's position. Once a day, the collected fingerprints and the predicted locations were sent to the centralized server. Moreover, we also allowed an active localization with feedback where the user could ask the system for the actual position and correct it in case of a wrong classification. These "active" fingerprints, the predicted location and the actual location, were also sent to the centralized server. This last stage also lasted 1 or 2 weeks.

After the 2–4 weeks evaluation, the users provided their comments about configuring and using this in-home monitoring application. We will focus only on the experiences they reported about mapping and collecting the active location fingerprints with feedback.

2.1.2 Smartwatch-Based Patient Monitoring

Similarly, we developed an Android Wear application for monitoring people at their homes and we also performed a preliminary feasibility study to determine whether nonobtrusive Wi-Fi fingerprinting with smartwatches was suitable for in-home monitoring in Spanish flats.

All scenarios used during the experimentation phase were urban flats, three of them located in the city of Castellón de la Plana and the other two in the city of Valencia. A total of five people participated in this study. The flats located in the same city are far enough to each other not to share any common AP.

In the calibration stage, the smartwatch asked the user to go to every place and, once there, to start the collecting fingerprints for that place (see Fig. 2) by just pressing a button on the screen and waiting during, more or less, 1 min.

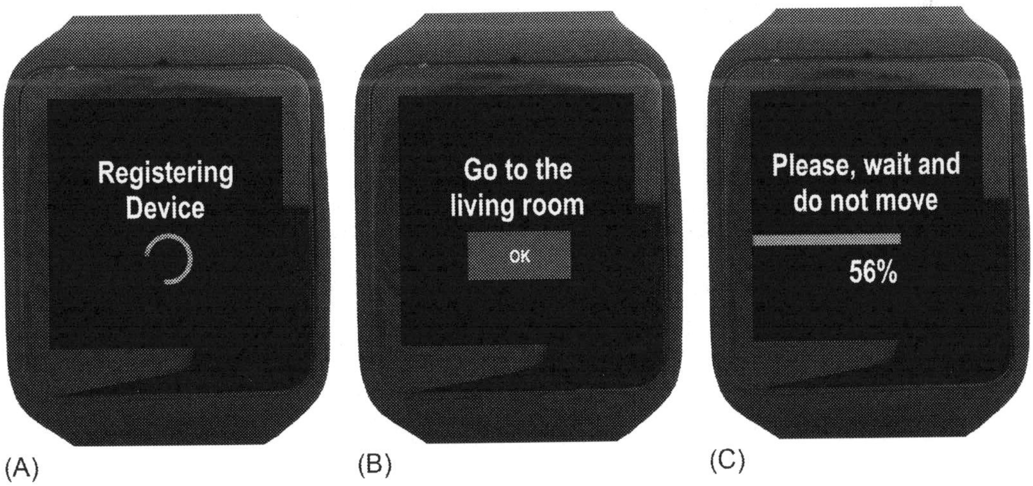

FIG. 2 Example of the Health Monitoring application at the mapping stage (smartwatch) when registering device (A), indicating a place to collect fingerprints (B) and collecting fingerprints at the selected place (C).

In the operational stage, a background process running in the wearable device collected one fingerprint per minute. Those fingerprints were stored in an internal database and sent to the server periodically. Moreover, two additional operations were allowed at this stage: (1) send new samples for recalibration and (2) force active localization to obtain the current position (feedback).

In contrast to the smartphone-based alternative presented before, the proposed monitoring system was less intrusive. After the 3 weeks evaluation, the users provided their comments about configuring and using this in-home monitoring application. As done before, we will focus only on the experiences they reported on mapping and collecting the active localization fingerprints with feedback.

2.2 Experiences and Lessons Learned

In-home data were collected to assess the feasibility of two different Wi-Fi-based monitoring systems. In most applications, like e-health and Aging-Assisted Living (AAL), only the location of the user at room level is needed.

2.2.1 Smartphone-Based Patient Monitoring

To perform the study with smartphones, the volunteers collected data in two stages. The data collected in the first and the second stages were used to train and test the solution, respectively. These data were collected in seven distinct scenarios (the volunteers' homes) and the stages' collection times were at least 1 week apart for all cases. Volunteers were asked to collect data at different hours throughout the day, to avoid that the usage patterns of neighbors' wireless networks to bias the measurements. This was not always possible due to work restrictions, so most data were taken at weekends (Table 1).

The first difference arose from the layout of the departments, whose area ranged from 76 to 114 m^2. Although all of them were one-floor flats, some of them had a square-like layout, and others had a rectangle-like layout.

The size and layout had a huge impact on the volunteers. In fact, the volunteer from scenario 1 reported that capturing one round of measurements in all the rooms was

Table 1 Main Characteristics of the Scenarios

Scenario	Area (m^2)	No. of APs	$\overline{APs}$	Time (s)	Training	Validation
1	114	127	15.78	12	5808	4313
2	76	103	14.90	5	1018	740
3	91	81	5.37	7.5	897	804
4	95	108	17.65	11	2019	1873
5	89	107	13.88	11	1900	1890
6	99	104	27.96	3.5	3627	2929
7	97	123	15.77	44	1960	1723
				Total	17,229	14,272

exhausting, whereas the user from scenario 3 collected multiple rounds of measurements without any complaint. Moreover, one complaint from volunteers was that the number of rounds of measurements was not defined. The previous facts explain the difference in captured fingerprints in the scenarios (more than 10,000 fingerprints in scenario 1 and $\approx$1800 in scenarios 2 and 3). According to the volunteers, they did not have any feedback about the quality of the collected fingerprints, so they could not determine when to finalize the training stage.

A reported issue related with the quality of the device used during the acquisition stage was the time each measure lasted. The time was 3.5 s on average in the best case for a high-end mobile phone, and 44 s for the worst device used in the acquisition stage. If the worst case is taken apart, the mean time for acquisition is about 10 s, which was a quite reasonable time for all users. The volunteers, specially those having a problematic device, suggested to add a timer to avoid being stuck at a position for a long time.

Finally, the battery drainage was a problem highlighted by a volunteer that took many measurements in weekdays after work. The battery status of the device was poor (a 2-year-old device) and a few mapping rounds could not be finished because the battery drained. The volunteer asked to implement a safe-energy mode or to avoid capturing fingerprints when the battery was below a safety threshold.

2.2.2 Smartwatch-Based Patient Monitoring

The experiments for the smartwatch-based monitoring were performed with the training and re-calibration modes (normal data collection and repeated collection, respectively, presented in Section 2.1.2) of the smartwatch application. Five volunteers participated in this study (see Table 2).

Four databases were created for each scenario. A total of 50 fingerprints per location (room) were considered in any of the 4 datasets, with each dataset differing from the others to cover different situations:

- Set 1: The user was standing up while collecting training fingerprints.
- Set 2: The user was moving around the room while collecting training fingerprints.
- Set 3: The user was standing up while collecting re-calibration fingerprints.
- Set 4: The user was moving around the room while collecting re-calibration fingerprints.

Table 2 Scenario characteristics

Scenario	Area (m^2)	No. of APs	Locations Mapped
1	120	33	Kitchen, office, living-room, bedroom
2	80	36	Kitchen, office, living-room, bathroom
3	90	27	Kitchen, office, living-room, bedroom
4	80	43	Kitchen, office, living-room, bedroom
5	62	23	Kitchen, office, living-room, bedroom

Sets 1 and 2 were collected during the first week of the experimental setup (calibration stage), whereas sets 3 and 4 were collected 2 weeks later (re-calibration at the operational stage).

A few users who participated in the experiments with smartphones also participated in the experiments with smartwatches, just to track the evolution of the procedure to collect fingerprints with a less invasive device.

First, the time required to capture the fingerprints at one location was not excessive according to the feedback provided by the volunteers than also participated in the previous experiment (with smartphones). Moreover, the fact of being guided was considered a better approach. However, some volunteers suggested to add a "skip location" button to avoid capturing fingerprint in a nonaccessible location (e.g., the bathroom when someone else is inside).

In general, the procedure to collect the data for this experiment was faster than for the previous one because: (1) people where guided about how, where, and how much measurements were to be captured; (2) the total number of fingerprints required was defined before starting the experiments (we avoid relying on the user to stop data capture) and (3) the smartwatch was an additional device provided for this experiment (the volunteers did not have to use their personal mobile device).

3 Lessons Learned at Very Large Scenarios

This section presents several efforts iteratively performed to apply an indoor positioning solution to the large scenario of a university campus. In a way similar to the previous section, details about each effort and experiences gathered with them are presented.

3.1 Calibration and Experimental Setup

There is a raising interest in providing applications and custom services inside shopping centers, airports and railway stations, and public institutions. Commercial IPS would be helpful for reaching a place of interest indoors, for example, a classroom in a university campus or the boarding gate in an airport. However, they can also be used for safe evacuation in case of an emergency disaster, or for monitoring and tracking patients in a hospital. These kinds of scenarios have common features: they are very large and heterogeneous scenarios.

We began to develop an Indoor and Outdoor Positioning System in 2013 (Torres-Sospedra et al., 2015). The main objectives were: improving mobility throughout the campus, obtaining high location accuracy indoors and outdoors, minimizing costs, being as less intrusive as possible, and developing as a smartphone-based application. Thus, we decided to develop a Wi-Fi fingerprinting technique for indoor positioning. We consider that our navigation application was a infrastructure-less solution since we did not install any Wi-Fi antenna for localization purposes and we only use the information gathered by smartphones to provide indoor location services.

Wi-Fi fingerprinting mapping at the university campus was done in six very differentiated stages, where different strategies and applications were carried out. In the first, second, and third stages, we collected the data that yielded the *UJIIndoorLoc* database as a result, including the public and private datasets. In the fourth stage, we mapped the building where the 17th AGILE Conference on Geographic Information Science was held. The fifth stage was done in parallel to an official Wi-Fi quality and coverage study at the university facilities. The last stage collected a year-long measurements at the university's library.

- The first stage consisted of collecting Wi-Fi fingerprints for the UJIIndoorLoc training/reference dataset. The fingerprints were taken at 933 reference points located in 3 different multifloor buildings (Universitat Jaume I). Although 18 people (mappers) participated in this process and installed the label-based mapping application on their phones, we divided the whole scenario into smaller areas. Each area was mapped by, at least, two different mappers with different Android devices (smartphone or tablet). The application interface was as simple as possible, and they only had to select their reference point using four combo boxes and click on a button in order to collect up to 10 consecutive Wi-Fi fingerprints (see Fig. 3A). Almost 20,000 valid fingerprints were collected.
- The second stage consisted of collecting Wi-Fi fingerprints for the UJIIndoorLoc public evaluation/testing dataset. The fingerprints were taken at any place inside the three previously mentioned buildings, that is, there was not any reference point. The scenario was divided into smaller areas as it was done for the first stage. The map-based mapping application collected one fingerprint and estimated the current position. Then, the user had to pin the actual current position on a map and click on a button to store the fingerprint, the estimated position, and the actual position (see Fig. 3B). This mapping procedure was simpler and more effective. More than 1000 valid fingerprints were collected at this stage by 12 people with different devices.
- The third stage consisted of collecting Wi-Fi fingerprints for the UJIIndoorLoc private evaluation/testing dataset. The fingerprints were also taken in any place inside the three previously mentioned buildings with another map-based application. Only six volunteers participated in this stage and installed a map-based mapping application on their phones. The interface was simpler and they only had to pin their real position and click on a button in order to collect up to 5 or 10 (depending on the user) consecutive Wi-Fi fingerprints (see Fig. 3C). More than 5100 fingerprints were collected at this stage.
- The fourth stage consisted of collecting samples in the Faculty of Law and Economics main building. We developed some applications for the 17th AGILE Conference on Geographic Information Science, which required indoor positioning techniques, so we mapped this building with precision in order to support the conference applications. At this stage, 18 people participated in collecting the Wi-Fi fingerprints. They installed another map-based mapping application on their phones, so they only had to pin

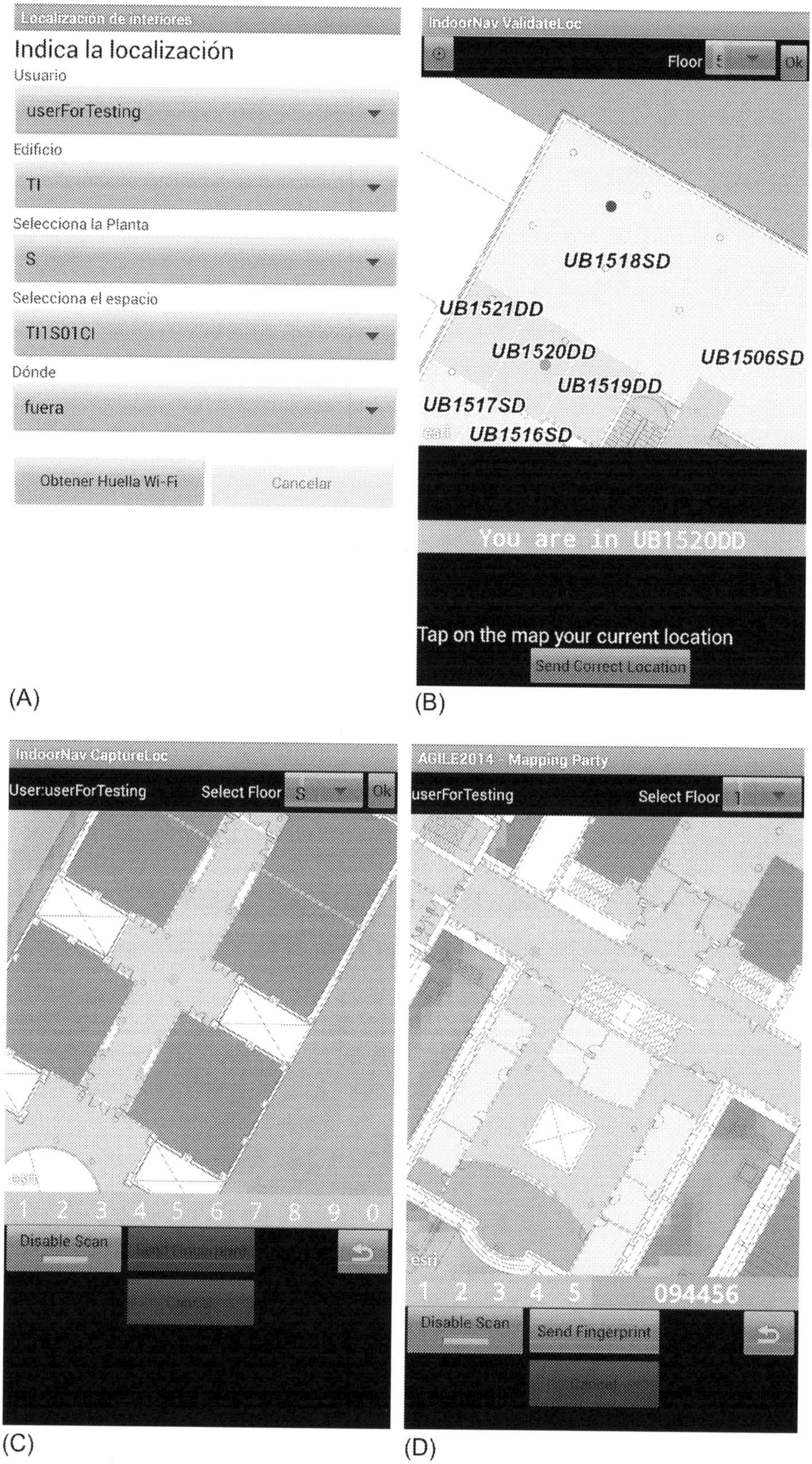

FIG. 3 Example of the UJI Monitoring applications for the first (A), second (B), third (C) and fourth (D) stages.

their real position and click on a button in order to collect up to five consecutive Wi-Fi fingerprints. In this application, we showed a timestamp code and we also introduced some warning messages to avoid errors (see Fig. 3D). In this case, we did not divide the whole scenario in smaller areas, but we divided it into routes. We calculated the most common routes among all the conference rooms and places and each route was mapped by, at least, two different people with different devices. Moreover, two users mapped the whole building (including stairs and the conference rooms) with higher precision. More than 12,000 fingerprints were collected by the 16 route-based volunteers and more than 9000 fingerprints were collected by 2 researchers with a background in Wi-Fi fingerprinting mapping.

- The fifth stage consisted in a full mapping of all the buildings located at the UJI's campus. We developed a Wi-Fi quality application that simultaneously run some Internet connectivity tests and recorded Wi-Fi fingerprints. This application was used once at every single facility (office, classroom, laboratory, among others) in the university to gather the information required to create Wi-Fi coverage and quality maps, and to collect up to 10 consecutive fingerprints (see Fig. 4). Only one

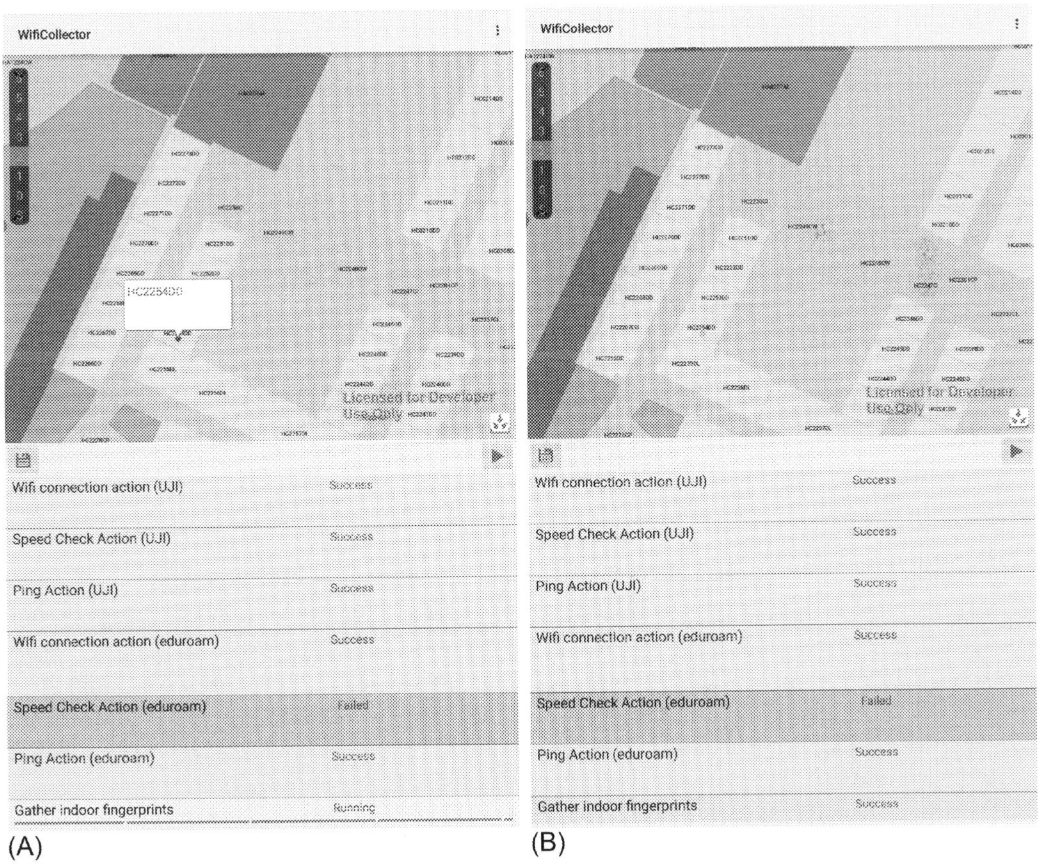

FIG. 4, CONT'D See legend on next page

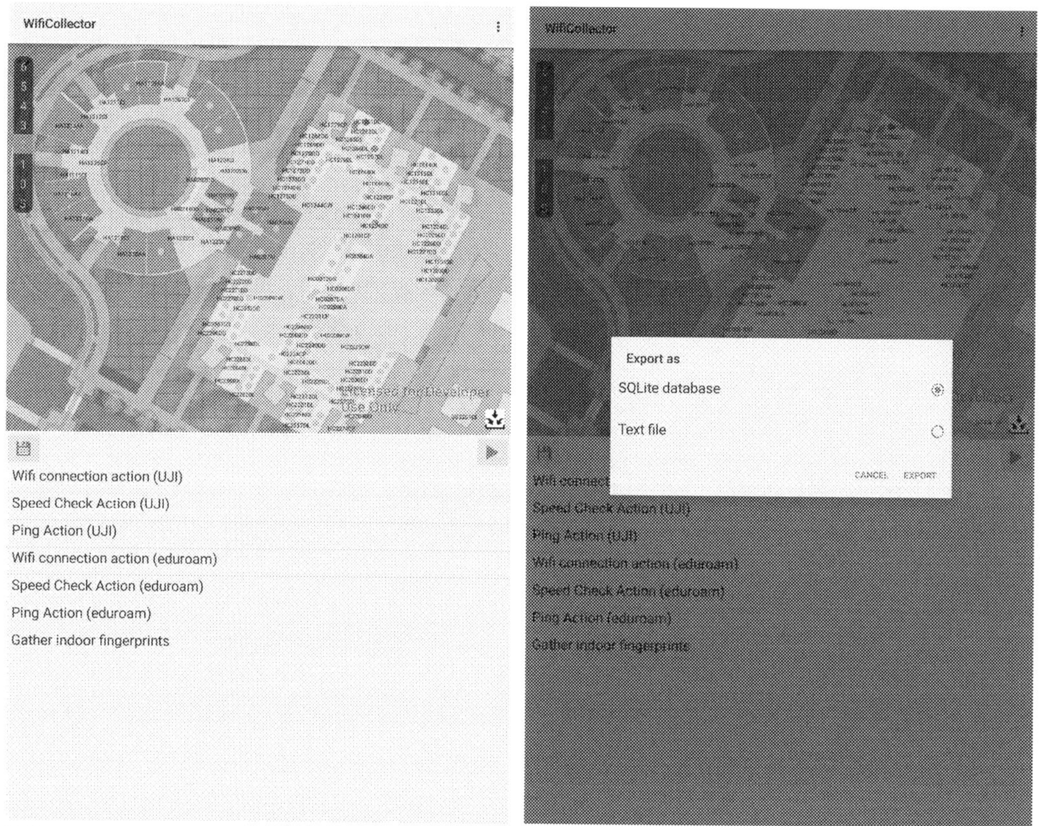

(C) (D)

FIG. 4 Example of the UJI Wi-Fi quality test application while gathering Wi-Fi fingerprints (A), after collecting all fingerprints (B), showing locations already mapped (C), exporting the database (D).

professional user mapper participated in this stage, using a Samsung Galaxy Tablet. More than 22,000 fingerprints were collected at this stage.

- The last stage created a long-term, dense mapping of mid-sized areas among the bookshelves of two floors at the UJI's library building. A trained professional performed the mapping using a Samsung Galaxy S3 smartphone and a new application. The application organized the signals measurement process around campaigns defined by an organizer and composed of ordered capture locations. Each capture location defined guidance aids for the mapper, that is, map's zoom and orientation, the current capture location, the direction that the mapper should face when collecting the batch of fingerprints, the locations already measured and those pending for collection (Fig. 5). The application let the mapper choose which available campaign to work with. The batch of fingerprints to be taken for each location contained six (the first one is later discarded) fingerprints. The mapping process did

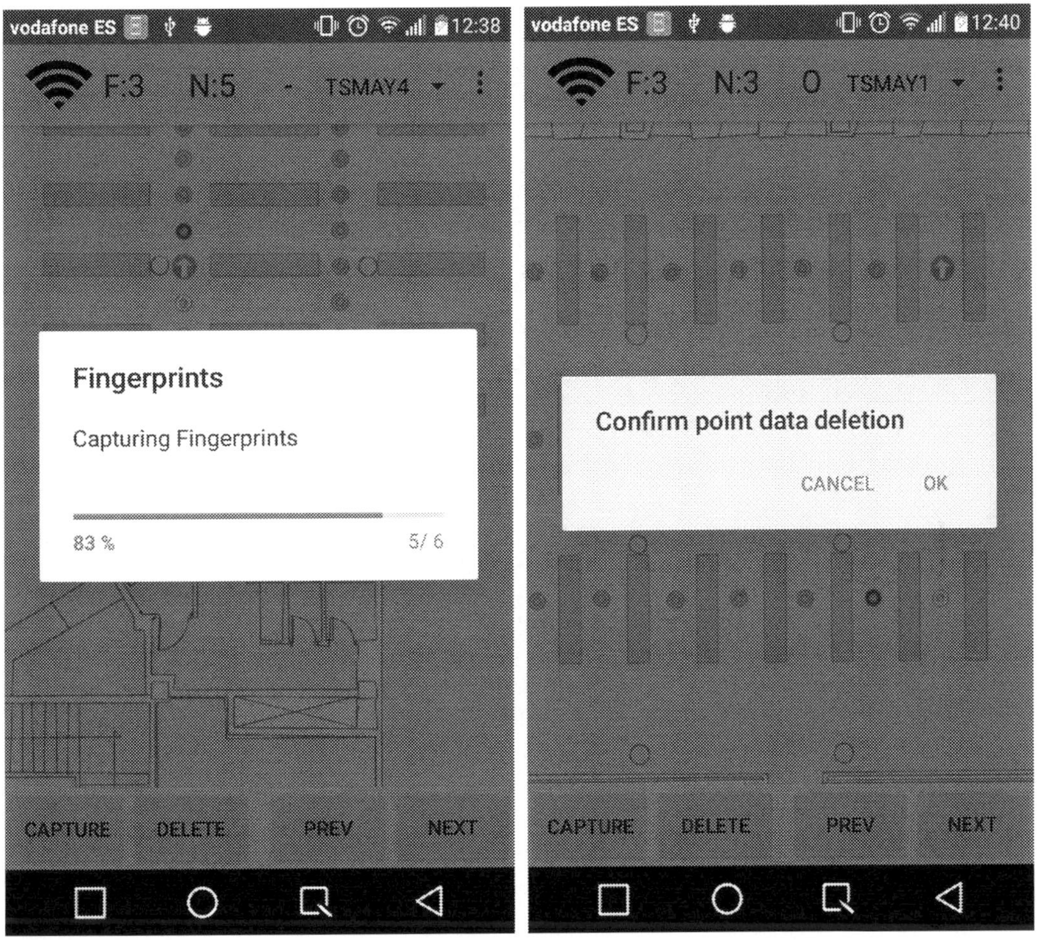

(A) (B)

FIG. 5 Example of the latest UJI Wi-Fi campaign-based application while capturing Wi-Fi fingerprints (A) and while deleting a database record (B).

not require an Internet connection. The number of fingerprints captured at this stage was 46,656.

After finishing each stage, the users reported the issues they had during the mapping procedure and they also reported their experiences about using the different mapping applications. We will focus on the experiences reported by a total of 28 people that participated in such mapping stages with different mapping applications. As far as we know, there are not any previous studies about mapping in a large scenario including such quantity of people, devices, and mapping strategies. Some of those 28 volunteers participated in more than one stage, and they were able to compare the procedures and applications used in one stage to the previous one, and provide their valuable feedback to us. The volunteers who had no previous experiences with Wi-Fi mapping also provided valuable information, as they were not biased by previous mapping approaches.

3.2 Experiences and Lessons Learned

Mapping the UJI's university campus was done in six well-differentiated stages. The first three stages corresponded to the generation of the public and private subsets of the *UJIIndoorLoc* database, and their fingerprints were collected in three different buildings at UJI's campus. The fourth stage was done to support the applications for a conference, whose venue was the main building of the Faculty of Law. The fifth stage was collected in parallel with a Wi-Fi quality evaluation in all the buildings of the UJI's Campus. The last stage collected measurements during a year at areas among bookshelves from the university's library.

Prior to the first stage, the *Indoor Positioning System* architect, the *Database* designer, and the *APP* developers tested the alpha and beta versions of the whole system. After some tests, they decided to release a label-based application to collect the fingerprints (see Fig. 3A). Due to the variability of Wi-Fi signals, 10 consecutive fingerprints were collected at each reference point.

3.2.1 First Stage

In the first stage, we provided a label-based application (Fig. 3A) to 18 volunteers in order to collect the training samples for the *UJIIndoorLoc* database. Moreover, we provided them with: (1) a list of the areas and reference points where they had to collect the Wi-Fi fingerprints and (2) a notebook to manually log this process and annotate the detected issues. In June 20, 2013, they collected almost 15,000 valid fingerprints, the 75% of the UJIIndoorLoc training samples. With this process, we collected most of the fingerprints with high diversity in approximately two-and-a-half hours.

After collecting all the training fingerprints, we revised all the collected fingerprints that the volunteers marked as issue. We manually fixed all the issues when possible, but we had to discard 750 fingerprints. The most common issues detected by the volunteers were:

- Wrong user. *The application crashed and I forgot to set my user. The fingerprints I collected with the default user are valid.*
- Wrong label about reference point. *I selected the XYZ label instead of YXZ at 12:34 a.m.*
- Wrong floor. *I selected the X floor instead of Y at 12:34 a.m.*
- Unknown reason. *The fingerprints I took from 12:34 a.m. to 12:45 a.m. are not valid.*

After the data correction, we matched the information provided by fingerprints with the manual logs provided by the volunteers. We detected that two volunteers did not report a few wrong label and wrong floor issues. So, we manually removed 125 more fingerprints from the database.

The volunteers reported that the office identifier was not appropriate to denote the reference points and, in some cases, it induced errors. So they suggested to use a map to improve the mapping task experiences.

At the end of this first stage, we realized that a label-based application was not appropriate for this kind of large-environment mapping because the volunteers felt uncomfortable during all the mapping procedure. Mapping 10 consecutive fingerprints at each reference point was very tedious according to the experiences reported by the volunteers. Despite the simplicity of the application interface, they had to stand about 1 or 2 min at each reference point in order to collect the fingerprints and log them on the notebook.

On the management side, a significant amount of time was spent inspecting fingerprints to assure the collection's quality. The fingerprints inspection included manually looking for inconsistencies in the temporal marks order, as well as considering the issues explicitly reported by the volunteers. We manually fixed about 21% of collected fingerprints, with 15% of them successfully edited and the remaining 6% removed. Afterward, a basic IPS was implemented using the 19,937 fingerprints collected at the first stage as reference data.

3.2.2 Second Stage

In the second stage, we developed a map-based mapping application (see Fig. 3B). This application collected one fingerprint, the IPS returned the estimated position and the user only had to select his/her current position if the estimation was not accurate. We introduced a map to avoid using labels in this large environment and it only collected 1 fingerprint (instead of 10) to perform a faster mapping. We provided this application to 12 volunteers, 2 of whom had not participated in the previous stage, to collect the fingerprints for the validation dataset (public test). Moreover, we provided them with: (1) a map with the areas (without explicit reference points) where they had to collect the Wi-Fi fingerprints and (2) a notebook to manually log this process and annotate the detected issues. On September 20, 2013, the volunteers collected almost 700 valid fingerprints. The 62.5% of the UJIIndoorLoc validation samples were taken in approximately 70 min.

The volunteers performed an initial mapping in which they simultaneously mapped the same area for a few minutes, and they were supported by the application developers. This initial mapping served as a training step on the application and the fingerprints were marked to be removed. According to the feedback provided by the volunteers, this initial fake mapping was useful to become familiar with the application.

After the initial mapping, the volunteers moved to the target collection areas. Although we improved the application interface, some issues were detected by users. We manually fixed all the issues when possible, but we had to discard 17 fingerprints. In this stage, the volunteers detected two issues:

- Wrong floor. *I did not realize that the system detected me in the wrong floor and I submitted my position with the wrong floor at 12:34.*
- Wrong location. *I wrongly established my position on the map at 12:34* or *I established my position on the map but I accidentally tapped another position when uploading the information at 12:34.*

Afterward, we matched the information provided by fingerprints with the manual logs provided by the volunteers. We only detected that two volunteers did not report one wrong location issue at this stage, so we removed the related fingerprints from the database. In fact, we detected more nonreported errors but they corresponded to the initial mapping, whose fingerprints had been previously marked as invalid and had not been considered for the final database.

At the end of the second stage we realized that using a map-based application is very useful for mapping since the visual information helps reducing errors. However, the negative point about this mapping is the density of fingerprints collected. We only collected an average of 0.90 fingerprints per minute and volunteer in this second stage, in contrast to the 5.55 fingerprints per minute and volunteer gathered at the first stage. Although the time required per capture has been the same in the two stages, 1 min approximately, the mapping procedure in the second stage has been more interactive and the volunteer felt more comfortable.

3.2.3 Third Stage

In the third stage, we combined some features from the first and second stages. This stage was done in two different periods of time, one in November 2013 and the other in March 2015. We improved the map-based application to collect 10 consecutive fingerprints for the first mapping period (see Fig. 3C), and we introduced two minor changes in the application for the second mapping period (see Fig. 3D). We provided the application to six volunteers, three of whom had not participated in the previous stages, to collect the fingerprints for the private test dataset. This dataset was used in an EvAAL-ETRI competition.

Each volunteer had to map, at least, one of the three buildings. Moreover, we only provided them with a notebook to manually log this process and annotate the detected issues. In November 2013 and March 2015, the volunteers collected (3779 + 1395) valid fingerprints for the private test dataset. Prior to each period, we trained the volunteers to use the application.

Although the mapping was successfully done without any important issues, the users suggested some minor changes to improve the mapping applications:

- The map size was not optimized for new versions of Android and it was too small under some configurations.
- Some devices, specially those supporting the 5 GHz Wi-Fi band, collected the fingerprints significantly slower than the others.

3.2.4 Fourth Stage

In the fourth stage, the scenario was the main building of the Faculty of Law and Economics, Universitat Jaume I. The main aim of this mapping stage was to collect the Wi-Fi fingerprints to support indoor positioning in the applications for the 17th AGILE Conference on Geographic Information Science. In this mapping stage, 2 experienced people performed a comprehensive mapping, whereas 16 volunteers focused in short common routes and in the main areas where the conference took place. The mapping was done during 4 different days in May 2014. The mapping application (see Fig. 3D) introduced minor changes with respect to the mapping application used in the third stage (first period), being the most remarkable ones: (1) the map fitted better on any display size, (2) the number of consecutive fingerprints was reduced to five, and (3) we introduced a capture identifier, so people could more easily log the fingerprints. These minor changes were also applied to the application used for the second mapping period of the third stage (March 2015). In essence, the mapping applications used for the third and fourth stages were very similar.

Although the route mapping was successfully done, the users who performed the comprehensive mapping reported that it was not possible to send the fingerprints to the server in a few areas out of 3G/4G and Wi-Fi stable coverage. Moreover, all volunteers had to use the 3G/4G Internet connection to send the fingerprints to the centralized server instead of the Wi-Fi. Due to the high density of Wi-Fi antennas and high presence of people in classrooms, some users reported issues with Wi-Fi connectivity in some areas. The problem was that the devices connected to the UJI's Wi-Fi networks tended to constantly disconnect from one AP and connect to another one with the strongest signal. This caused severe delays while collecting and sending the fingerprints to the server, so we decided to announce volunteers to quit using Wi-Fi and switch to 3G/4G connections.

The volunteers who performed the route-based mapping felt comfortable with the collection approach. The comprehensive mapping took approximately 7 h, with an average of 16 fingerprints per minute, for the first experienced mapper and about 9 h, with an average of 7.5 fingerprints per minute, for the second experienced mapper. In some places, the first experienced mapper collected more than 20 fingerprints per minute. Both experienced users reported that the comprehensive mapping was very demanding due to the size of the environment and the comprehensive mapping strategy.

Only one volunteer reported severe errors and his fingerprints were not considered in the fourth stage. He received a phone call and after 5 min he got lost in the building. He did not know how many fingerprints had collected and the location of the last one, and he finally decided to quit the mapping process. We realized that we should have included in the application information about the fingerprints that the user, and other users, had collected to avoid these kind of situations.

3.2.5 Fifth Stage

With the experience gained in the fourth stage, we completely rebuilt the mapping application in order to be a multipurpose application (see Fig. 4). Although it allowed Wi-Fi tests and gathering Wi-Fi fingerprints, it could easily be extended to perform any test and collect data from any sensor embedded in the device. We highly improved the user interface to make it adaptable to a high diversity of devices and screen sizes; we added user configuration parameters and new functionalities:

- Maps were embedded in the application and they were simplified to improve the user experience in mobile devices. We realized that a simple but effective interface reduces the issues in the mapping process and improves the quality and size of the reference databases (see Fig. 4A and B).
- The number of consecutive Wi-Fi fingerprints or a timeout could be set to avoid time issues in those devices whose Wi-Fi refresh rate is very low.
- We showed the information of the previous gathered data in the map as red and green bullets (see the gray bullets in the black and white screenshot showed in Fig. 4C). The user knew where mapping had been done, as well as the status, success (green/light-gray bullets) or failure (red/dark-gray bullets), of each individual capture.
- The mapping application stored all gathered data in an internal database: quality tests, Wi-Fi fingerprints, and the information provided by other sensors. The user decided when to send this information to a centralized server (see Fig. 4D). Therefore, we avoided those cases in which severe communication problems, mainly due to low or null coverage, did not allow to register the fingerprints in the centralized server.

In the fifth stage, all the buildings of the UJI's Campus were mapped by a single professional person, who was explicitly hired to perform this task. Due to the coverage problems detected in previous stages, the new application collected the Wi-Fi fingerprints offline and they were registered in the centralized server when all the Campus was already mapped. The application was set to capture between 5 to 10 fingerprints per observation. This fifth mapping lasted 24 working days, starting on April 20, 2015 and ending on May 20, 2015. A total of 2700 observations (Wi-Fi tests) were made and 25,000 fingerprints were collected at 2300 different locations during this stage. Performing the quality tests and gathering the Wi-Fi fingerprints took, approximately, 2 min per each location (office, classroom, laboratory, among other spaces). In particular, 4.5 fingerprints per minute were taken with this professional mapping strategy.

Although the mapping was successfully done without any important issues, the professional user suggested some minor changes to improve the mapping applications:

- More detailed maps were required. He collected the fingerprints approximately at the center of each classroom or office. However, it was not easy to select his position with high precision in the map since the maps did not included the furniture in classrooms. The professional user suggested the inclusion of an additional layer with furniture found in the classrooms and the library.
- Include predefined reference points instead of manually selecting them in the map, since most of the measurements were collected at the geometric center of the surveyed classrooms.
- Add the feature to remove a set of measurements. Although the systems warn about the user's position, he had to triple check his position before taking the measurements, since there was no option to remove the captured measurements.

Moreover, the professional mapper suggested us to include a functionality to add predefined reference points with detailed instructions in the application. According to his experience, nonexpert mappers (without knowledge of radio signal propagation) may suffer stress while taking the measurements because they do not know if they are in the appropriate place to collect the fingerprint. In such case, detailed instructions (reference points, orientation, among other features) must be provided to them. In other words, we might not rely the responsibility of selecting the reference points on nonexperts.

3.2.6 Sixth Stage

This stage demonstrated that incorporating the mapping instructions to the application makes the mapping process less cumbersome, more accurate, and faster. The users no longer have to carry hard copy instructions to follow a well-planned mapping campaign. They just need to focus on the map and instructions from the application. They may finish one campaign and then switch to another one, and upload the campaigns' data at the time of their choosing.

In the mapping process with a map-based application, mappers may make two types of errors: (1) place themselves at wrong locations, and (2) wrongly indicate their current location on the application's map. In this stage, the second type of error is no longer present because users are not required to indicate their current floor or tap their location on the application's map. The application's design reduces the likeliness of the first type of error by:

- map's zoom and orientation adjustments at each location;
- indications of the facing direction at each location;
- actions for traversing the list of locations and deleting the captures that the user may consider erroneous; and
- indication of the point that follows the current one in the capture process.

The collection in this stage was performed by a professional mapper. With the new approach, almost the mapping time employed by the mapper was devoted to fingerprint capture. For example, for a campaign totalizing 96 fingerprint batches, each batch composed of 6 fingerprints, the mean time to complete it was 50 min. As the mean time for capturing one batch was 30 s, a mean of only 2 min was used for displacements, orientation, and other tasks. Among the factors influencing the huge time utilization are follows:

- Capture locations corresponding to the same floor were close to each other.
- The user did not have to check hard copy instructions nor to indicate the current capture location or the floor to the application.
- An indication of the following location to capture was provided, which let mappers use the batch capture time in spatially locating themselves on the following destination and help them to stay focused on the task.
- The termination of the batch capture process was announced by sound and vibration actions in order to regain the mapper's attention.

The new mapping approach provided notable results in terms of error avoidance and time utilization. The application improvements were accompanied by the automation of other tasks performed by the campaign organizers, like a tool for campaign definition, captured data validation (amount, order, and direction of collected data points), and data transformation. The new approach has consolidated the idea of providing a platform for the smartphone-based citizen-science data collection.

In general, mapping in large scenarios has been a complex task. Several factors have to be properly balanced:

- size of the environment;
- number and kind of people involved in mapping;
- density of fingerprints;
- distance between reference points; and
- distribution of reference points.

Other large environments may add new challenges to those we have faced in our university, for example, difficulties for mapping at very crowded places (e.g., in shopping centers during sales season), and the need of performing the same mapping at several times of the day due to environment variations (e.g., check-in areas in airports may be very crowded at some times and completely empty at others, which influences the measured Wi-Fi signal intensities). In general, the mapping process requires previous careful considerations of the target environment.

4 General Experiences

This chapter has shown the lessons learned from mapping large environments for Wi-Fi fingerprinting. Despite two very different contexts have been considered (in-home monitoring and pedestrian navigation inside a campus), the experiences have been positive in both of them. On the one hand, the in-home monitoring is interesting since many homes have been considered, and thus it can be seen as a large area composed by many independent small areas. On the other hand, the campus' size resembles a small city or a neighborhood of a big city, which is of interest for real deployments of indoor positioning in smart cities. In conclusion, the experience gained from both environments was positive and allowed us to prepare increasingly better mapping campaigns.

For in-home monitoring with smartphones, using different devices means different Wi-Fi hardware that can bias the feedback from users. The time required to collect a single fingerprint varied depending on the device. Moreover, giving freedom to users was seen positive by some volunteers, which collected the fingerprints at their own will without any intrusion. However, a few volunteers stated that they could not realize when additional an fingerprint capture was not required since they did not have a metric to know the quality of collected data. Moreover, battery drainage was highlighted as the major concern of the users.

The second strategy guided mapping with smartwatches was better suited for volunteers since they were totally guided and the number of fingerprints required to have an effective system was much lower. Minor details were pointed out by the volunteers to be improved in successive versions of the smartwatch in-home monitoring application.

The mapping strategy for collecting the *UJIIndoorLoc* database (public and private datasets) was very demanding and positive. We collected a realistic database for large multibuilding multifloor environments, which has supported all our developments in Indoor Positioning and Indoor Navigation. Moreover, it has public access through the *UCI Machine Learning Repository (University of California, Irvine, USA)*. Furthermore, this database has been used to evaluate the IPSs that participated in the EvAAL-ETRI Competition, which was organized in conjunction with the 2015 Indoor Positioning and Indoor Navigation Conference (October 13–16, Banff, Canada) (Potortì et al., 2015). Competitors and attendees reported that meaningful comparisons are possible with this huge database. Feedback provided by external undergrad students, external researchers, and, even, companies suggest that this database, and others, may serve as de facto standard to fairly compare different IPSs.

The strategy with 2 comprehensive and 16 route-based mapping was also very positive. Although reviewing this mapping was less demanding because severe issues were not reported, two people had to map the environment for some hours. The database was used to support the official applications for the 17th AGILE conference on Geographic Information Science. The feedback provided by the attendees who used the applications was positive because the applications were useful to attendees and they supported wayfinding to the rooms and other points of interest where the conference events took place.

The professional mapping experiences have shown us that mapping a large scenario, such as university campus, and making a dense mapping at a mid-sized scenario over a long time, are very demanding tasks. A basic mapping covering all the campus required 1 month. The dense, long-term mapping required at least 6 h a month during 12 months. Mapping for just one purpose may not compensate the efforts if the cost of a professional mapping is considered. However, professional mapping also sheds light on the interest of integrating different tests and observations about the environment, and a reliable mapping can provide insights into (long-term) signals variability that would be otherwise untrustworthy. Developing a multipurpose application that collects information from diverse sensors and runs different test may be valuable. The campaign-driven application's value lies not only in improving the collection experience and reducing the errors, but also in its potentials for crowdsourced collection usage.

Assisting the mapper when collecting samples is also useful, since the mapper has only to go the places shown in the map. According to the data we have collected, having the responsibility of selecting the references points might require additional time and might produce stress in the people who collects the data.

Finally, we can state that the process of generating a reference database for Wi-Fi-based fingerprinting is very hard and demanding. Moreover, it requires constant refinement in order to avoid errors. Label-based mapping applications are appropriate for label-based positioning, such as in-home monitoring, when the number of labels is low. However, it introduces some errors during the ground truth generation if the environment is large and the number of labels is also large (a building with multiple offices or reference points). For big environments, such as a university campus, offline mapping applications are a suitable solution to generate the ground truth. Moreover, the users reported better feedback when an extensive mapping process was complemented with route-based mapping. In general, it seems that crowdsourcing will be well-established in the future, so future work will also be focused on these kinds of methodologies to generate and keep up-to-date the reference data for fingerprinting.

Acknowledgments

Parts of this work were funded in the frame of the Spanish Ministry of Economy and Competitiveness through projects TIN2015-70202-P and TEC2015-71426-REDT. Germán M. Mendoza-Silva gratefully acknowledges funding from grant PREDOC/2016/55 by Universitat Jaume I.

References

Ayadi, M., Torjemen, N., Tabbane, S., 2015. Two-dimensional deterministic propagation models approach and comparison with calibrated empirical models. IEEE Trans. Wirel. Commun. 14 (10), 5714–5722. ISSN 1536-1276. https://doi.org/10.1109/TWC.2015.2442572.

Berkvens, R., Weyn, M., Peremans, H., 2016. Position error and entropy of probabilistic Wi-Fi fingerprinting in the UJIIndoorLoc dataset. In: 2016 International Conference on Indoor Positioning and Indoor Navigation (IPIN), pp. 1–6.

Čabarkapa, D., Grujić, I., Pavlović, P., 2015. Comparative analysis of the Bluetooth low-energy indoor positioning systems. In: 2015 12th International Conference on Telecommunication in Modern Satellite, Cable and Broadcasting Services (TELSIKS), pp. 76–79.

Chiou, Y.S., Wang, C.L., Yeh, S.C., 2010. An adaptive location estimator using tracking algorithms for indoor WLANs. Wirel. Netw. 16 (7), 1987–2012. ISSN 1572-8196. https://doi.org/10.1007/s11276-010-0240-8.

Deasy, T.P., Scanlon, W.G., 2007. Simulation or measurement: the effect of radio map creation on indoor WLAN-based localisation accuracy. Wirel. Pers. Commun. 42 (4), 563–573. ISSN 1572-834X. https://doi.org/10.1007/s11277-006-9211-x.

El-Kafrawy, K., Youssef, M., El-Keyi, A., Naguib, A., 2010. Propagation modeling for accurate indoor WLAN RSS-based localization. In: 2010 IEEE 72nd Vehicular Technology Conference, Fall, pp. 1–5.

Ezpeleta, S., Claver, J.M., Pérez-Solano, J.J., Martí, J.V., 2015. RF-based location using interpolation functions to reduce fingerprint mapping. Sensors 15 (10), 27322–27340. ISSN 1424-8220. https://doi.org/10.3390/s151027322. Available from: http://www.mdpi.com/1424-8220/15/10/27322.

Guimarães, V.G., Castro, L., Carneiro, S., Monteiro, M., Rocha, T., Barandas, M., Machado, J., Vasconcelos, M., Gamboa, H., Elias, D., 2016. A motion tracking solution for indoor localization using smartphones. In: 2016 International Conference on Indoor Positioning and Indoor Navigation (IPIN), pp. 1–8.

He, S., Chan, S.H.G., 2016. Wi-Fi fingerprint-based indoor positioning: recent advances and comparisons. IEEE Commun. Surv. Tutorials 18 (1), 466–490. ISSN 1553-877X. https://doi.org/10.1109/COMST.2015.2464084.

Hernández, N., Ocaña, M., Alonso, J.M., Kim, E., 2017. Continuous space estimation: increasing WiFi-based indoor localization resolution without increasing the site-survey effort. Sensors 17 (1). ISSN 1424-8220. https://doi.org/10.3390/s17010147. Available from: http://www.mdpi.com/1424-8220/17/1/147.

Hossain, A.K.M.M., Soh, W.-S., 2015. A survey of calibration-free indoor positioning systems. Comput. Commun. 66, 1–13. ISSN 0140-3664. https://doi.org/10.1016/j.comcom.2015.03.001. Available from: http://www.sciencedirect.com/science/article/pii/S0140366415001115.

Li, W., Wei, D., Yuan, H., Ouyang, G., 2016. A novel method of WiFi fingerprint positioning using spatial multi-points matching. In: 2016 International Conference on Indoor Positioning and Indoor Navigation (IPIN), pp. 1–8.

Liu, W., Chen, Y., Xiong, Y., Sun, L., Zhu, H., 2014. Optimization of sampling cell size for fingerprint positioning. Int. J. Distrib. Sens. Netw. 10 (9), 273801. https://doi.org/10.1155/2014/273801.

Liu, W., Fu, X., Deng, Z., Xu, L., Jiao, J., 2016. Smallest enclosing circle-based fingerprint clustering and modified-WKNN matching algorithm for indoor positioning. In: 2016 International Conference on Indoor Positioning and Indoor Navigation (IPIN), pp. 1–6.

Markets&Markets, 2014. Indoor location market by positioning systems, maps and navigation, location based analytics, location based services, monitoring and emergency services. worldwide market forecasts and analysis (2014–2019). Available from: http://www.researchandmarkets.com/reports/2570920.

Marques, N., Meneses, F., Moreira, A., 2012. Combining similarity functions and majority rules for multi-building, multi-floor, WiFi positioning. In: 2012 International Conference on Indoor Positioning and Indoor Navigation (IPIN), pp. 1–9.

Marti, J.V., Sales, J., Marin, R., Jimenez-Ruiz, E., 2012. Localization of mobile sensors and actuators for intervention in low-visibility conditions: the zigbee fingerprinting approach. Int. J. Distrib. Sens. Netw. 8 (8), 951213. https://doi.org/10.1155/2012/951213.

Mathisen, A., Sørensen, S.K., Stisen, A., Blunck, H., Grønbæk, K., 2016. A comparative analysis of indoor WiFi positioning at a large building complex. In: 2016 International Conference on Indoor Positioning and Indoor Navigation (IPIN), pp. 1–8.

Mizmizi, M., Reggiani, L., 2016. Design of RSSI based fingerprinting with reduced quantization measures. In: 2016 International Conference on Indoor Positioning and Indoor Navigation (IPIN), pp. 1–6.

Potortì, F., Barsocchi, P., Girolami, M., Torres-Sospedra, J., Montoliu, R., 2015. Evaluating indoor localization solutions in large environments through competitive benchmarking: the EvAAL-ETRI competition. In: Proceedings of the Sixth Conference on Indoor Positioning and Indoor Navigation.

Raspopoulos, M., Laoudias, C., Kanaris, L., Kokkinis, A., Panayiotou, C.G., Stavrou, S., 2012. 3D ray tracing for device-independent fingerprint-based positioning in WLANs. In: 2012 9th Workshop on Positioning, Navigation and Communication, pp. 109–113.

Torres-Sospedra, J., Montoliu, R., Martínez-Usó, A., Avariento, J.P., Arnau, T.J., Benedito-Bordonau, M., Huerta, J., 2014. UJIIndoorLoc: a new multi-building and multi-floor database for WLAN fingerprint-based indoor localization problems. In: Proceedings of the Fifth Conference on Indoor Positioning and Indoor Navigation, pp. 261–270.

Torres-Sospedra, J., Avariento, J., Rambla, D., Montoliu, R., Casteleyn, S., Benedito-Bordonau, M., Gould, M., Huerta, J., 2015. Enhancing integrated indoor/outdoor mobility in a smart campus. Int. J. Geogr. Inf. Sci. 29 (11), 1955–1968. https://doi.org/10.1080/13658816.2015.1049541.

Zhang, M., Pei, L., Deng, X., 2016a. GraphSLAM-based crowdsourcing framework for indoor Wi-Fi fingerprinting. In: 2016 Fourth International Conference on Ubiquitous Positioning, Indoor Navigation and Location Based Services (UPINLBS), pp. 61–67.

Zhang, W., Hua, X., Yu, K., Qiu, W., Zhang, S., 2016b. Domain clustering based WiFi indoor positioning algorithm. In: 2016 International Conference on Indoor Positioning and Indoor Navigation (IPIN), pp. 1–5.

4

Radio Maps for Fingerprinting in Indoor Positioning

Filipe Meneses, Adriano Moreira, António Costa, Maria João Nicolau
ALGORITMI RESEARCH CENTER, UNIVERSITY OF MINHO, GUIMARÃES, PORTUGAL

1 Introduction

Knowledge of own absolute position, or relative position regarding other known humans, objects, or places, has always been necessary for human activities ever since. With today's technology, that can be achieved continuously, almost in real time, with the help of positioning devices increasingly accurate and increasingly cheaper. Positioning systems are available on vehicles, mobile phones, personal computers, and many other smart devices that can be carried around, giving support for new applications or new features in existing ones. Localization requires the existence of a map and a coordinate system. Navigation requires the knowledges of possible traveling paths between two or more points in the map (Zhang et al., 2017). But *mapping* and *navigation* are just two examples of a new set of a variety of location-based applications (LBA) for humans nowadays. Other important examples include emergency assistance, first responders aid, social interactions, event dissemination, localized marketing, nearby places, and augmented reality.

In 3GPP (2008), the 3GPP forum presents a clear definition of LBA and also a taxonomy for them. According to 3GPP (2008) an LBA is an application software processing location information or utilizing it in some way. The location information can be provided by a user, detected by the user equipment (UE) or by the network. Mapping and navigation are location application examples. 3GPP also identifies and standardizes the type of services that the user, or in more clear terms the Location Client, may need, in the perspective of a global network operator. They can be categorized in four distinct categories. First one is called *Commercial Services* or *added value services*. With the provided terminal location, many useful information can be provided or acquired by the user for that location. This includes nearby places and events, marketing of products and services, etc. The second category is *Internal Services*, which are important to the network operation, like assisted handovers. The third category is *Emergency Services*, which can help emergency providers locate the terminal on an emergency call. This type of service is vital and may be

Geographical and Fingerprinting Data for Positioning and Navigation Systems. https://doi.org/10.1016/B978-0-12-813189-3.00004-6

mandatory. Finally, *the lawful intercept* for legally required services. The last two categories are considered mandatory in many countries.

Outdoors, location information can be obtained by Global Navigation Satellite System (GNSS) available worldwide (Kahvecı, 2017). GNSSs are based on two components: the satellites on space and the terminal receivers on earth. The user terminal computes its own location based on received time differences between signals from different visible satellites. This type of outdoor localization system became popular in the 1970s, with the advent of the US Global Positioning System (GPS). GPS was first developed to support military operations, but soon it was clear its enormous potential for civil usage. Besides the US GPS system, GNSS includes the Russian GLONASS and the European Union GALILEO systems.

Satellite-based positioning systems are, however, usually not available for indoor positioning systems, and also in some dense urban scenarios (Locubiche-Serra et al., 2016), due to high attenuation of signals and nonline-of-sight (NLOS) propagation. In those scenarios, carrier-to-noise ratio (C/N0) is typically below 20 dB Hz (Locubiche-Serra et al., 2016). GNSS position systems need therefore to be complemented, even outdoors, and various approaches have been proposed to solve those problems. One possible way is to use global wireless networks, available in those scenarios as a network supported positioning system. Cellular networks like 4G and 5G/LTE can provide simple cell coverage positioning methods. But also more complex methods based on radio signal measurements that can be processed to estimate the location. According to Peral-Rosado et al. (2012), new long-term evolution (LTE) specification (3GPP, 2008) provides network-based mechanisms to compute terminal location. The method can only be used by the network, and not by user terminals, because it uses the difference in the arrival times of downlink radio signals from multiple base stations to compute the user position, and so it may be provided as service by network providers.

Similar methodologies can also be explored in more universal WLAN networks like IEE802.11x Wi-Fi, due to their popularity, low cost, and global availability in mobile phones and smart devices. Radio signal methodologies for localization are surveyed in Liu et al. (2007) and Yassin et al. (2017). Not all of them can be applied with good accuracy to Wi-Fi signals, but some of them provide very good results, specially the ones based on the *scene analysis* methodology. In Yassin et al. (2017) the authors identify measuring principles and positioning algorithms, organizing them by techniques into three major categories: *proximity, triangulation,* and *scene analysis.* Proximity uses relative position to a nearby reference point (RP) like an Access Point to provide a less accurate and descriptive position information. Triangulation uses geometric properties of triangles to compute location. It has two derivations: *lateration* and *angulation.* Lateration-based techniques include (i) Time of Arrival (TOA) where distance is assumed to be proportional to the measured time; (ii) Time Difference of Arrival (TDOA) where a relative measure of time between multiple signal arrivals is used instead of the absolute time of arrival; (iii) Receiver Signal Strength (RSS) attenuation, which assumes a theoretical model for signal strength attenuation with distance, that can be used to compute distance from signal sources; (iv) Round-trip time of flight (RTOF), that uses time measured from the transmitter to the receiver and back again; and (v) Phase of Arrival (POA), that uses received signal phase or carrier phase to estimate the range. Regarding *angulation* techniques, receiver use at least two RPs and two measured angles of arrival to compute position.

A complete different type of techniques is based on scene analysis (Liu et al., 2007). This type of algorithms first collects features (called fingerprints) of signals on a scenario at a set of RPs, and then later uses online measurements to find the *closest* offline collected fingerprints. This technique is called Location Fingerprint and is based on the assumption that the features observed and measured are location dependent. Localization is done in two phases. In the first phase, called the offline phase, a *FingerPrint Map* or radio map has to be constructed for the scenario. On the second phase, called online phase, the user device uses the same features (measured signal properties) as an input to a localization algorithm that estimates the location. RADAR (Bahl and Padmanabhan, 2000) is considered as the pioneer work on this type of location systems using Wi-Fi networks. Since then, extensive research has been done on Wi-Fi Fingerprinting-based systems, published in specific publications and special conferences like (IPIN, 2018).[1]

Fingerprint Maps, or radio maps, are a crucial component of those systems and the main focus of this chapter. They can be constructed for Wi-Fi networks, but also for many other radio frequency (RF) alternative technologies, as described in next sections.

2 Radio Maps for Different Radio Technologies

A *radio map* is defined by Kjaergaard (2007) as a model of network characteristics in a deployment area. It is used to estimate a position, using a *localization algorithm*. The localization algorithm, sometimes also called *estimation method*, uses the information collected and stored in the *radio map* in a deterministic or probabilistic way to predict positions. Fig. 1 (from He and Chan, 2016) shows the relation between those components. In order to construct the radio map, a selected number of points, designated as *reference points*, must be identified in the scenario. Each RP must be sampled to construct the

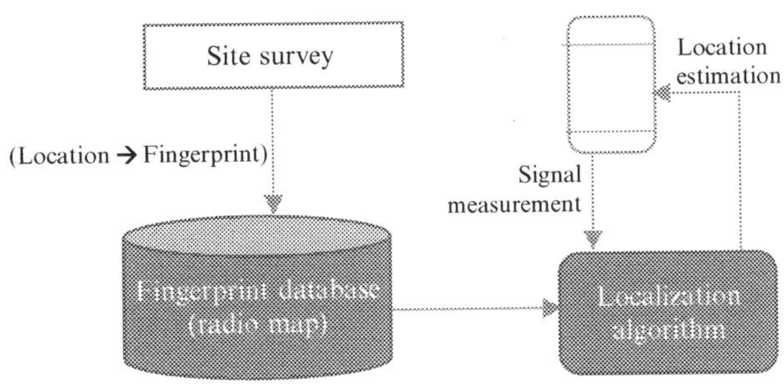

FIG. 1 Radio map and Localization Algorithm relationship. *(From He, S., Chan, S.H.G., 2016. Wi-Fi fingerprint-based indoor positioning: recent advances and comparisons. IEEE Commun. Surv. Tutorials 18 (1), 466–490.)*

[1] See http://ipin2018.ifsttar.fr/

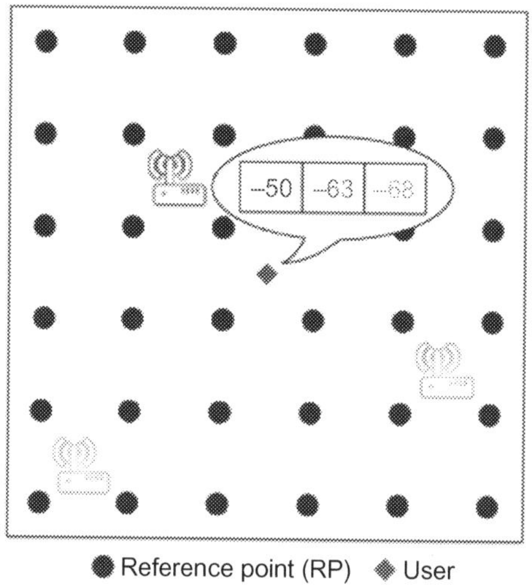

FIG. 2 Scenario area with sampled referenced points (He and Chan, 2016).

location fingerprint, as shown in Fig. 2. The *radio map* can now be defined as a *(Key, Value)* relationship between each *reference point* and its *location fingerprint*.

In RADAR (Bahl and Padmanabhan, 2000), several important conclusions regarding Wi-Fi radio maps were presented. Authors used WaveLAN NIC interfaces to collect both *signal strength* (SS) and *signal-to-noise ratio* (SNR), both they soon concluded that SS feature was a stronger location function than SNR. SS is measured in units of dBm while SNR is expressed in dB. A signal strength of s watts is equivalent to $10 * \log_{10}(s/0.001)$ dBm. A signal strength of s watts and a noise power on n watts give an SNR of $10 * \log_{10}(s/n)$ dB. They also concluded that collected values varied in at least 5 dBm according to the user orientation so, at each (x, y) position in the map they also collected samples in four distinct directions d (*north*, *south*, *east*, and *west*). The authors collected multiple samples at each *reference point* and each direction, and then they merged them into a single record for each location, computing statistics like mean and standard deviation. The first Wi-Fi fingerprint map (Bahl and Padmanabhan, 2000) was therefore a set of (x, y, d, ss_i) records, where (x, y, d) is the key and (ss_i) the value associated, for each Access Point i found in the scenario. Besides user orientation, RADAR also studied other important factors in radio map construction, like (i) the number of RPs in the scenario; (ii) the number of collected samples per RP3; and (iii) different localization algorithms. They concluded that only a small amount of samples are needed per RP and few points are needed if physical locations are uniformly distributed by the floor (the results were not much different with 40 or 70 points in a 22.5×43.5 m area).

In Kjaergaard (2007) an important taxonomy for radio location fingerprinting is proposed, based on a survey of 51 published papers and 30 different Wi-Fi fingerprinting systems. This taxonomy is focused mainly on fingerprinting and not on general systems.

The paper starts by identifying the taxonomic units that are relevant to the problem of localization, pointing out that only four of them really distinguish RF fingerprint systems: (i) the *scale*, regarding the size of the deployment area for the system (a city, a campus, a building, a floor); (ii) the *output*, or the type of information that the system returns back to the user (description of the place in words or coordinates in a map); (iii) the *roles* in the system, if it is infrastructure based or not, based on the user terminal equipment with or without network support; and (iv) the *measurements*, referring to the types of measured network characteristics.

Regarding measurements, for all surveyed works, only a few signal properties were identified, besides the mandatory *Base Station Identifier*. The first two are the same already used by RADAR (Bahl and Padmanabhan, 2000), currently used by other systems: the *Receiver Signal Strength* (measured in dBm) and the *Signal-to-Noise Ratio* (measured in dB). Some works used combinations with other metrics like *Link Quality Indicator* (collected by radios to measure link quality), *Power Level* (of sender), and *Response Rate* (the frequency of received measurements).

But perhaps the best contribution of Kjaergaard (2007) regarding radio maps is the classification system used. Authors first distinguish between *empirical* and *model-based* maps. Empirical maps are obtained from measurements only, while model-based use signal propagation models (direct path or ray tracing) to help in radio map construction. Regarding representation, radio maps can be represented *empirically* or *probabilistically*, aligned with the estimation method (localization algorithm) to use with the map. An empirical representation keeps a single value for each RP, like RADAR, while a probabilistic representation uses probabilistic distributions for each point. In both cases, *outliers* can be previously removed, and values *aggregated* or *interpolated*. Interpolation is used to augment the map with extra RPs using some interpolation function. Aggregation can be done either by using a simple *mean* function or by using a Gaussian distribution fit function.

2.1 Deterministic Radio Maps

In Kaemarungsi and Krishnamurthy (2012), authors analyze in detail the receiver signal strength indication (RSSI) that is used to build the great majority of deterministic indoor location radio maps. The goal is to do extensive data analysis around the presumably unique relationship between an RSSI value of a WLAN and an indoor location. Since RSSI values can be viewed as sensor data that refer to indoor positions, the characteristics of RSSI should be carefully studied. There is extensive knowledge of signals but in a communication's perspective, not specifically for fingerprinting. Authors enumerate a set of factors that can influence the statistics of an RSSI fingerprint and focus on five of them: (i) make of the hardware card (different well-known card makers were considered); (ii) time of measure (time of day and day of week); (iii) period of measurement (second minute and hour); (iv) interference (cochannel and adjacent radio channel); and (v) building environment (corridor, small office, large hall). These are basically two types of factors: hardware and environment.

Regarding different cards, there is a clear conclusion that the differences in hardware cannot be ignored. The mapping between the actual RF energy value and the RSSI value varies from vendor to vendor. Although, while IEEE recommends a range of 0–255 for RSSI, many vendors assume a range between 0 and a Max_RSSI. Each vendor has its own range and accuracy. Regarding environmental factors, the authors show that a mean value stays stationary for long periods of time and usually only changes when there are changes in furniture or in human presence and movement. But the standard deviation changes with long periods of time, like different hours of the day. Due to this time dependency, the fingerprint collection should be done at different periods of a day. Other conclusion is that the interference by cochannel signals does not have any strong correlation. Signals from different APs may be considered independent. The most influential effect on the performance of location fingerprinting is the standard deviation of RSSI. The second most influential effect is time dependency, and the third is the quality of WLAN card used. The least influential effect is the cochannel interference or the correlation among signals from multiple Access Points (AP) using the same frequency. The study reported in Kaemarungsi and Krishnamurthy (2012) also confirms that the mean value used by RADAR (Bahl and Padmanabhan, 2000) can represent a fingerprint; however, the distance between two locations does not translate well to distance between two fingerprints. Therefore, simple Euclidean distance will perform worse than probabilistic methods or Mahalanobis distance. These conclusions are useful to determine the procedure for map construction. And according to the taxonomy presented in Kjaergaard (2007), radio map construction should address *spatial variations, time variations, equipment,* and *collecting agent.*

2.2 Bluetooth Low Energy Radio Maps

In Rahman (2017), a system for locating shoppers in a large wholesale shopping store indoor environment, using Bluetooth Low Energy (BLE) beacons, is presented. The system uses RSSI readings from multiple beacons, measured asynchronously using commercial devices. There are not much studies addressing BLE fingerprinting, compared with Wi-Fi fingerprinting, partially because it is a recent technology not yet widely available, that requires extra cost and complexity in deployment. Authors in Rahman (2017) used 136 beacons on a 6000 m^2 area. The Android devices used, record Beacon Identifier, RSSI, and timing data. Data were collected by walking continuously at a steady pace along pathways. Authors presented a mapping scheme that represents the environment as a graph traversed by the user. This is used to constrain the search space of the localization algorithm. Three localization methods were used: nearest beacon, averaged beacon-pair range, and particle filter-based tracking. Particle filter methods outperformed the others significantly.

2.3 Other Technologies: FM and AM Radio Maps

In Chen et al. (2013), authors propose a robust indoor fingerprinting system using frequency modulation (FM) broadcast radio signals. With an experimental setup implemented in different buildings in United States, they demonstrate that FM broadcast signal strength (RSS) can be used to achieve room-level indoor localization with similar or better accuracy than Wi-Fi fingerprinting. FM broadcast radio signals have lower frequency and are less susceptible to multipath, fading, and human presence. They also penetrate very well in buildings. Authors show that accuracy can be further improved with more than five signals available. And since errors are not correlated with Wi-Fi obtained errors, both techniques can be combined to improve results. FM signals range from 88 to 108 MHz and are stronger than Wi-Fi. They cover areas of hundreds of kilometers and FM towers are also separated by hundreds of kilometers. This means that the variation of RSS experienced in the received signals is not significant in nearby locations. To overcome this difficulty, the RSS values are augmented with extra information collected from physical layer, like multipath indications. The extra values allow to distinguish between similar values of RSS. As in other fingerprinting systems, the system uses an offline and an online phases. In the offline phase, the system recorded FM signal strengths from 32 distinct radio stations, in parallel with normal Wi-Fi fingerprints. The radio map is composed of records with the following structure (r_i, s_i, m_i, f_i) for each RP and radio station i, where r_i is the RSS measured value, s_i is the SNR value, m_i is the multipath physical indication, and f_i is the frequency. As the authors point out in the paper, the most challenging aspect in this approach is how to engineer the setup, due to the required extra hardware.

In Rahman (2017), authors designed a system to use Amplitude Modulation (AM) radio broadcast signals. They argue that AM signals provide extensive coverage in urban environments and that receivers consume minimum power. A radio map was constructed using only RSSI values captured using a real-time spectrum analyzer for a total of eight AM broadcasting channels. Authors conclude, based on results obtained in one experimental setup, that accuracy results are very similar to the ones obtained using FM fingerprinting.

3 Building and Updating Radio Maps

Generically, two wireless devices can communicate if they are within the coverage area. That is, the devices can communicate if the radio signal propagates and the receiver is within the propagation area. A coverage map indicates the service area of a certain emitter. The coverage area depends on a number of factors including the transmission power, RF, orography, buildings, and other constructions and the sensitivity of the receiver. For long-range systems, the coverage area is frequently defined based on computerized models that estimate the signal propagation.

Many indoor positioning systems are built based on the principle that each location has a different radio signature, that is, the radio waves/signals that can be detected in a location are different from another location and the difference is enough to differentiate locations that are nearby. Radio waves propagate from one point to another being affected by several phenomena, including attenuation, reflection, absorption, etc.

A Wi-Fi radio map includes, for each location, the MAC address of the observed AP and the corresponding signal level. Optionally it may include information like the Service Set IDentifier (SSID) and the frequency channel.

3.1 Build a Radio Map

Building a radio map, a task often referred to as calibration or survey, is a time-consuming and labor intensive task. Several dimensions influence the time necessary to build a radio map, including the overall size of the building, the building structure (number and size of the interior divisions, like rooms, corridors, etc., and kind of materials used in the construction), the density of the RPs, and the number of samples collected in each RP.

To build a radio map means to use a specific application or tool and move along the building to collect radio data in different places. In a building with large rooms, the locations for the radio survey are often defined based mostly on the area, having the number of scanning locations defined by the density of the map to be built. Often the scanning locations are defined over a grid, collecting data in equidistant locations. In buildings with small rooms, the locations can be also defined considering a grid of points but adjusting it in a way that ensures that at least one RP is surveyed.

The propagation of a wireless signal is influenced by several factors including the building wall materials. A wall made of glass has low influence on a wireless signal while one built with large stones may completely block a signal. Furniture and other appliances installed nearby or between the sender and the receiver may also influence the signal level. Such scenarios may lead to the need for a more dense radio map, collecting data in more locations.

To build a radio map for very large building such as hospitals, universities, shopping centers, airports can be a challenging task since the number of scanning locations for an average dense radio map can easily grow to hundreds of locations. An additional challenge is to link the radio data to a place inside a building since there is no standard to represent the space or to represent the indoor maps.

3.2 Crowdsourcing

Crowdsourcing is an interesting solution to build Wi-Fi radio maps for large buildings. Instead of having a person or team collecting data all around the building, the idea is enroll the final users in this process and have them collaboratively contributing to the creation of a radio map. This way, big building and facilities can be surveyed more quickly, since the end users would help in the process of creating a radio map, avoiding the time-consuming task of doing it room by room (often several times in the same room if it is a larger one). Another advantage is that rooms with more people will get more scanning and thus the radio map coverage will increase faster in the more crowded areas, bootstrapping the quality of the maps for largely crowded areas.

To build a radio map it is necessary to link the collected radio data to a specific location inside a building. Unfortunately, there is no universal standard for interior maps.

CIMLoc (Zhang et al., 2014) and GraphSLAM (Zhang et al., 2016) are two crowdsourced solutions: CIMLoc is an example of a crowdsource solution to build indoor maps and GraphSLAM combine inertial sensor-based user motion measurement and sensed Wi-Fi signals to build a crowdsourced Wi-Fi fingerprint radio map.

Some authors propose to use techniques based on Simultaneous Location and Mapping (SLAM) and based on Pedestrian Dead Reckoning (PDR) to continuously update the radio maps. Lim et al. (2013) proposes a solution where the APs with weak signal strengths should have more chances to be updated into the radio map using scoring function. Ma et al. (2017) proposes to combine PDR trajectory matching with floor maps to crowdsource Wi-Fi radio maps.

Many other systems try to combine data from several sensors and from multiple users to improve the accuracy of indoor positioning systems. The AcMU system introduced in Wu et al. (2018) exploits the static behavior of mobile devices, using it to collect new fingerprints when the device is static at specific location. Liu et al. (2015) describes a system that uses the motion sensor of a smartphone to build a radio fingerprint map in a short time while Zhao et al. (2018) proposes a crowdsourcing and multisource fusion-based fingerprint sensing to replace the traditional site survey approach.

3.3 A Crowdsourcing Solution to Build a Radio Map

This section describes a crowdsourcing solution to build Wi-Fi radio maps in large buildings. The Where@UM app (Where@UM, 2018)[2] is an Android application based on a social network concept that uses an indoor positioning system based on Wi-Fi fingerprinting. In particular, the aim was to try to create a complete Wi-Fi radio map for all the buildings of university campi and simultaneously deploy a large-scale indoor positioning.

The University of Minho has two large campi and a set of other buildings spread around two towns, including students' dorms, old historical building, and offices. This is an example of a very large institution where creating and maintaining a complete Wi-Fi radio map updated is a very demanding task. Since a big number of persons, including students, faculty staff, and employees, use the buildings everyday, it was decided to try to engage those persons to crowdsource the Wi-Fi radio map.

The Where@UM user can become a friend of other users, being able to see his/her friends' location and share with them its own location, facilitating real-life encounters. Users are motivated to use this app since it was specifically tailored to the university. The primary target is the students, creating a way to easily find their friends inside the campi. Additionally, the app also allows sending text messages to the friends. The messages are forward through a server that delivers the messages as soon as the friends are online. By integrating a social network concept, the Where@UM tries to make the user experience better and more attractive, increasing the number of users and thus widening the mapped area.

[2]See http://where.dsi.uminho.pt/. Accessed on July 2018.

The app uses a network base positioning engine that receives Wi-Fi fingerprints and provides the location estimation based on an existing Wi-Fi radio map. The update frequency of the user' location is configurable. Default value is 5 min, which does not compromise the mobile device battery. The application retrieves the Wi-Fi fingerprint from the mobile device wireless interface and submits it to a server that processes and provides the corresponding estimated location.

The server uses a preexisting Wi-Fi fingerprint database to compute the device location. If the fingerprint is not similar enough to any of the fingerprints that exist in the database then the user location is reported to be unknown. On these cases, the user is prompt to manually provide his own location using the application interface. The same interface may be used to correct the location if it has been poorly estimated. On both cases, the user' new defined location is linked to the collected Wi-Fi fingerprint enriching the Wi-Fi radio map.

The lack of a universal space model for indoor positioning continues to be a problem. The Where@UM app was tailored to the university campi and thus the authors created a fixed hierarchical symbolic space model that includes four levels: area, building, floor, and room/space. For the first three levels, the user can select only between a predefined set of values. The University has eight locations in two different towns that constitute the top-level "area." In each of these locations, it is possible to find one or more buildings (layer two of the symbolic space model) and each building has a predefined number of floors. The last level is open, allowing the user to define the name or ID of the rooms that exist on each floor. To minimize the chances of creating more than one name for the same room the app shows all the existing rooms' names before allowing adding a new one.

Outside the University premises, each place is defined by the fields' country, city, address, and place. The app uses the Foursquare API to get a list of nearby places that suggests to the user. This prefiled information helps some of the users that for chance are in fact in one of the nearby places identified by Foursquare and it helps others because it makes easy to clearly identify a part of the address (e.g., the street name).

Wi-Fi fingerprints are stored linked to positions. When a user claims to be in a specific place (because the server was not able to estimate the position or because it was poorly estimated), the new data are stored into a secondary database and it is not assumed immediately as being trustable. When creating a radio map from a crowdsource application, it is necessary to consider that not all data are correct. Some users are honest and provide data that is correct but accidentally they can also contribute with wrong information. Some other users may be malicious and provide, intentionally, wrong information aiming to hinder the system. The wrong information will contribute to a low-quality radio map that will influence the quality of the indoor positioning system. Several problems were identified: wrong annotation of a fingerprint (eventually later providing the correct information); different users giving different names to the same place; a user providing wrong information to try to fake his current location or to try to attack the system.

To solve the quality problem it is important to assess the quality of the contribution. Fingerprints collected for the same place should be similar, that is, they should have a similar set of AP. Similarity allows defining a Place Discrimination metric that measures the similarity between all fingerprints associated with a place. Based on the Place Discrimination metric, it is possible to calculate a Contribution Credibility (for each new fingerprint): new contributions that are not similar to previous ones should not be considered. The Contribution Credibility of the most recent fingerprints allows creating a user' reputation metric. If a user has continuously low reputation then its data should not be considered at all, since he is probably just trying to hinder the system.

Radio map degrades over the time. As time goes by new APs are installed everywhere while existing ones can be moved or turned off. Additionally, changes in the environment (e.g., repositioning or adding or removing furniture) also influences the radio environment. Aging the radio maps, by removing the older records while new ones are added, leads to a more updated solution that will improve the indoor positioning systems.

4 Wi-Fi Radio Map Density

The first step toward indoor positioning using Wi-Fi fingerprinting is to build a radio map for the intended positioning area, which contains a number of samples corresponding to the various places where the location service is to be arranged. Collecting these samples usually represents a difficult and time-consuming task. A relevant aspect, which must be taken into account, is the location and determination of the number of RPs, that is, the locations where the samples for the Wi-Fi radio map will be collected. The RPs should be arranged throughout the environment and depend on the size and characteristics of the physical space itself. At each of RP, several measurements of RSS values are taken, processed and stored on the radio map. This is the most time-consuming and expensive process in the development of Wi-Fi fingerprinting localization systems.

In addition, the maintenance of the Wi-Fi radio map is also a difficult task. Any changes in the layout of the furniture, for example, can affect the validity of information stored on the radio map and consequently the accuracy of the estimates that are given by the location system, that is the localization environment is dynamic and it may be necessary to rebuild the Wi-Fi radio map very often or at least when some significant environmental changes are made.

To mitigate the large amount of time needed to build the Wi-Fi radio map, the main goals are to reduce the time spent at each sampling point and the number of points needed to obtain the Wi-Fi radio map. This includes not only the initial construction of the initial radio map, but also its updating over time. Several studies suggest that adding more references points with the interpolation method to the radio map can save time and improve localization accuracy. According to this strategy, Received Signal Strength values are measured in some positions and the rest of fingerprints are calculated using the values measured and different interpolation methods. In this way, since it is only necessary to take samples at some of the positions, the time needed for radio map creation is reduced. Other studies suggest the use of propagation models to predict signal strengths or even hybrid models that require only a few RSS samples being the rest of the radio map estimated using an RSS prediction algorithm based on propagation models.

4.1 Radio Map Construction Using Interpolation

Interpolation is a mathematical method to estimate the value of a function at a certain point using other available values of this function at different points. Spatial interpolation is a particular case of interpolation, which assumes that the available values of the function are continuous over space. This allows the estimation of unknown function values, at any location within the available values boundary. Another assumption is that the data are spatially dependent, in other words the values closer together are more likely to be similar than the values farther apart.

As a first approach, the interpolation can be performed combining in some manner the known function values. Since the known points closer to the new estimated function value should have a greater influence in the interpolation, the known values have to be combined by using a weight function where weights are chosen to be larger for nearby values than for more distant ones. For example, weighting could be chosen as a simple inverse function of the distance.

4.1.1 Inverse Distance Weighting Interpolation

The inverse distance weighting (IDW) interpolation (Kuo and Tseng, 2011) is a deterministic, nonlinear interpolation technique where the weight function is a simple inverse power function in $\Re$ expressed as: $w(x) = x^{-a}$, where the constant a is a positive value. The final expression of the interpolation function using these weights is:

$$y(x_0) = \frac{\sum_j w(d_j) \cdot y(x_j)}{\sum_j w(d_j)} \qquad (1)$$

being $y(x_j)$ the set of known values of the function, and d_j the distance between these points (x_j) and the new estimated point (x_0)

In the example presented in Fig. 3, the circles represent known sample points and the square is an unknown point for which we would like to estimate a value $(y(x_0))$. The distances between the sample points and the unknown point are shown in black text, while the attribute values of the sample points are shown inside the circles. Using Eq. (1) and the inverse distance-squared relationship to establish the weights, we obtain 52, 19 for the unknown point. As already mentioned, the distance affects the influence of a point on the estimated value in the equation, nearer points have a significantly greater effect on the estimated value than more distant points.

This type of interpolation method may be used to estimate an unknown RSS value in a predefined location using effective measured values of RSS values in other known locations. In order to get a good estimate, the algorithm can be enhanced. For example, when distance is larger than a specific value, the measurement on that reference location can be ignored.

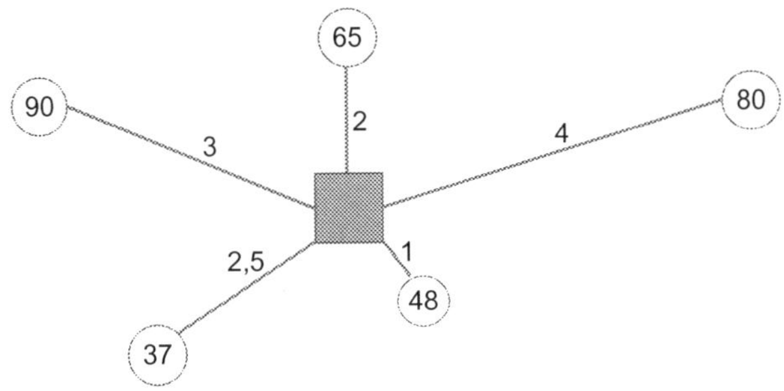

$$w(x) = x^{-2} = \frac{1}{x^2}$$

$$y(x_0) = \frac{\frac{1}{1^2}.48+\frac{1}{2^2}.65+\frac{1}{2{,}5^2}.37+\frac{1}{3^2}.90+\frac{1}{4^2}.80}{\frac{1}{1^2}+\frac{1}{2^2}+\frac{1}{2{,}5^2}+\frac{1}{3^2}+\frac{1}{4^2}} = 52.19$$

FIG. 3 Example of using the IDW interpolation method.

4.1.2 Radial Basis Function Interpolation

A radial basis function (RBF) is a real function whose value depends only on a distance from some point called origin (Krumm and Platt, 2003). These basis functions are radially symmetric around the origin and decline toward zero as we move away. Some examples of RBFs calculated at a point s in $\Re^2$ are:

- The Euclidean distance linear basis function: $f(s) = \| s \|$
- The multiquadratic function: $f(s) = \sqrt{1+ \| s \|^2}$
- The inverse multiquadratic function: $f(s) = 1/\sqrt{1+ \| s \|^2}$
- The thin plate spline function: $f(s) = \| s \|^2 \log(\| s \|)$

Using this interpolation method, the origin of each of the basis functions is placed at every position where a known function value is available and then the unknown values are estimated using a weighted combination of all the RBFs used. For example, suppose that you are using three RBFs, the value of each RBF at the prediction location can be taken from $f1(si)$, $f2(si)$, and $f3(si)$, which simply depend on the distance from each data location si. The predictor is formed by taking the weighted average $w1 \cdot f1(si) + w2 \cdot f2(si) + w3 \cdot f3(si)$.

The calculation of the weights w_j is made in such a way that ensures the equality between the interpolation results and the initial known values at the origins of the RBFs. The calculation of the weights can be carried out solving the following system of linear equations:

$$yi(si) = \sum_{j=1}^{m} wjf \cdot j(si), \quad i = 1, \ldots, n \tag{2}$$

where $yi(si)$ is the set of known values used in the interpolation, si are the points where the known values were taken, wj are the weights, and $fj(si)$ are the RBFs, each one centered at a different si point.

When comparing RBF to IDW, IDW cannot estimate above maximum or below minimum measured values, which is not the case when using RBF.

In Krumm and Platt (2003) a new location algorithm is presented to work in spite of missing calibration data based on RBFs. It takes a set of signal strengths from known locations in a building and builds an interpolation function giving (x, y) as a function of signal strength. The authors evaluated this new algorithm on one floor of a building with 118 rooms. The location error was 3.75 m using, on average, one RP every 19.5 m^2 of floor area.

4.1.3 Kriging Interpolation

Like IWD and RBF interpolation methods, the basic idea of kriging interpolation (Binghao et al., 2005; Zhao et al., 2016) is to estimate the value of a function at a given point using the weighted average of the known values of the function in the neighborhood of the point. However, kriging interpolation considers both the distance and the degree of variation between known data points when estimating values. It is based on statistical models that include the statistical relationships among the known points, also called auto-correlation.

Kriging uses a formula similar to the one used by IDW to derive a prediction at an unmeasured location.

$$\hat{Z}(x_0) = \sum_{i=1}^{N} \lambda_i Z(x_i) \tag{3}$$

where $Z(x_i)$ is the measured value at the ith location, λ_i is an unknown weight for the measured value at the ith location, x_0 is the prediction location, and N is the number of measured values.

In IDW, the weight, λ_i, depends only on the distance to the prediction location, the measured values closest to the unmeasured locations have the highest weights. However, with the kriging method, the weights are based not only on the distance between the measured points and the prediction location but also on the overall spatial arrangement of the measured points. Kriging weights come from a semivariogram that should be developed by looking at the spatial nature of the data.

In Binghao et al. (2005) a method based on kriging for obtaining the Wi-Fi fingerprint radio map has been presented. It utilizes the spatial correlation of measurements to generate the database during the offline phase. An experiment was carried out and the results indicate that the proposed method does work efficiently. On the basis of results obtained, only 1/4 or 1/8 of the number of RPs are needed using the proposed method compared with other methods that do not take into account spatial correlation. Another method based on universal kriging interpolation is presented in Zhang et al. (2016). With just 28 observation points the authors claim that it was possible to achieve an average error of 1265 m and the proposed system can be compared with other indoor positioning methods with 112 observation points.

4.1.4 Other Interpolation Methods

Besides IWD, RBF and kriging other methods have been proposed to build Wi-Fi fingerprinting radio maps. In Lee and Han (2012) a new interpolation method based on higher-order Voronoi tessellation is presented. Unlike other interpolation methods, the proposed method adopts the log-distance path-loss model and takes into account the signal fading caused by walls and obstacles. The authors claim that this method achieves better accuracy than other conventional methods such as RBF and IDW through experiments with two infrastructures.

4.2 Radio Map Construction Using Propagation Models

An alternative to radio maps made of a set of collected samples, or a filtered subset of those samples, is to build a model representing the expected signal strength of the radio signals transmitted by each AP across the operational area. These models are particularly useful when probabilistic methods are used to estimate the position as both the mean and standard deviation of the signal strength can be represented in the radio map. One advantage of model-based radio maps is that a much sparser set of RPs can be used, therefore reducing the time and effort associated with the site survey. Liu et al. (2014) proposed a method to determine the optimum cell size that balances the effort in collecting data in the offline phase with the positioning accuracy in the online phase.

The most common propagation model used to build a radio map is the log-distance path-loss (LDPL) model, which is expressed as follows:

$$PL = PL_0 + \gamma * \log_{10} \frac{d}{d_0} \tag{4}$$

where PL represents the total path loss, PL_0 the path loss at reference distance d_0, γ the (environment-specific) path-loss exponent, and d the distance from the transmitter. According to this model path loss varies exponentially with distance. The advantage of using this model is that it eliminates the need of taking RSS measurements at the cost of decreased localization accuracy. In Lim et al. (2006) a localization algorithm for building a zero-configuration indoor localization system is presented. To allow unmodified, off-the-shelf APs to be used the authors propose the use of Wi-Fi sniffers at known locations. These sniffers measure the RSS from the various APs and use the LDPL model to construct the Radio Map.

Yiu et al. (2017) recently reported on the comparison of traditional radio maps with parametric and nonparametric regression models. In their work, they concluded that nonparametric regression models based on Gaussian process perform better that traditional parametric models by providing smaller mean errors. These authors also concluded that their proposed GP model-based radio map is not significantly affected by a large reduction on the number of visible APs, and that it can be created from a small fraction of the entire set of RAW samples without a significant degradation, one feature that contributes to reduce the time needed to collect the samples across the operational area.

5 Radio Maps Filtering

Estimating the position of a device based on fingerprinting, whether using a probabilistic or deterministic method, benefits from the availability of a dense radio map. This trend results from the nature of the fingerprinting method: the more samples the radio map includes, collected at each RP, the better that location is characterized and the easiest is to distinguish it from other locations. A denser grid of RPs is also assumed to contribute to a better accuracy. Moreover, the higher the number of AP covering the space, the better each location can be characterized. However, very dense radio maps are associated with some known problems. In the first place, building very dense radio maps by collecting samples over a dense grid of RPs is a tedious and time-consuming task. Second, if the number of AP is large, the grid of RPs is very dense, and a large number of samples is collected at each RP, the time taken to estimate the position of a device might turn too large since the operational fingerprint (the fingerprint collected during the online phase) needs to be compared against a too large number of samples represented by a large number of radio strength readings. Third, and not less important, some previous works demonstrated that a smaller and higher quality radio map leads to more accurate position estimates than a larger radio map that includes all the collected samples.

5.1 Radio Map Density and Positioning Performance

It is widely accepted that large radio maps contribute to improve the positioning accuracy and/or precision in estimating the user's location (Kim et al., 2013). Fig. 4 illustrates this trend, obtained from a simulated system using synthetic radio samples.

One interesting conclusion obtained from this simple simulation is that increasing the density of the radio map (distance between adjacent RPs) provides performance improvements in terms of mean positioning error, but only to some extent. Above a certain

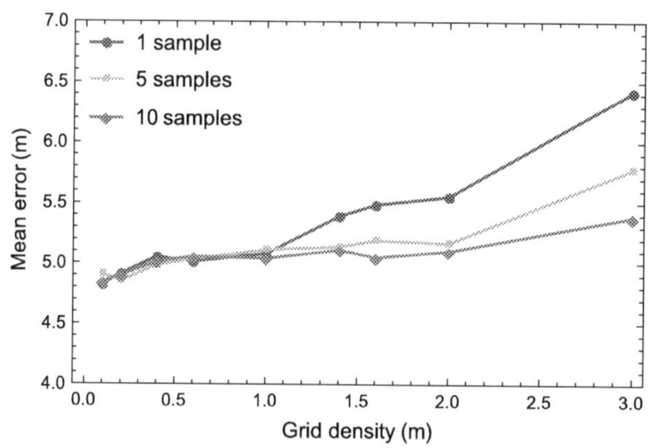

FIG. 4 Mean positioning error as a function of the grid density and number of samples per reference point (noise standard deviation = 4 dBm).

point, no significant improvements are obtained by using a denser radio map. Given that the time required to estimate the position increases linearly with the number of RPs in the radio map (and with the number of samples per RP in some cases), above a certain point the error improvements do not compensate the time penalty. Therefore, even if a large number of samples is collected to create a dense radio map, some potential computational time improvements are expected if the RAW radio map is preprocessed before being used during the online phase.

5.2 AP Selection

One approach to optimize the radio map is based on selecting only a subset of all the APs that are observed across the operating area. AP selection can be performed offline or during the online positioning phase. When performed offline, the RAW radio map, made of all the samples collected at all RPs and including signal strength measurements from all visible APs, is filtered to result into an optimized radio map that include only the APs that best discriminate among different locations. On the other hand, online AP selection is performed at the positioning phase based on each newly measured sample.

5.2.1 Offline AP Selection

Several research teams addressed the problem of AP selection aiming at reducing the computation effort and/or the energy required to obtain a position estimate. The work described in Chen et al. (2006) is focused on reducing the energy consumption on battery-powered devices by optimizing the computational cost associated to estimate their position. The proposed solution is based on an offline AP selection method, where the discriminative ability of each AP is assessed using information theory. The authors propose a *InfoGain* metric to rank the APs that best contribute to distinguish between different locations. Then, only the top *k* APs are retained in the radio map. The authors also propose to divide the operational area into clusters of RPs, and to further reduce the number of APs describing each cluster in the radio map by using again the *InfoGain* metric to select the best APs. The samples in each cluster are then used to create a model to be used by a decision tree algorithm in the online phase. The final result is a radio map made of a set of models, one model per cluster. With this solution, the authors claim to obtain better positioning accuracy when compared with other methods (e.g., the *MaxMean* method reported in Youssef et al., 2003) while also reducing the computational cost and energy consumption of the client device. However, the computational cost gains achieved by reducing the number of APs in the radio map are obtained at the cost of a slight degradation on the positioning accuracy.

A similar solution, also based on the information gain of each AP, is described in Deng et al. (2011). The authors propose a method where the discriminant ability of each AP is measured considering all the APs, and not individually, named Joint Location Information Gain. The provided experimental results show that this solution outperforms previous approaches reported by other teams, and also that a slightly better accuracy is achievable by using only a subset of all the APs.

The work from Lin et al. (2014) includes a Group Discriminant AP selection method where the importance of a subset of all the APs is determined based on the risk function from support vector machines. The results reported in Lin et al. (2014) demonstrate the benefits of this criteria when compared to AP selection methods where the relevance of each AP is assessed independently from the contributions of the other APs to the positioning accuracy.

A later, more comprehensive, study reported in Laitinen and Lohan (2016) compares several offline AP selection approaches combined with three positioning methods. Experimental results, obtained from data collected in three multistorey buildings, show that the *maxRSS* AP selection method (a method similar to *MaxMean*) performs better with the three positioning methods, and is more consistent across the three buildings. These results suggest that benchmarking AP selection methods should not be based on data from a single building floor, as is the case with the works described earlier (Chen et al., 2006; Deng et al., 2011). This study also evaluates the impact of the grid size in the positioning error, showing that both the fingerprinting and path-loss-based methods perform considerably well even for large grid sizes, such as 10 m. This is consistent with the result shown in Fig. 4, especially if multiple samples are collected at each RP.

5.2.2 Online AP Selection

In Feng et al. (2012), three online AP selection schemes are evaluated as part of a positioning system based on Compressive Sensing theory. The three schemes, namely strongest APs, Fisher criterion, and random combination, are used to select a subset of all the observed APs for each new position estimate. An experiment, conducted on a single floor of a building at the University of Toronto, reveals the superiority of the Fisher criterion. However, all the three schemes show to provide better results with a subset of all the APs than when all the APs are used, meaning that some APs can actually degrade the positioning performance. This is in contrast with most of the other AP selection solutions, where reducing the number of APs often results into a degradation in the position accuracy, even if slight.

Another AP selection method that determines the best APs in the online phase has been proposed by Zou et al. (2015) . Their approach is to select the best APs by measuring the discriminative ability of each AP taking into consideration all the APs using a mutual information metric. In their paper, the authors claim that this approach is capable of better adapting to changes in the environment after the calibration (offline) phase, and also that it outperforms other AP selection methods, namely the offline *MaxMean* and *InfoGain* methods. However, this method requires that many samples be collected at the same location during the online phase (the authors used 100 samples in their experiments), which limits the approach to stationary devices. Moreover, the mean error (positions were estimated using the weighted k-nearest neighbor method) increases if too many APs are removed. In their experiments, the authors were able to reduce the number of APs from 16 to 8 with no degradation on the mean error. The gains in the computational load obtained from the reduction on the number of APs were not discussed or measured.

5.3 Samples Filtering

All samples in a radio map are affected by noise, interference, multipath fading, and shadowing that occurs while the RSSI values are being measured at each RP. It happens that some samples are more affected than others, meaning that some samples collected at a given RP represent that location better than other samples collected at the same location. Therefore, some potential gain can be obtained by filtering out samples that degrade the positioning performance. Quite often, a simple method to minimize the impact of low-quality samples is used where all the samples collected at the same RP are averaged to create a single aggregated sample. An alternative is to create a single sample from the maximum signal strength value observed from each AP.

A solution, where the RAW radio map is filtered to preserve only a subset of the collected samples, has been proposed by Kim et al. (2013). In their work, clustering is used to segment the samples collected at each RP into a set of clusters. Then, a representative sample is selected from each one of the clusters as the one more similar to all the other samples in the cluster. The result is a much smaller radio map (5.4 samples, on average, per RP against 130 samples originally collected), obtained at the cost of just a small degradation of the positioning precision (from 0.88 to 0.76). This method contributes mainly to the scalability of the radio map.

A similar approach has been proposed in Eisa et al. (2013). In this work, simplification of radio maps aims mainly at reducing the time needed to compute each position estimate. The proposed approach includes a reduction on the total number of samples and a reduction on the number of APs. Filtering of samples is achieved by computing the estimated position for each sample in the radio map using all the other samples. Position estimation is done by using the J48 classification algorithm. All samples that are not classified into the correct room are discarded from the radio map. Filtering of APs is performed by computing a set of statistical features for each AP, such as the number of distinct RSSI values, percentage of missing RSSI values, and overall standard deviation of the RSSI values across all samples. Then, by applying a set of rules with proper threshold values, some APs are retained while others are removed from the radio map. Experimental results obtained by applying these methods to real-world data showed to reduce the processing time by more than 50% while improving the precision in detecting the correct room by more than 3%.

6 Standards

In recent years, the expectations about the penetration of location-based services (LBS) and Real-Time Location Systems (RTLS) have grown considerably, and indoor positioning and tracking technologies are a fundamental building block of such systems. However, a limited number of indoor positioning systems are actually being used around the world and use of LBS indoors is far from being ubiquitous as is the use of cellular and wireless local area networks. Within the wide range of positioning technologies, systems based on Wi-Fi fingerprinting have been extensively studied due to its potential to exploit already deployed radio-based communication infrastructures, and also due to the ubiquity of powerful mobile devices that can be used to run software modules capable of estimating their position/location without any additional hardware. However, even solutions based on Wi-Fi fingerprinting are just now starting to be used in real-world contexts. In the following we discuss some of the causes for such a slow deployment rate by arguing that standards, or the lack of them, might have a significant role in this process.

6.1 Fundamental Building Blocks of an Indoor Positioning and Tracking System

As discussed earlier, we focus this analysis on indoor positioning systems based on Wi-Fi fingerprinting. However, similar considerations also apply to many other techniques, such as those based on cellular and FM radio fingerprinting, BLE proximity, magnetic field fingerprinting, and, to some extent, to inertial navigation systems.

The fundamental building blocks of an indoor positioning system based on fingerprinting are depicted in Fig. 5. On this system, we assume that a person carrying a mobile device is using an LBS that needs to know its position within the operating area, including the 2D position, floor and building.

Depending of the specific solution, the position of the mobile device can be estimated at the mobile device or at a network component, or even by a combination of local and remote computations. In any case, a radio map, previously created for the operational area,

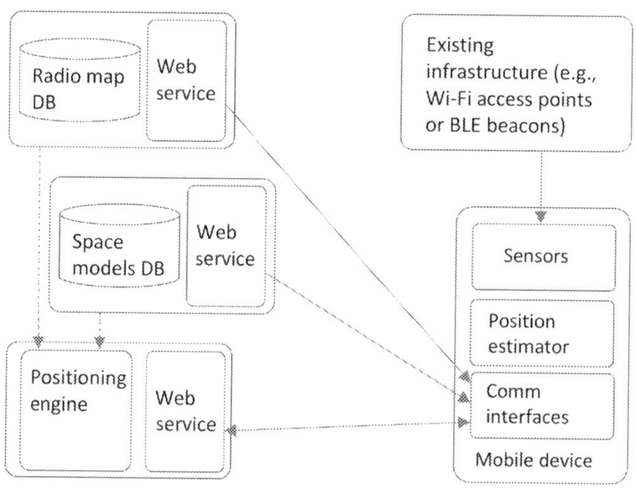

FIG. 5 Anatomy of an indoor positioning system.

must be made available to the components performing the position estimations. These estimations rely on the comparison of the data collected locally at the mobile device[3] through the existing sensors with data stored in the radio map.

The estimation process might also rely on the geometry of the space, for example, when particle filters are used or when fingerprinting is combined with inertial navigation. Information about the geometry of the operational area is also required for the users' interface, such as to support indoor navigation. Additionally, the topology of buildings might also be required to compute shortest paths in indoor navigation or safety applications. All that data might be stored on a remote Space Models component.

6.2 The Need for Standards

Currently, each particular indoor positioning solution uses its own proprietary architecture, data formats, access protocols, and estimation methods. We envision a future where standards play a significant role in promoting interoperability among different systems, in contributing to the wide adoption of indoor positioning systems and LBS, and in creating a larger market that can benefit from large economies of scale. Such a vision is illustrated through the following scenario.

> *After landing at a foreign airport, Claire opens the MyIPS app in her smartphone to find the best route to the baggage claim area. After leaving the security area, Claire uses the same app to find the car rental desk of the company where she previously booked a car. In her way to the hotel, Claire stops at a shopping mall to look for some flowers to offer to her friend on this evening diner. Upon entering the mall's underground car park, MyIPS helps her in spotting an empty place to park the car. As Claire is using MyIPS for a long time, she quickly finds two flower shops in the mall. MyIPS also helps her in easily finding her way back to the car.*

This scenario looks like the script of any advertising promoting an indoor positioning system. However, it is far from what can be achieved today as it encompasses many practical challenges, as described in the following sections.

6.3 Automatic Discovery Protocols

The implementation of the scenario described earlier calls for the capability of the MyIPs app to discover local services, at each one of the visited premises, from where to retrieve local radio maps, local floor maps, and other data such as the location of AP or BLE beacons, and/or local positioning engines able to estimate the position of Claire's smartphone. No such discovery protocols exist today, and very few steps have been given toward that direction. Among them is the work described in Sousa et al. (2014)

[3]In some solutions, the sensors are part of the infrastructure. These cases are not considered here as they require the deployment of specific hardware.

that proposes the use of the Domain Name System (DNS) to enable a mobile device to retrieve its location from a DNS server. Other related contributions are described in Tschofenig et al. (2006) and Maass (1998). In Tschofenig et al. (2006), the Geopriv protocol is proposed, enabling the exchange of location information over the Internet by exploiting some capabilities of the Dynamic Host Configuration Protocol. Maass (1998) proposes a solution to facilitate the development of LBS and applications based on the use of a directory service (X.500). While these efforts contribute to a solution for automatically discovering of local localization/positioning services, no complete solution exist, not to mention its wide acceptance.

6.4 Radio Maps: Formats and Protocols

Fingerprinting-based positioning systems rely on the availability of radio maps for each area visited by Claire, in the scenario described earlier. A global and distributed positioning system would benefit from the possibility of accessing, or downloading, local radio maps upon entering a certain operational area. In addition to the need to discover a local server from where to retrieve the radio map, an access protocol and a normalized format for representing the radio map are required. Accessing local radio maps could be easily solved through a predefined protocol implemented over REST to communicate to a local web service. On the other hand, security (e.g., access control) issues should also be addressed in order to control when and who have access to local radio maps.

With more or less variations, radio maps are widely accepted as being represented by a set of vectors, each one of them made of a collection of radio Received Signal Strength (RSS) measurements from each one of the visible AP and of the geometric representation of the Reference Point where it was obtained (e.g., a pair of coordinates, floor and building identifiers). Therefore, it would not be difficult to define a standard for the representation of radio maps. On the other hand, representing maps about magnetic field signatures, images, of BLE beacons requires a more comprehensive solution. Unfortunately, no such universal solution exists.

6.5 Floor Maps and Other Space Models

Floor maps are an essential part in many LBS (e.g., a map showing the location of the flower shops in the scenario described earlier). In addition, a description of the geometry and topology of the space is also fundamental for some indoor positioning and tracking techniques and a key requirement for indoor navigation. In a universal indoor positioning solution, mobile devices should be able to access local servers providing this kind of space models. Similarly to radio maps, as described in the previous section, a global solution depends on the wide acceptance of access protocols and data formats.

Some companies and research groups have been addressing the problem of creating floor maps for use in indoor positioning and LBS. Among them are the efforts of companies such as Google,[4] Microsoft,[5] Apple,[6] MapsPeople,[7] IndoorAtlas,[8] Cartogram,[9] MazeMap,[10] and the OpenStreetMaps (OSM) indoorOSM project.[11] This current situation illustrates how the indoor positioning market is being take seriously due to its potential for big business, with major players trying to find their market share, while also highlighting how difficult will be to develop a standard for indoor space models.

Moreover, creating space models for indoor spaces is difficult and costly. For large buildings, and in particular for old buildings, creating floor maps requires a considerable manual effort. In order to automate this task, or at least minimize the amount of human labor required, many research teams have been proposing methods and software tools to create space models. The reader is referred to the works reported in Stahl and Haupert (2006), Schäfer et al. (2011), Xuan et al. (2010), Zhang et al. (2014), Philipp et al. (2014), and Pintore et al. (2016) for set of proposals related to the task of building floor maps.

One of the major steps toward the creation of a standard for indoor maps has been given with the indoorGML[12] proposal from OGC (Open Geospatial Consortium) (Kim et al., 2014). However, this standard has been criticized for being too complicated and very difficult to implement, not being widely used yet. Other related standards are CityGML and IFC, whose characteristics are discussed by Chen and Clarke (2017).

6.6 Remote Positioning Engines

As shown in Fig. 5, the position/location of a mobile device can be estimated at the mobile device, with advantages in terms of privacy, at a remote positioning engine, with advantages in terms of energy consumption of the mobile device, or through a hybrid solution involving both local and remote computations. Using remote positioning engines specially deployed to serve a specific area (e.g., a positioning engine for a shopping mall), and accessible over a local area network contribute to shorter delays and can benefit from the specific configuration of the deployed infrastructure. As an example, consider a positioning solution where the coarse position of the mobile device is estimated from the Wi-Fi network side, while the more accurate position is estimated at the mobile device. This functionality requires the definition of a protocol to enable the communication between mobile devices and remote positioning engines, as well as standard formats for the representation of data such as fingerprints, magnetic signatures, and position/location representations. Such a specification might even extend to the standard representation of the data collected from the sensors embedded in smartphones and other mobile devices.

[4]See https://www.google.com/maps/about/partners/indoormaps/.

[5]See http://www.bing.com/maps/.

[6]See http://www.theverge.com/2015/11/2/9657304/apple-indoor-mapping-survey-app.

[7]See http://mapspeople.com.

[8]See https://www.indooratlas.com.

[9]See http://www.indoormaps.com.

[10]See http://www.mazemap.com.

[11]See https://wiki.openstreetmap.org/wiki/Indoor_Mapping.

[12]See http://www.indoorgml.net.

This level of detail would, eventually, enable the negotiation between the mobile device and the remote positioning engine about the available sensors on the client side and the available features at the server side.

6.7 Standardization Initiatives

Besides the standardization projects referred previously, very few initiatives are currently working on standards for indoor positioning, tracking, and navigation. Among them is an ad hoc group recently created within the context of the IPIN conference[13] that is addressing some of the issues and challenges associated with standards for IPIN (indoor position and indoor navigation).

Another initiative has been coordinated by the National Institute of Standards and Technology of the US Department of Commerce, especially in what concerns testing and evaluation of localization and tracking systems.[14] One of the outcomes of this initiative is the ISO/IEC 18305 Standard published in 2016.[15]

None of the above referenced standards or standardization projects addresses the representation of radio maps, which is not that strange given that the market for IPIN is still dominated by proprietary solutions.

7 Conclusion

An indoor positioning system based on fingerprinting is made of several elements. Building and updating a radio map is a challenging task but a fundamental one for many indoor positioning systems. The quality of a radio map may influence largely the accuracy of the position estimates, leading to the need of having dense and up to date maps to achieve the best results.

Crowdsourcing, by enrolling the final users, has been proposed a solution for large buildings where the number of scanning locations for an average dense map can grow to hundreds of locations. Crowdsourcing is a solution not just to build radio maps in large buildings but also to update them. As expected, the quality of the contributions must be accessed to avoid degrading the maps quality by accepting data from malicious users.

Different interpolation methods and propagation models can be applied to increase the density of radio maps. Denser maps are expected to lead to solutions that are more accurate but it can increase the computational power necessary to estimate the position. On the other hand, filtering the radio map, by selecting the appropriate APs and/or selecting the best samples, may lead also to a better map (improved accuracy and/or faster processing time). The wide number of constraints that may influence the interpolation methods, propagation models, and filtering techniques shows that this should continue to be a research area for some years to come.

[13]See http://ipin-conference.org.

[14]See https://perfloc.nist.gov.

[15]See https://www.iso.org/standard/62090.html.

The creation of standards, promoted by some international organizations, is fundamental to boot the deployment of large-scale ubiquitous indoor positioning solutions. Universal radio map formats and protocols for indoor positioning service discovery, along with indoor floor maps and space models, are currently still missing.

References

3GPP, 2008. Functional stage 2 description of Location Services (LCS) in GERAN. No. 43.059.

Bahl, P., Padmanabhan, V.N., 2000. RADAR: an in-building RF-based user location and tracking system. In: Proceedings IEEE INFOCOM 2000. Nineteenth Annual Joint Conference of the IEEE Computer and Communications Societies (Cat. No. 00CH37064), vol. 2, pp. 775–784.

Binghao, L., Wang, Y., Lee, H.K., Dempster, A.W., Rizos, C., 2005. A new method for yielding a database of location fingerprints in WLAN. IEE Proc. Commun. 152 (5), 580–586.

Chen, J., Clarke, K., 2017. Modelling standards and file formats for indoor mapping. In: Proceedings of the 3rd International Conference on Geographical Information Systems Theory, Applications and Management, vol. 1. GISTAM, pp. 268–275.

Chen, Y., Yang, Q., Yin, J., Chai, X., 2006. Power-efficient access-point selection for indoor location estimation. IEEE Trans. Knowl. Data Eng. 18 (7), 877–888.

Chen, Y., et al., 2013. Indoor localization using FM signals. IEEE Trans. Mob. Comput. 12 (8), 1502–1517.

Deng, Z., Ma, L., Xu, Y., 2011. Intelligent AP selection for indoor positioning in wireless local area network. In: 2011 6th International ICST Conference Communications and Networking in China (CHINACOM). IEEE, pp. 257–261.

Eisa, S., Peixoto, J., Meneses, F., Moreira, A., 2013. Removing useless APS and fingerprints from WiFi indoor positioning radio maps. In: 2013 International Conference in Indoor Positioning and Indoor Navigation (IPIN). IEEE, pp. 1–7.

Feng, C., Au, W.S.A., Valaee, S., Tan, Z., 2012. Received-signal-strength-based indoor positioning using compressive sensing. IEEE Trans. Mobile Comput. 11 (12), 1983–1993.

He, S., Chan, S.H.G., 2016. Wi-Fi fingerprint-based indoor positioning: recent advances and comparisons. IEEE Commun. Surv. Tutorials 18 (1), 466–490.

Kaemarungsi, K., Krishnamurthy, P., 2012. Analysis of WLAN's received signal strength indication for indoor location fingerprinting. Pervasive Mob. Comput. 8 (2), 292–316.

Kahvecı, M., 2017. Recent advances in GNSS: technical and legal aspects. In: 2017 8th International Conference on Recent Advances in Space Technologies (RAST), Istanbul, pp. 501–505.

Kim, Y., Chon, Y., Ji, M., Park, S., Cha, H., 2013. Scalable and consistent radio map management scheme for participatory sensing-based Wi-Fi fingerprinting. In: Proceedings 3rd Int'l Workshop Mobile Sensing (IWMS), pp. 14–18.

Kim, J.S., Yoo, S.J., Li, K.J., 2014. Integrating IndoorGML and CityGML for indoor space. In: International Symposium on Web and Wireless Geographical Information Systems. Springer, Berlin, Heidelberg, pp. 184–196.

Kjaergaard, M.B., 2007. A taxonomy for radio location fingerprinting. In: International Symposium on Location Context, pp. 139–156.

Krumm, J., Platt, J., 2003. Minimizing calibration effort for an indoor 802.11 device location measurement system. Tech. Rep. MSRTR-2003-82. Microsoft Research.

Kuo, S.P., Tseng, Y.C., 2011. Discriminant minimization search for large scale RF-based localization systems. IEEE Trans. Mobile Comput. 10 (2), 291–304.

Laitinen, E., Lohan, E.S., 2016. On the choice of access point selection criterion and other position estimation characteristics for WLAN-based indoor positioning. Sensors 16 (5), 737.

Lee, M., Han, D., 2012. Voronoi tessellation based interpolation method for Wi-Fi radio map construction. IEEE Commun. Lett. 16, 404–407.

Lim, H., Kung, L.C., Hou, J.C., Luo, H., 2006. Zero-configuration, robust indoor localization: theory and experimentation. In: Proceedings IEEE INFOCOM 2006. 25th IEEE International Conference on Computer Communications, Barcelona, Spain, pp. 1–12.

Lim, J.S., Jang, W.H., Yoon, G.W., Han, D.S., 2013. Radio map update automation for Wi-Fi positioning systems. IEEE Commun. Lett. 17 (4), 693–696.

Lin, T.N., Fang, S.H., Tseng, W.H., Lee, C.W., Hsieh, J.W., 2014. A group-discrimination-based access point selection for WLAN fingerprinting localization. IEEE Trans. Veh. Technol. 63 (8), 3967–3976.

Liu, H., Member, S., Darabi, H., Banerjee, P., Liu, J., 2007. Survey of wireless indoor positioning techniques and systems. IEEE Trans. Syst. Man Cybern. C (Appl. Rev.) 37 (6), 1067–1080.

Liu, W., Chen, Y., Xiong, Y., Sun, L., Zhu, H., 2014. Optimization of sampling cell size for fingerprint positioning. Int. J. Distrib. Sens. Netw. 10 (9), 273801.

Liu, H.H., Liao, C.W., Lo, W.H., 2015. The fast collection of radio fingerprint for Wi-Fi-based indoor positioning system. In: 2015 11th International Conference on Heterogeneous Networking for Quality, Reliability, Security and Robustness (QSHINE), Taipei, pp. 427–432.

Locubiche-Serra, S., López-Salcedo, J.A., Seco-Granados, G., 2016. Sensitivity of projection-based near-far mitigation techniques in high-sensitivity GNSS software receivers. In: 2016 International Conference on Indoor Position and Indoor Navigation (IPIN), pp. 4–7.

Ma, L., Fan, Y., Xu, Y., Cui, Y., 2017. Pedestrian dead reckoning trajectory matching method for radio map crowdsourcing building in WiFi indoor positioning system. In: 2017 IEEE International Conference on Communications (ICC), Paris, pp. 1–6.

Maass, H., 1998. Location-aware mobile applications based on directory services. Mob. Netw. Appl. 3 (2), 157–173.

Peral-Rosado, J.A.D., López-Salcedo, J.A., Zanier, F., Criscim, M., 2012. Achievable localization accuracy of the positioning reference signal of 3GPP LTE. In: 2012 International Conference Localization GNSS. ICL-GNSS.

Philipp, D., Baier, P., Dibak, C., Durr, F., Rothermel, K., Becker, S., Peter, M., Fritsch, D., 2014. MapGenie: grammar-enhanced indoor map construction from crowd-sourced data. In: 2014 IEEE International Conference Pervasive Computing and Communications (PerCom). IEEE, pp. 139–147.

Pintore, G., Garro, V., Ganovelli, F., Gobbetti, E., Agus, M., 2016. Omnidirectional image capture on mobile devices for fast automatic generation of 2.5 D indoor maps. In: 2016 IEEE Winter Conference Applications of Computer Vision (WACV). IEEE, pp. 1–9.

Rahman, M.M., Moghtadaiee, V., Dempster, A.G., 2017. Design of fingerprinting technique for indoor localization using AM radio signals. 2017 International Conference on Indoor Positioning and Indoor Navigation (IPIN), Sapporo, pp. 1–7.

Schäfer, M., Knapp, C., Chakraborty, S., 2011. Automatic generation of topological indoor maps for real-time map-based localization and tracking. In: 2011 International Conference Indoor Positioning and Indoor Navigation (IPIN). IEEE, pp. 1–8.

Sousa, A., Costa, A., Santos, A., Meneses, F., Nicolau, M.J., 2014. Using DNS to establish a localization service. In: 2014 International Conference Indoor Positioning and Indoor Navigation (IPIN). IEEE, pp. 385–392.

Stahl, C., Haupert, J., 2006. Taking location modelling to new levels: a map modelling toolkit for intelligent environments. In: International Symposium on Location and Context Awareness. Springer, Berlin, Heidelberg, pp. 74–85.

Tschofenig, H., Schulzrinne, H., Newton, A., Peterson, J., Mankin, A., 2006. The IETF Geopriv and presence architecture focusing on location privacy. In: Position paper at W3C Workshop on Languages for Privacy Policy Negotiation and Semantics-Driven Enforcement, Ispra, Italy.

Wu, C., Yang, Z., Xiao, C., 2018. Automatic radio map adaptation for indoor localization using smartphones. IEEE Trans. Mob. Comput. 17 (3), 517–528.

Xuan, Y., Sengupta, R., Fallah, Y., 2010. Crowd sourcing indoor maps with mobile sensors. In: International Conference on Mobile and Ubiquitous Systems: Computing, Networking, and Services. Springer, Berlin, Heidelberg, pp. 125–136.

Yassin, A., et al., 2017. Recent advances in indoor localization: a survey on theoretical approaches and applications. IEEE Commun. Surv. Tutorials 19 (2), 1327–1346.

Yiu, S., Dashti, M., Claussen, H., Perez-Cruz, F., 2017. Wireless RSSI fingerprinting localization. Signal Process. 131, 235–244.

Youssef, M.A., Agrawala, A., Shankar, A.U., 2003. WLAN location determination via clustering and probability distributions. Pervasive computing and communications, 2003 (PerCom 2003). In: Proceedings of the First IEEE International Conference. IEEE, pp. 143–150.

Zhang, X., Jin, Y., Tan, H.X., Soh, W.S., 2014. CIMLoc: a crowdsourcing indoor digital map construction system for localization. In: 2014 IEEE Ninth International Conference on Intelligent Sensors, Sensor Networks and Information Processing (ISSNIP), Singapore, pp. 1–6.

Zhang, M., Pei, L., Deng, X., 2016. GraphSLAM-based crowdsourcing framework for indoor Wi-Fi fingerprinting. In: 2016 Fourth International Conference on Ubiquitous Positioning, Indoor Navigation and Location Based Services (UPINLBS), Shanghai, pp. 61–67.

Zhang, H., Lu, H., Zheng, S., Wu, E., 2017. Unified navigation graph model of indoor space and outdoor space. In: 2017 International Conference Indoor Positioning and Indoor Navigation (IPIN). Work-in-Progress paper.

Zhao, H., Huang, B., Jia, B., 2016. Applying Kriging interpolation for WiFi fingerprinting based indoor positioning systems. In: 2016 IEEE Wireless Communications and Networking Conference, Doha, pp. 1–6.

Zhao, W., Han, S., Hu, R., Meng, W., Jia, Z., 2018. Crowdsourcing and multi-source fusion based fingerprint sensing in smartphone localization. IEEE Sensors J. 1–1. https://doi.org/10.1109/JSEN.2018.2805335.

Zou, H., Luo, Y., Lu, X., Jiang, H., Xie, L., 2015. A mutual information based online access point selection strategy for WiFi indoor localization. In: 2015 IEEE International Conference on Automation Science and Engineering (CASE). IEEE, pp. 180–185.

Further Reading

Google, 2018. Google Maps Interior. Available from: https://www.google.com/maps/about/partners/indoormaps/. Accessed on March 2018.

Matos, D., Moreira, A., Meneses, F., 2014. Wi-Fi fingerprint similarity in collaborative radio maps for indoor positioning. In: Proceedings of 6º Simpósio de Informática (INForum 2014), Porto, Portugal, pp. 184–194.

Microsoft, 2018. Path Guide. Available from: https://mspg.azurewebsites.net/. Accessed on March 2018.

Moreira, A., Meneses, F., 2015. Where@UM—dependable organic radio maps. In: Proceedings of the 2015 International Conference on Indoor Positioning and Indoor Navigation (IPIN), Banff, Alberta, Canada.

Patel, N., 2014. Taking core location indoors. In: Apple's World Wide Developer Conference, San Francisco, CA.

5

Crowdsourced Indoor Mapping

A.K.M. Mahtab Hossain

DEPARTMENT OF COMPUTING AND INFORMATION SYSTEMS, UNIVERSITY OF GREENWICH, LONDON, UNITED KINGDOM

1 Introduction

While global positioning system (GPS) solved the outdoor localization problem quite successfully, it is not as successful indoors. This is due to the fact that GPS signals generally require an unobstructed line-of-sight (LOS) view from receiver to satellite, but cannot penetrate most construction materials. Also, indoor location-based service (LBS) applications require finer granularity and precision of localization accuracy that GPS is unable to offer under such constraints (Hossain and Soh, 2015). Besides the usual applications of resource tracking to finding the nearest store or distribution of electronic coupons within close proximity, LBSs have tremendous prospects in terms of business intelligence (BI) applications. Brickstream, an in-store analytics-related software company, published a survey result in 2014 collating the transcripts of interviews of 124 retail executives across the United States, Europe, Asia, and South America who emphasized the benefits of using data collected from in-store consumers for their businesses (Harkins, 2014). The executives identified operations, merchandizing, and profits as functions of being aware of what is happening in the store, for example, counting the customers (71% agreed), the value of adopted technology (68% preferred Wi-Fi), mobile/contact-less payment (of interest to 58% respondents), etc. Around 88% of retailers acknowledged the necessity to deploy technologies in-store, and the importance of various mobile device-driven applications to attract customers. Indoor positioning based on a customer's carried mobile device's sensor readings is a prerequisite to perform such consumer-driven data analysis. On top of it, availability of indoor maps has been treated as a natural assumption of such indoor localization research (Bahl and Padmanabhan, 2000; Youssef and Agrawala, 2005; Hossain and Soh, 2015). Only recently, it has been realized that the access to these maps especially for public or commercial buildings requires lengthy negotiations with the owners or operators. This is identified as a major hindrance toward ubiquitous availability of LBSs' indoors.

The newer emerging calibration-free indoor localization techniques succeed the labor intensive indoor fingerprinting solutions eliminating its pre-deployment site survey component of radio-map creation. Just like the calibration-free techniques, indoor floor

Geographical and Fingerprinting Data for Positioning and Navigation Systems. https://doi.org/10.1016/B978-0-12-813189-3.00005-8

layout conceptualization research also aims for automated, and implicit user participation by collecting their traces through the carried smartphone sensor measurements. Most of the commercial products based on calibration-free indoor localization use and process the crowdsourced inertial sensor measurements from their smartphones. For example, Navigine (2017) uses Wi-Fi signals together with the smartphone's inertial sensors for navigation with a claimed accuracy of 1–2 m whereas Navisens (2017) only uses the inertial sensor measurements. indoo.rs (2017) follows similar techniques as Simultaneous Localization and Mapping (SLAM) (Durrant-Whyte and Bailey, 2006) that relies on dead-reckoning (Constandache et al., 2010), sensor fusion, and filtering algorithm for indoor navigation purpose. They utilize crowdsourced data, and provide software as a service (SaaS) for incorporating their location service inside the various location-based applications.

Google Maps, Apple Maps, Bing Maps, etc., can provide street maps, and navigation direction outdoors with the help of GPS. However, the indoor counterparts were not as successful. Manual adding/editing to maintain up-to-date indoor floor maps undertaken by Google Maps did not offer a feasible and scalable solution. Only major airports, museums, and other business locations that are partnered with Google have been addressed (Google, 2017a). On the contrary, the crowdsourced approach's basic idea is to automate the modeling of floor plans by collating the pedestrian traces perceived from their smartphone sensors. Although these approaches are promising utilizing the already existing infrastructure, for example, Wi-Fi networks, and off-the-shelf hardware existing inside user smartphones, they either require a large number of motion traces or warrant the active collaboration of users collecting traces, or help of a number of sensor fusion (e.g., camera) in order to refine the traces. Considering the fact that the crowdsourced data acquired are from different sensors, and even coming from different persons' devices, they may be noisy, uncertain, and incomplete. The challenges of processing these data and the subsequent adoption of algorithms to come up with a realistic indoor map are investigated in this chapter. Subsequently, a comparative discussion based on the findings is provided by pointing out a few future research directions.

This chapter is organized as follows. We explain the existing popular outdoor Web map systems and its apparent shortcomings when applied indoors in Section 2. Next, we discuss a few emerging crowdsourced indoor map construction techniques, and some relevant calibration-free indoor localization methods in Section 3. In Section 4, we point out some crowdsourced indoor floor layout construction-related challenges, and a future research directions. Finally, we conclude in Section 5.

2 Some Existing Crowdsourced Outdoor Map Systems

An accurate Web map system is necessary to bridge the gap between the real world (i.e., what we see) and the online world (i.e., the interpretation of our surroundings). While majority of the map systems are dedicated to provide accurate location and navigation information outdoors, very few have ventured into the domain of indoor map dynamics. This is fundamental to enable location-based BI applications for various consumer-centric industry like retail store marketing and management. The ubiquity of smartphone, and its equipped sensors' rich technology, and cost effectiveness opened the door for harnessing the crowd's intelligence into such map systems. The popular Web map systems, such as Google Maps and Apple Maps to some extent acknowledge this opportunity. They had laid out certain projects, for example, Tango (2017) and TryRating (2017), respectively, to incorporate crowdsourcing for conceptualization of accurate map systems. In this section, some popular Web map systems that incorporate crowdsourcing to some extent are discussed, and a comparative analysis is provided based on the discussion.

2.1 Google Maps

Google Maps operates by means of a request/response protocol for providing various LBSs. The request from a client device (e.g., smartphone) contains its current location (if the device is equipped with a GPS receiver), Wi-Fi access points (APs), and cellular tower profile information. The response from Google servers includes the current geographical position of the client device, together with the locations of the Wi-Fi APs, and cellular towers—the profile of which were included inside the request message. This facilitates subsequent faster determination of the client device's position to enable LBSs. In addition to providing visualization of real-time traffic information, Google Maps also offers other services such as point of interest (POI) search, route planning, street view, geocoding, mass transit system's status and navigation information, etc. (Google, 2017b).

Crowdsourcing was initially considered unimportant by Google Maps during the build of US proprietary database, which was constructed by acquisition of maps from the authoritative or trusted state, regional and city sources. However, crowdsensing is then used as the main agent for the revision of the map database worldwide, which has recently been discontinued (Neis et al., 2012). Google also encourages public building owners to submit their indoor maps in order to bring them under their services. Even though Google Maps APIs are free for a number of use cases, its full functionality access requires a licensed version. Google's Tango envisioned crowdsensing to enable indoor navigation, conceptualization of indoor 3D maps, and interesting "Augmented Reality" (AR) applications (Tango, 2017). The ubiquity of mobile devices where over 1.4 billion new smartphones were sold in 2016 alone, and the advancement, and cost-effectiveness of its equipped sensor technologies motivated Google to invest heavily in such research (Lee, 2017). Google launched Tango hardware and software that equips mobile phone carried by a user with motion tracking, depth sensing, and area learning capabilities.

2.2 OpenStreetMap

OpenStreetMap (OSM) is one of the most utilized and analyzed Volunteered Geographic Information (VGI) platforms that generated keen interests among researchers and practitioners (Arsanjani et al., 2015). The OSM project was conceptualized in 2004 with the vision of creating an editable map of the whole world through crowd participation. Volunteers collect the map data via performing systematic surveys using inexpensive portable satellite navigation devices such as a GPS receiver equipped hand-held or a digital camera, which is then uploaded into the OSM database. The registered diverse set of trusted volunteers contribute to the overall data collection procedure, and editing of the uploaded maps by taking advantage of their local area knowledge. The crowdsourced map data is made available to the general public under the Open Database license (ODbL). As a result, researchers have attempted to adopt OSM tools more compared to their commercial alternatives over the years in order to create indoor maps of railway stations, hospitals, public library, etc., via volunteer participation (OpenStreetMap, 2017).

2.3 MapQuest

MapQuest uses a standard client-server architecture where the user's client requests for a particular map, and the MapQuest servers respond with the resultant map within a web page. Introduced in 1996, it quickly became the premier online map provider before ultimately falling behind Google Maps around 2009 (Peterson, 2014). Today, MapQuest provides street-level details, and navigation planning only for some selected countries. MapQuest-based iOS and Android mobile app also provide other features such as POI search, voice-guided navigation, real-time traffic information, etc. (MapQuest, 2017). In 2010, MapQuest tried to venture into crowdsourced mechanism of creating or manipulating maps by using OSM of Section 2.2 for few of its services but with limited success (Peterson, 2014).

2.4 Waze

The largest community-based free traffic and navigation application (Waze, 2017) utilizes floating car data (FCD) obtained from the driver's or passenger's smartphones in order to generate real-time traffic information similar to Google Maps. Even though the main purpose of the Waze application is to facilitate user's vehicle navigation, it also gives users the provisions for manipulating maps (e.g., adding new roads, reporting hazards and potholes). Just like Google Maps, Waze works upon the simple request/response protocol principle where the smartphone client application sends periodic messages to the Waze server with its current position acquired through the phone's GPS receiver. Subsequently, the Waze server returns the navigation plan together with the traffic information along the planned route in a response message. Since Waze is a community-driven approach, it generally requires the user to register before using the app. Once the user logs into the system, an appropriate Waze server ID is returned with a cookie, which is used to differentiate the individual user session for subsequent request messages. Waze incorporates crowdsourced sensor data in a limited manner for outdoor navigation map manipulation purpose only, but does not address the indoor map construction-related challenges.

2.5 Others

Not all the existing map services try to incorporate crowdsourcing as part of their operating principles. The four popular map systems discussed earlier try to include crowdsourcing aspects of map building in some ways. However, there are a few other alternatives as well such as Apple Maps, Microsoft's Bing Maps, HERE, etc., which have treated users only as consumers so far; they do not include them in building or refining the maps.

Apple Maps (Apple, 2017)—the default map system of iOS offers similar services like Google Maps such as navigation, POI search, route planning, mass transit system's status, navigation information, etc. TomTom Maps is the primary provider of Apple's map data. Apple is reportedly trying to involve freelance users to refine its map data through a program called "TryRating," where they are paid a nominal incentive for verifying POI or other searches made with Apple Maps.

HERE, formerly known as "Nokia Here," concentrates on collecting road map data to offer navigation, traffic information, and location solutions (HERE, 2017). It is the primary provider of LBSs to Microsoft Bing, and Facebook. HERE also works in the autonomous vehicle industry by providing its HD map data to manufacturers such as Alpine, Mercedes, Garmin, Hyundai, Pioneer, Volkswagen, and Toyota for testing their driverless vehicles. HERE recently partnered with Mobileye that harnesses crowdsourcing by collecting geometry and landmarks around the user's driving path through camera-equipped vehicles in order to maintain real-time map information.

The API of Microsoft's Bing Maps enables a wide range of applications to include its functionality like navigation, real-time traffic, POI search, geocoding, etc., through a variety of licenses (Bing, 2017). Some of the Microsoft products such as SharePoint, Excel, and Office 365 Pro Plus offer them out of the box. Bing Maps collects its road data from HERE, and various country-specific partners, and its imagery data generally comes from their own team.

2.6 Discussions

The popular Web map systems' knowledge base is dependent upon the service provider's associated commercial partners—not necessarily on its users, for example, Google Maps (in most cases) and Apple Maps as can be seen from Table 1. OSM is the only exception where its operating principle is completely based upon user participation and trusted volunteers' local knowledge. Another community-driven system, Waze is a traffic and navigation application that is targeted toward a niche area (i.e., vehicles only). In general, the crowdsourced data-set arising from the smartphone sensors carried by the people are still largely unexplored by such map systems. If they are used effectively, it will not only enhance their existing system's performance, but also can give rise to new interesting business cases. The big players such as Google Maps, and Apple Maps realized its importance, and announced innovative crowdsensing-based projects such as Tango (2017) and TryRating (2017), respectively. Another important observation is that—the attempts to bring indoor environments under these map systems were largely unsuccessful.

Table 1 Comparison of Some Existing Popular Web Map Systems

	Map Data Providers	Map Environment	Signals and Sensors	Crowdsourcing	License	Supported Apps
Google Maps	Federal, state, regional, user contributions, and other partners	Indoor and outdoor	Satellite (GPS), Cell-based, Wi-Fi, Tango: Inertial sensors, Camera	Yes; Google's Tango, device: user smartphones, apps: indoor 3D maps, AR applications, etc.	Proprietary	Google Earth, BMW, and Tesla navigation
OpenStreetMap	Users, open data	Indoor and outdoor	Satellite (GPS), Wi-Fi, Camera	Yes; add, edit, and refinement of maps by users	ODbL	Foursquare, Craiglist, Wikipedia, World Bank
MapQuest	TomTom, OpenStreetMap, and other partners	Outdoor	Satellite (GPS)	Attempted in the past by incorporating OSM	Proprietary	N/A
Waze	Users	Outdoor (for vehicles)	Satellite (GPS)	Yes; seamless integration of user smartphone's position data	Free	Waze Carpool
Apple Maps	TomTom and other partners	Outdoor	Satellite (GPS), Wi-Fi	Reported to venture into it recently through paid freelancers	Proprietary	iOS LBS apps
HERE	Naviteq (Nokia)	Outdoor	Satellite (GPS), Cell-based, Wi-Fi	Yes; via camera-equipped vehicles	Proprietary	Mercedes-Benz, Alpine car systems
Bing Maps	HERE, country-specific partners	Outdoor	Satellite (GPS), Cell-based, Wi-Fi	No	Proprietary	Windows OS LBS apps

One of the prerequisite of incorporating user participation data into such map systems is the availability of their position information. Using this position data as a reference, services such as traffic and navigation information can be correctly correlated with it. While position in an outdoor environment can be successfully resolved by GPS; it could not be the overwhelming solution indoors since GPS signals are blocked by most buildings. As a result, conceptualizing indoor maps through crowdsourcing is likely to pose additional challenges compared to its outdoor map systems counterpart in terms of unavailability of GPS signals, indoor environment complexity, lack of intelligent incorporation of crowdsourced rich sensor data, development of efficient mapping algorithms and architecture, etc.

3 Indoor Map Systems' Research

Indoor localization refers to the technique of obtaining location/position information of a device (or of a person carrying the device) indoors with the help of a set of reference nodes within a predefined space (Hossain and Soh, 2015). This space is characterized by indoor environment maps that are generally assumed to be available by the localization researchers. However, this assumption may not hold for many scenarios, especially for public buildings such as shopping malls, airports, museums, hospitals, etc. Furthermore, it is common for such environments to go through frequent changes or rearrangements, therefore, a up-to-date mechanism of ensuring the correct detailed indoor maps is also necessary.

GPS enables a user carrying a GPS receiver to pinpoint its location using the signals from the GPS satellites (i.e., reference nodes). A well-known observation is that GPS performs poorly in urban environments where buildings block GPS signals (especially indoor) (Hossain and Soh, 2015). An alternative cost-effective solution compared to GPS could well be through the possible use of the available resources (e.g., user smartphone sensors) and infrastructure (e.g., Wi-Fi networks) existing in an indoor environment. Traditionally, the fingerprinting approach is deemed appropriate in such scenarios where the surveyor laboriously collects signal signatures over the localization area that are annotated with the locations where they are captured. They are then stored inside the database as a ⟨location, signal fingerprint⟩ tuple. Subsequently, some well-known machine learning algorithms such as Maximum Likelihood Estimator (MLE) or Nearest Neighbor (NN) are applied in order to locate a user.

There is a newer family of emerging calibration-free techniques that try to relieve the pre-deployment woes of the laborious signal signatures collection phase of the fingerprinting solutions discussed earlier. Many work of this family just assume the availability of a floor-plan of an indoor environment. They feel the map knowledge is required to consume any LBSs (Rai et al., 2012); so it is only natural to assume its availability. A few map construction tools can also be accessed from the robotics literature (Shin et al., 2012). However, the inherent assumption of indoor map availability may have resulted in skewed optimistic conclusions for such indoor localization research. The conceptualization of the indoor maps rather than having it as a precondition for offering LBSs is therefore drawing attention from the indoor localization research community. In this section, we discuss both the robotics, and calibration-free indoor localization perspective of map construction, and the associated challenges. The relevant sensor technologies are also identified together with the well-known algorithms that are adopted.

3.1 Simultaneous Localization and Mapping

SLAM is the computational problem in robotics navigation and mapping where it constructs and updates the map of an unknown environment, and simultaneously locates the robot's position within it (Durrant-Whyte and Bailey, 2006). Various filtering algorithms, for example, Kalman filter, Particle filter, etc. (Welch and Bishop, 1995; Montemerlo et al., 2002), and odometry or dead-reckoning (Constandache et al., 2010) techniques utilizing the inertial sensor measurements are adopted in order to build the map, and for localization purpose. The filtering techniques generally consist of two different phases—the prediction step (i.e., the state transition model through odometry) and the update step, which takes into account the sensor measurements for correction of the prediction step. Odometry navigation method's fundamental idea is the integration of incremental motion (i.e., wheel revolution information over time). Inertial sensors' purpose is to explore the properties of inertia, for example, sense the change in angular motion (gyroscope), and change in linear motion (accelerometer). GPS, landmark, magnetic compass, active beacons (Wi-Fi, sonar, etc.), and computer vision (e.g., camera) sensor measurements can be considered during the update step of the filtering techniques to refine or correctly estimate the state.

Traditional SLAMs generally require custom-made inertial measurement units (IMUs), for example, FootSLAM (Robertson et al., 2009) and ActionSLAM (Hardegger et al., 2012) use foot-mounted and body-mounted IMUs, respectively, to track a user's motion. PlaceSLAM (Robertson et al., 2010) is an improvement over FootSLAM by explicitly involving user participation for refining the interpretation of the physical surroundings. Unlike traditional SLAMs, SmartSLAM (Shin et al., 2012) uses smartphone's inertial sensors, and Wi-Fi modules to observe the device's movement and environment, and thereby construct the indoor map. The on-going incorporation of computer vision technologies enriched the SLAM approaches, which largely concentrated on robot navigation via changing direction when faced with an obstacle. The application of drone or micro aerial robot facilitates the creation of indoor 3D maps through the use of IMU, camera, laser scanner with deflective mirrors (Shen et al., 2012; Intel, 2017). SLAM has also been used as a viable commercial tool especially in autonomous vehicle industry, major Web map systems such as Google and Apple navigation, etc. iRobot which is MIT institution's brainchild already developed SLAM-based product in the form of a vacuuming robot—so did a few other industries such as Samsung, Dyson 360 Eye, Robo.com, etc. Wi-Fi SLAM (Ferris et al., 2007), which was later acquired by Apple, and SmartSLAM (Shin et al., 2012) attempted to utilize the sensor readings that are normally available inside the user carried smartphones or mobile devices. Hence, crowdsensing can potentially be part of such SLAM variants for indoor map construction, while most of the other SLAM techniques or even commercial solutions require additional customized sensors such as sonar or laser sensors, mechanical parts (wheels or wings) of robots or drone, etc.

3.2 Calibration-Free Indoor Positioning System

SurroundSense (Azizyan et al., 2009) opened the door for various ambient sensors (e.g., sound, light, color), and Wi-Fi, accelerometer, to be collectively used for localization through user participation, and subsequently, others followed suit (Chintalapudi et al., 2010; Rai et al., 2012; Wang et al., 2012). Most of them assume these sensors' availability in the mobile phones carried by the crowdsourcing users which might be unreasonable. However, modern smartphones are equipped with a few inertial sensors, for example, accelerometer, gyroscope, and other additional sensors such as Wi-Fi, compass, camera, etc. In this section, we first briefly discuss a few calibration-free localization techniques which can be used for indoor map conceptualization, and then also explain a few crowdsensing-based indoor map construction research.

TIX

Crowdsourced Wi-Fi received signal strength (RSS) measurements are utilized for Triangular Interpolation and eXtrapolation (TIX)'s localization purpose (Gwon and Jain, 2004). TIX does not require the indoor floor plan map to operate; however, it needs the Wi-Fi APs' location information with respect to the indoor map. The distance of the client device to each AP is approximated via its perceived RSS from them, and also through the inter-AP RSS measurements. Then the TIX algorithm is applied (Gwon and Jain, 2004). Any other lateration-based algorithms (e.g., Trilateration) could work as well. Assuming RSS (log-scale) decays linearly with distance, the Wi-Fi RSS measurements are enough to localize the client device without the need of a map when the APs' location information are known. TIX's reported localization accuracy was not great though; it was only 5.4 m inside an office setting of size 1020 m^2 with zero calibration effort.

SDM

Just like TIX, signal-distance map technique, SDM (Lim et al., 2010) also utilizes only the online RSS measurements for localization. Periodic inter-AP RSS measurements facilitate to calibrate RSSs in the spatiotemporal domain. To map the relationship between these RSSs and the inter-AP distances, a truncated singular value decomposition technique is applied. No floor plan map is necessary for its operation but the locations of the APs with respect to the indoor map are required. Its reported localization accuracy is within 3 m inside a small building, and it claims to perform better than TIX.

EZ

The Wi-Fi RSS measurements are implicitly reported by EZ clients that typically run inside a user's hand-held device. The log-distance path loss is used to model the RSS measurements where its unknown parameters are resolved using Genetic Algorithm (Chintalapudi et al., 2010). EZ does not require the knowledge of indoor map layout or even the AP's location and its transmit power information. The only assumption is the availability of occasional location fix, for example, GPS lock at entrance, or other uncluttered places.

UNLOC

UnLoc (Wang et al., 2012), an unsupervised indoor localization scheme, follows similar principle as the SLAM techniques discussed previously. It utilizes dead reckoning (Constandache et al., 2010) for tracking a user while recalibrating its estimate whenever it encounters a landmark—the location of which is known. The landmarks can be identified through the equipped inertial sensor measurements of the client device where they exhibit distinct signatures indoors (e.g., elevators, escalators, entrances, etc.). UnLoc is purely a crowdsensing-based technique since these landmarks are also identified in an automated manner through the user device sensor measurements (i.e., no a priori knowledge is necessary).

WALKIE-MARKIE

Walkie-Markie (Shen et al., 2013) leverages crowdsourced user traces to generate indoor pathway maps without any a priori knowledge of the propagation characteristics, and floor plan map. It uses Wi-Fi RSS trend (increasing or decreasing) as a location fingerprint in an indoor environment. These fingerprints (i.e., landmarks where the RSS trend tripping point occur) are then placed inside the 2D plane by a graph embedding algorithm incorporating user trajectories. Subsequently, a user is colocated with a landmark when he/she passes it, and dead reckoning (Constandache et al., 2010) is utilized in order to localize him/her in between two such landmarks.

The global indoor location market has seen a rapid growth over the years, and is expected to grow even more. To enable these LBSs, the availability of an indoor floor plan is mandatory. While the outdoor Web map systems discussed in Section 2 gather map/street data from various sources such as federal database or even volunteer participation, indoor floor plan acquisition is deemed to be more challenging. There are multiple challenges, which pose an obstacle to acquire an indoor floor plan: (i) in an multi-tenant public/commercial buildings, different sections might be managed by different entities; therefore, no central authority may be liable for the whole area's map, (ii) it requires effort to maintain an up-to-date map since it is quite common for indoor premises to go through frequent changes or rearrangements, and (iii) manual adding/editing of indoor floor plans is not a scalable solution which was attempted in the past (e.g., Google Maps) but with limited success. The indoor floor plan construction through crowdsensing has drawn attention recently mainly due to: (i) the ubiquity of smartphones and (ii) the rich technology and cost-effective sensor technologies equipped inside the modern smartphones. In the following, we discuss a few such research that are targeted toward conceptualizing indoor floor plan through user participation utilizing their smartphone sensor measurements.

CROWDINSIDE

CrowdInside (Alzantot and Youssef, 2012) pioneers the indoor map conceptualization research utilizing only user smartphone sensor measurements. It operates following client/server architecture where the smartphone client records sensor measurements, and periodically sends them to the server which in turn collates and processes them to come up with the floor plan layout. The user motion traces take the form of ⟨timestamp, location, Wi-Fi measurements⟩ tuple where dead-reckoning approach is followed for their generation together with error resitting fix through known absolute landmarks. CrowdInside can provide both overall and detailed floor plan layout where the overall floor plan only shows the occupancy map based on the user motion traces. The detailed one can identify rooms/corridors, and define the boundaries among them. To provide such details, the motion traces are divided into segments by taking into account the events such as sharp turn, inactivity tracking identified through the smartphone inertial sensor measurements. These segments are classified as rooms and corridors, and then a density-based clustering algorithm is used to merge "similar" segments together, and define boundaries/connectivity between them.

MAPGENIE

MapGenie (Philipp et al., 2014) infers indoor floor plan layout of a building from user smartphone's inertial sensor measurements, and its exterior structural information which can be obtained via OSM of Section 2.2. From the user motion traces, first, the hallways are detected, and their skeletons are approximated under certain constraints inferred from the exterior structural information of the building. Thereafter, the room geometry is estimated where room segments correspond to the maximum-length continuous sequences of motion traces without overlapping the hallway skeleton. Finally, the grammar-based room-layout generation corrects the trace-based indoor model's inaccuracies.

JIGSAW

Jigsaw combines mobile computing with computer vision technologies, and uses probabilistic and optimization methods to conceptualize indoor floor layout (Gao et al., 2014). Computer vision enables to obtain rich features and detailed information of a user's surroundings, while inertial sensors of mobile devices provide coarser information of the indoor environment but at lower computational complexity. First, some absolute landmarks' (entrance, elevators, etc.) placement and orientation information are inferred from crowdsourced photos and smartphone inertial sensor measurements. Subsequently, this map is augmented by including walls for external hallways, its structure, and room shapes.

IMOON

iMoon's operation is based on client/server architecture where the smartphone client collects imagery (i.e., photos), inertial sensor (i.e., accelerometer, gyroscope), and Wi-Fi RSS measurements, and then sends them to the server (Dong et al., 2015). The iMoon server builds the 3D floor plan utilizing the crowdsourced photos applying Structure from Motion (SfM) techniques (Stockman and Shapiro, 2001), and incorporating user trajectories perceived via the inertial sensor measurements, and application of dead-reckoning method. The Wi-Fi measurements serve the purpose of geolocation through a typical location fingerprinting technique (Bahl and Padmanabhan, 2000; Youssef and Agrawala, 2005). Localizing a client device is resolved via a two-step process where a query photo together with the Wi-Fi measurements are submitted. First, k-nearest neighbor (k-NN) algorithm is applied to acquire a set of approximate locations based on Wi-Fi fingerprinting. Second, iMoon server selects the partitions corresponding to these coarse locations, and searches the photos within the partition space to find the query photo's best match. This two-step process improves the computational complexity of iMoon localization. The features of this client query are added to the overall knowledge base under certain constraints to facilitate crowdsensing approach of constructing and maintaining an up-to-date database.

3.3 Discussions

It is apparent from Table 2 that the earlier calibration-free indoor localization research (TIX, SDM, or even EZ) required occasional fix from known locations for their operation while the more recent ones (UnLoc, Walkie-Markie) tend to focus on crowdsourcing approach of inferring such fixes. They also tend to incorporate modern smartphone's equipped inertial sensor such as accelerometer, gyroscope measurements more effectively. Although none of the discussed calibration-free localization research requires floor plan knowledge; the crowdsensing operating principle of UnLoc and Walkie-Markie will put them in a better position for adaptation for constructing such maps if required.

Table 2's list of crowdsourced indoor map conceptualization techniques make use of inertial sensors. Their sole usage was not as successful as the ones where they are used in conjunction with vision sensors (e.g., camera) (Gao et al., 2014; Dong et al., 2015). However, the photo captures require explicit participation on the users' parts, which arguably undermine the true spirit of crowdsourcing approach. All the listed crowdsourcing based floor plan construction approaches such as CrowdInside, MapGenie, Jigsaw, iMoon, follow client/server architecture where a data acquisition client runs inside a user's smartphone. The main floor plan construction-related computing intensive tasks are performed at the server. While this division of tasks between a client and the server is practical from energy efficiency perspective, early calibration-free localization methods attempted pure client-based approach ensuring security/privacy but they could only provide coarser accuracy (e.g., TIX/SDM). On the contrary, EZ, UnLoc, Walkie-Markie ensured better localization accuracy but at the expense of client being tracked transparently at the server.

Table 2 Comparison of Some Map-Free Indoor Localization, and Crowdsourced Indoor Map Conceptualization Research

	Need Location Fix	User Participation	Sensors	Accuracy and Precision	Security and Privacy	Architecture
TIX/SDM	Yes; from APs with known locations	N/A	Wi-Fi	Average error ~5.4 m (TIX) and ~3 m (SDM)	Yes	Client-based
EZ	Yes; occasional fix, e.g., GPS lock near entrance	Implicit	Wi-Fi	2 m inside a building of size 486 m^2; 7 m inside a large building (12,600 m^2)	No	Location computed at server
UnLoc	No; only one landmark's location needed during bootstrapping	Implicit	Inertial, Wi-Fi	Median error ~1.69 m across three different indoor setups (largest being 4000 m^2)	No	Location computed at server
Walkie-Markie	No; only one landmark's location needed during bootstrapping	Implicit	Inertial, Wi-Fi	Average error ~1.65 m inside a medium-sized office floor (3600 m^2)	No	Location computed at server
SmartSLAM	No	Implicit	Inertial, Wi-Fi	Average error 10 ± 5.73 m using raw mobility traces	N/A	Client/server
Wi-Fi SLAM	No; but Isomap without finer resolution required	Implicit	Wi-Fi	Average error 3.97 ± 0.59 m along $250 \sim 500$ m traces	N/A	Client/server
CrowdInside	No	Implicit	Inertial, Wi-Fi	Not reported	No	Client/server
MapGenie	No	Implicit	Inertial	72% rooms are found with at most 0.37 m average error	No	Client/server
Jigsaw	No	Explicit	Inertial, Wi-Fi, Camera	Average error 0.61–1.80 m; 80% of the hallway size is correctly identified	No	Client/server
iMoon	No	Explicit	Inertial, Wi-Fi, Camera	More than 90% measurements points are correctly localized	No	Client/server

4 Research Challenges of Crowdsourced Indoor Floor Plan Construction

4.1 Quality of Crowdsourced Data

The volume of user smartphone originated crowdsensed data has increased significantly over the years as a result of its widespread availability and affordability. There is a surge in utilizing the smartphone equipped technology in order to understand consumer behavior from a retailer perspective as well. Gathering such information would facilitate interesting BI applications such as devising personalized marketing strategy, and ensuring efficient store management functions. For example, becoming aware of the route a customer has taken can play a significant role in arranging product displays, thereby enhancing a customer's shopping experience on one hand, and also helping to maximize sales, on the other.

One of the challenges for the incorporation of crowdsourced data for indoor floor layout conceptualization is to ensure the localization algorithm's accuracy and precision since the more accurate and precise the indoor location information is; the more meaningfully the position annotated crowdsourced data can be used. Consequently, the challenges associated with the calibration-free indoor localization techniques, for example, acquiring location fix, additional sensors requirement, available infrastructure, algorithmic complexity, etc., which are discussed in Hossain and Soh (2015), have impact in this family of research as well. The crowdsourced mechanism of conceptualizing indoor floor layout mandates implicit user participation as can be seen from Table 2. Therefore, their challenges are quite different that are identified for explicit user feedback-related research, for example, sentiment analysis of social media, quantifying trust-worthiness of an individual user, etc. The main challenge here is to correctly model the user surroundings through the usage of their devices' sensor measurements of *good* quality, and adopted localization and mapping algorithm's efficiency.

4.2 Implications of Internet of Things (IoT) Devices' Equipped Sensors

The inertial sensor (accelerometer, gyroscope, compass) measurements are incorporated in all the crowdsourced indoor floor plan construction-related research discussed in Section 3, such as CrowdInside, MapGenie, Jigsaw, iMoon. Additionally, Jigsaw and iMoon utilize camera sensor measurements, and apply computer vision technology in conjunction with them. The adopted inertial sensor measurements traditionally constitute a fundamental component of dead-reckoning techniques commonly seen in SLAMs. Through the use of camera sensors, the geometric features of the surroundings are captured which in turn refine the inertial sensors' motion traces by identifying the known landmarks. Both Jigsaw and iMoon reported finer accuracy and precision compared to CrowdInside and MapGenie which were purely inertial sensor-oriented research. This outlines the importance of combining computer vision technologies with mobile technologies.

Alternatively, it also necessitates more sophisticated algorithms compared to the ones that have been used so far regarding inertial sensors-based solutions. All the user IoT devices, for example, smartphones are generally equipped with the inertial sensors and camera that are utilized in such research. However, the quality of the sensor measurements (e.g., camera resolution) may have impact on the outcomes of the algorithms which need to consider these factors in more detail.

4.3 Dimension of the Floor Plan Layout

By definition, a 2D map is two-dimensional, which is the flat representation of the surroundings, whereas a 3D map is three-dimensional, which depicts the surroundings' length, width, and height. Depending on the applications, both variants can be useful, for example, a 2D map is commonly used for navigation using only street names and numbers, whereas a 3D map may be needed when the depth information is also required such as Google Map's street view. The inertial sensor-based research can only construct a 2D map, whereas iMoon's output is a 3D model, which is only possible because of the use of camera sensor and computer vision techniques. This may come at the expense of higher computational complexity, and additional latency incurred for accessing the LBSs. All these factors should be weighed carefully against the provided LBSs' needs, for example, whether they can be met by a 2D or 3D map which subsequently have impact on the technologies adopted, and the algorithm choices.

4.4 Type of Architecture

The crowdsourced indoor floor layout construction technique generally operates upon client/server architecture as can be from Table 2. The data acquisition component resides inside the smartphone client which delivers the collected sensor data to the server, which then conceptualize the indoor map using them. To reduce the communication overhead, the smartphone client might include additional component for preprocessing the data in order to ensure its quality before sending it to the server. However, this may have adverse impact on the resource constrained IoT devices in terms of energy efficiency. Therefore, a balance between certain performance metrics should be explored in designing the architecture, and the delegation of various tasks between the smartphone client and the server. In addition to that, the number of sensor data that needs to be communicated may increase the communication overhead again, and also add to the server's computational complexity. An ideal solution would be to use as little sensor data as possible but still able to model the surroundings correctly. This warrants more sophisticated algorithms to be adopted which leaves a plenty of research opportunity in terms of the architecture chosen, and the division of client and server functionalities.

4.5 Privacy and Security

Security corresponds to system resilience toward attacks from adversaries, and privacy ensures the confidentiality of system data. Traditionally, the client-based indoor localization where the client computes its location itself by using the available infrastructure beacons was preferred in terms of privacy. This is due to the fact that the client's location information will then be private. It may not need to be communicated with the localization infrastructure to access the available LBSs. There are alternative measures to ensure anonymity of a client device, for example, MAC randomization for the infrastructure-based solutions as well. Accessing location-based data at the server or infrastructure is a prerequisite for the crowdsourced approach of indoor map construction as can be seen from Table 2. Ensuring client anonymity at the server is more relevant compared to opting for a client-based solution for such systems. Furthermore, the research that strive to combine computer vision techniques with mobile computing such as iMoon, Jigsaw would require photos to be taken inside the areas of interest. This not only poses a question mark whether they are truly a crowdsourced approach but also gives rise to various privacy- or security-related issues especially for a commercial or public building setting.

5 Conclusion

In this chapter, we reviewed the existing popular outdoor Web map systems, and discussed the degree to which crowdsourcing was incorporated inside them. Their inapplicability inside an indoor environment was outlined with rationale. Subsequently, a few crowdsourced indoor map construction techniques, and their characteristics were discussed. The similarity of their working principle with some calibration-free indoor localization techniques was pointed out, and a comparative discussion is provided based on a few performance metrics. Finally, the challenges associated with the crowdsourced approach of indoor floor layout conceptualization were discussed, and a few future research directions have been identified.

References

Alzantot, M., Youssef, M., 2012. Crowdinside: automatic construction of indoor floorplans. In: Proceedings of the 20th International Conference on Advances in Geographic Information Systems. SIGSPATIAL '12. ACM, New York, NY, pp. 99–108.

Apple, 2017. iOS—Maps. Available from: https://www.apple.com/ios/maps/ (Accessed July 2018).

Arsanjani, J.J., Zipf, A., Mooney, P., Helbich, M., 2015. OpenStreetMap in GIScience: Experiences, Research, and Applications. Springer Publishing Company, Inc, Berlin.

Azizyan, M., Constandache, I., Roy Choudhury, R., 2009. Surroundsense: mobile phone localization via ambience fingerprinting. In: Proc. of ACM. MobiCom'09, pp. 261–272.

Bahl, P., Padmanabhan, V.N., 2000. RADAR: an in-building RF-based user location and tracking system. In: Proc. IEEE INFOCOM, Tel Aviv, Israel, pp. 775–784.

Bing, 2017. Bing Maps—directions, trip planning, traffic cameras & more. Available from: https://www.bing.com/maps (Accessed July 2018).

Chintalapudi, K., Padmanabha Iyer, A., Padmanabhan, V.N., 2010. Indoor localization without the pain. In: Proc. of ACM. MobiCom'10, pp. 173–184.

Constandache, I., Choudhury, R.R., Rhee, I., 2010. Towards mobile phone localization without war-driving. In: Proc. of IEEE INFOCOM, pp. 1–9.

Dong, J., Xiao, Y., Noreikis, M., Ou, Z., Ylä-Jääski, A., 2015. imoon: using smartphones for image-based indoor navigation. In: Proceedings of the 13th ACM Conference on Embedded Networked Sensor Systems, SenSys '15. ACM, New York, NY, pp. 85–97.

Durrant-Whyte, H., Bailey, T., 2006. Simultaneous localization and mapping: part I. IEEE Robot. Autom. Mag. 13 (2), 99–110.

Ferris, B., Fox, D., Lawrence, N., 2007. WiFi-SLAM using Gaussian process latent variable models. In: Proc. of the 20th International Joint Conference on Artificial Intelligence, pp. 2480–2485.

Gao, R., Zhao, M., Ye, T., Ye, F., Wang, Y., Bian, K., Wang, T., Li, X., 2014. Jigsaw: indoor floor plan reconstruction via mobile crowdsensing. In: Proceedings of the 20th Annual International Conference on Mobile Computing and Networking, MobiCom '14. ACM, New York, NY, pp. 249–260.

Google, 2017a. Indoor Maps—About—Google Maps. Available from: https://www.google.com/intl/en_uk/maps/about/partners/indoormaps/ (Accessed July 2018).

Google, 2017b. Map—Google Maps. Available from: https://maps.google.com/ (Accessed July 2018).

Gwon, Y., Jain, R., 2004. Error characteristics and calibration-free techniques for wireless LAN-based location estimation. In: Proc. of the Second ACM MobiWac. New York, NY, pp. 2–9.

Hardegger, M., Roggen, D., Mazilu, S., Troster, G., 2012. ActionSLAM: using location-related actions as landmarks in pedestrian SLAM. In: Proc. of Indoor Positioning and Indoor Navigation (IPIN), pp. 1–10.

Harkins, H., 2014. Real Analytics: What's In-Store Brickstream Global Retailer Survey, Georgia, USA.

HERE, 2017. HERE offers you accurate location-driven data solutions. Available from: https://here.com/en (Accessed July 2018).

Hossain, A.K.M.M., Soh, W.S., 2015. A survey of calibration-free indoor positioning systems. Comput. Commun. 66, 1–13. https://doi.org/10.1016/j.comcom.2015.03.001.

indoo.rs, 2017. Indoo.rs | Accurate real-time indoor localization. Available from: https://indoo.rs/ (Accessed July 2018).

Intel, 2017. Intel Euclid™ Community. Available from: http://www.euclidcommunity.intel.com/ (Accessed July 2018).

Lee, J., 2017. 4-1: Invited paper: mobile AR in your pocket with Google Tango. SID Symp. Dig. Tech. Pap. 48 (1), 17–18. https://doi.org/10.1002/sdtp.11563.

Lim, H., Kung, L.C., Hou, J.C., Luo, H., 2010. Zero-configuration indoor localization over IEEE 802.11 wireless infrastructure. Wirel. Netw. 16 (2), 405–420.

MapQuest, 2017. Official MapQuest—maps, driving directions, live traffic. Available from: https://www.mapquest.com/ (Accessed July 2018).

Montemerlo, M., Thrun, S., Koller, D., Wegbreit, B., 2002. Fastslam: a factored solution to the simultaneous localization and mapping problem. In: Proceedings of the AAAI National Conference on Artificial Intelligence, pp. 593–598.

Navigine, 2017. Navigine | Universal Platform for Navigation. Available from: http://www.navigine.com/ (Accessed July 2018).

Navisens, 2017. Navisens | Location Aware. Available from: http://www.navisens.com/ (Accessed July 2018).

Neis, P., Zielstra, D., Zipf, A., 2012. The street network evolution of crowdsourced maps: openstreetmap in Germany 2007–2011. Futur. Internet 4, 1–21.

OpenStreetMap, 2017. OpenStreetMap. Available from: https://www.openstreetmap.org (Accessed July 2018).

Peterson, M.P., 2014. Mapping in the Cloud. The Guildford Press, New York, USA.

Philipp, D., Baier, P., Dibak, C., Durr, F., Rothermel, K., Becker, S., Peter, M., Fritsch, D., 2014. MapGenie: grammar-enhanced indoor map construction from crowd-sourced data. In: 2014 IEEE International Conference on Pervasive Computing and Communications (PerCom), pp. 139–147.

Rai, A., Chintalapudi, K.K., Padmanabhan, V.N., Sen, R., 2012. Zee: zero-effort crowdsourcing for indoor localization. In: Proc. of ACM MobiCom'12, pp. 293–304.

Robertson, P., Angermann, M., Krach, B., 2009. Simultaneous localization and mapping for pedestrians using only foot-mounted inertial sensors. In: Proc. of ACM Ubicomp, pp. 93–96.

Robertson, P., Angermann, M., Khider, M., 2010. Improving simultaneous localization and mapping for pedestrian navigation and automatic mapping of buildings by using online human-based feature labeling. In: IEEE/ION Position Location and Navigation Symposium (PLANS), pp. 365–374.

Shen, S., Michael, N., Kumar, V., 2012. Autonomous indoor 3D exploration with a micro-aerial vehicle. In: IEEE International Conference on Robotics and Automation, pp. 9–15.

Shen, G., Chen, Z., Zhang, P., Moscibroda, T., Zhang, Y., 2013. Walkie-Markie: indoor pathway mapping made easy. In: Proceedings of the 10th USENIX Conference on Networked Systems Design and Implementation, Berkeley, CA, pp. 85–98.

Shin, H., Chon, Y., Cha, H., 2012. Unsupervised construction of an indoor floor plan using a smartphone. IEEE Trans. Syst. Man Cybern. C (Appl. Rev.) 42 (6), 889–898.

Stockman, G., Shapiro, L.G., 2001. Computer Vision, first ed. Prentice Hall, Upper Saddle River, NJ.

Tango, 2017. Tango—Google. Available from: https://get.google.com/tango/ (Accessed July 2018).

TryRating, 2017. Tryrating—Apple. Available from: https://tryrating.com/ (Accessed July 2018).

Wang, H., Sen, S., Elgohary, A., Farid, M., Youssef, M., Choudhury, R.R., 2012. No need to war-drive: unsupervised indoor localization. In: Proc. of ACM. MobiSys'12, pp. 197–210.

Waze, 2017. Free Community-Based Mapping, Traffic & Navigation App. Available from: https://www.waze.com/ (Accessed July 2018).

Welch, G., Bishop, G., 1995. An Introduction to the Kalman Filter University of North Carolina at Chapel Hill, Chapel Hill, NC.

Youssef, M., Agrawala, A., 2005. The Horus WLAN location determination system. In: Proc. of ACM MobiSys, pp. 205–218.

6

Radio Fingerprinting-Based Indoor Localization: Overcoming Practical Challenges

Thomas Burgess

INDOO.RS GMBH, WIEN, AUSTRIA

1 Introduction

Awareness and demand for location-based services (LBS) has grown with the ever-increasing ubiquitousness of Global Navigation Satellite System (GNSS) enabled devices. Current generation LBS drives multitudes of novel technological developments in location and context aware computing over a wide range of fields (ABI Research, 2015). This has also lead to increasing demand for LBS in environments without access to reliable GNSS, such as urban and indoor environments. Hence, many alternate approaches to providing accurate urban and indoor positioning have been developed (Torres-Solis et al., 2010). The range of applicable approaches is significantly reduced by focusing on solutions able to work on unmodified smartphones along with cheap and easy to install infrastructure. Radio fingerprinting localization (Kjærgaard, 2007) is one of the most common approaches able to overcome these restrictions.

The remainder of this chapter provides further motivation for radio fingerprinting in Section 1.1, introduces the basic methodology and assumptions in Section 1.2, and elaborates on how to overcome some of the technical challenges in Section 2. This overview is written from the point of view of a practical commercial use.[1]

1.1 Motivation

With direct line of sight to enough satellites GNSS can enable accurate location awareness at low cost to a vast range of mobile devices. However, in urban and indoor environments large infrastructure such as buildings can disrupt the signal path. In the extreme case of urban canyons only a narrow strip of sky is directly visible at street level. These disruptions

[1] Based on several years of experience deploying radio fingerprinting solutions at indoo.rs GmbH.

Geographical and Fingerprinting Data for Positioning and Navigation Systems. https://doi.org/10.1016/B978-0-12-813189-3.00008-3

can lead to complete signal loss or to signal distortion through reflection and refraction which invalidates the line of sight assumptions in the localization. Furthermore, there may be multiple signal paths of similar strengths for the same source (Braasch, 2017). Together these effects dramatically impair the applicability of satellite-based methods in such environments.

Therefore, some alternate approach must be taken, such as Ultra Wide Band (UWB) radio (Ingram et al., 2004), Radio-Frequency IDentification (RFID) (Bekkali et al., 2007; Gikas et al., 2016), optical solutions (Mautz and Tilch, 2011; Kuo et al., 2014), ultrasound (Hazas and Hopper, 2006), or dead reckoning (Jimenez et al., 2009). However, for a solution to be widely adopted it is required to work on unmodified end-user mobile devices such as smartphones. In this sense the aforementioned methods have severe shortcomings: UWB radio is unavailable to most devices, RFID methods are range limited on smartphones, optical and ultrasound methods work poorly with pocketed devices, and dead reckoning solutions with fixed sensor placement require phones to be predictably coupled to their owners.

This situation leaves solutions using the motion sensors and radio receivers commonly available on smartphones. Some possible radio solutions are impeded by requiring expensive or hard to install custom hardware. Additionally, many methods are sensitive to strong multipath and fading effects in indoor environments. Together, this reduces viability of multilateration, Angle Of Arrival (AOA) (Rong and Sichitiu, 2006), and Time Difference Of Arrival (TDOA) (Liu et al., 2007) methods. Moreover, methods operating standalone on the device are preferable, as maintaining an uninterrupted remote server connection is not always possible. Server side solutions also introduce additional latency from round trip time and increased battery drain due to data transfers, which may significantly degrade user navigation experience.

Radio fingerprinting with on-board radios can overcome all of these obstacles. With enough visible signal sources the method is robust to noise in any particular input. Depending on the device, different sources can be used, such as FM radio (Chen et al., 2012), GSM (Otsason et al., 2005), Wi-Fi (Al Nuaimi and Kamel, 2011), and Bluetooth Low Energy (BLE) (Faragher and Harle, 2015). In practice, the most common fingerprinting sources are Wi-Fi and BLE, which both offer unique and complementary features. Wi-Fi is supported on a multitude of devices, is ubiquitously installed, and has 2.4 and 5 GHz radio bands with different propagation properties. Meanwhile, BLE offers iBeacon and Eddystone devices that can be made very small, cheap and easy to install, and still have good localization performance (Zhao et al., 2014). Unlike most Wi-Fi installations, BLE setups are often optimized for localization performance. Typical BLE scan rates are faster than for Wi-Fi leading to more responsive localization. Another important difference is that on Apple iOS devices only the BLE Application Programming Interface is available to developers. In some cases it is possible to use both sources at the same time, however, as the antenna commonly is shared this may lead to poor performance. With this in mind, the remainder of the chapter is restricted to a discussion of radio fingerprinting on mobile devices with Wi-Fi and BLE receivers.

It should be noted that commonly absolute radio fingerprinting positioning is augmented by fusing it with estimates based on on-board motion sensors, for instance, using Pedestrian Dead Reckoning (PDR) (Ettlinger et al., 2017; Pratama et al., 2012) and Kalman filters (Burgess and Dong, 2016), or by exploiting local disturbances to the geomagnetic field (Li et al., 2016; Guo et al., 2014).

1.2 Radio Fingerprint Localization Assumptions

At its core, radio fingerprinting relies on premeasured radio reference maps with point measurements. Each reference point contains a set of measured received signal strength indicator (RSSI) readings (fingerprints) from identifiable sources. Positions are estimated by comparing a new RSSI measurement to the radio map, and computing a position from the most similar reference points. This approach makes several basic assumptions:

Environment No major change has taken place in the radio environment since the creation of the radio map.

Transmitters Transmitters are fixed in place, transmit at a constant signal strength, and they are uniquely identifiable.

Reference point Reference points see enough visible signals with enough variations to provide a location unique hierarchy.

Receivers Radio receiver characteristics of locating devices and mapping devices are similar.

As long as these assumptions hold, relatively good localization is possible with residuals in the range of 2–5 m. With infrastructures that obfuscate transmitter ID or modulates transmission power, localization becomes inaccurate or unfeasible. Over time small changes to the radio environment are inevitable, and subsequently the localization accuracy will deteriorate unless the radio map is updated. These changes can have many causes, for example, architectural modifications such as the addition of walls or partitions, installation of doors or windows; interior changes such as displacement of tables or white boards; or changes in the radio infrastructure through aging or replacement of beacons, or new sources of interference such as microwaves or Wi-Fi infrastructure. However, if sufficient sources are visible, it can be several years until such random environment changes are larger than the inherent noise in the RSSI measurements. Certain enterprise systems sometimes randomize transmitted Basic Service Set IDentifier (BSSID) (unique transmitter identifier), share it between multiple physical devices, transmit multiple virtual devices from the same physical unit, or dynamically throttle the transmission strength. Furthermore, mobile access points and beacons are not uncommon and cannot be assumed to stay in place for any extended period of time. The presence of such unreliable transmitters likely will degrade localization accuracy, and should be filtered both from observations and the radio map. For opportunistic approaches using whatever signal sources are available filtering is essential for consistent results. The reference points should be dense enough so that neighboring points are similar but not identical, furthermore it is preferable that the reference point density is uniform throughout the building. Finally, the receiving (locating) devices should give comparable results to the devices used in mapping, and not be too sensitive to device posture, for example, hand held, placed in pocket, carried in hand bag, placed on desk, etc.

2 Fingerprinting Challenges

As described in Section 2.1, the similarity between an observation and the reference points can be estimated by comparing the signal sets. Special care has to be taken to ensure that signals are comparable, and that missing signals due to packet collisions, or weak signal strength are handled correctly. Once similarities to all reference points are estimated, they should be used to infer the position of the observations. There are many ways to do this, as outlined in Section 2.2. Which combination of similarity and estimation is most suitable may depend on environment, devices in use, and overall requirements of a solution. Furthermore, differences in device receiver characteristics must be taken into account, as follows in Section 2.3. Finally, as shown in Section 2.4, a radio map has to be available and up-to-date with the physical conditions at the time of localization.

2.1 Fingerprint Point Similarity

A radio fingerprint $\vec{x}$ is a set of observable signals under some given conditions. In a space equipped with N transmitters a fingerprint can be expressed as the vector

$$\vec{x} = \{x_0, \ldots, x_N\}, \tag{1}$$

where x_i is the RSSI obtained for transmitter i. It should be noted that x_i does not have to be a single reading, it can be a time series of observed RSSI or a set of summary statistics such as mean μ_i, standard deviation σ_i, and number of readings n_i. In principle, these vectors are sparse and may not have an entry for each of the N possible transmitters. This can be due to signal loss from announcement package collisions during measurements, or the RSSI is below the sensitivity threshold of the measuring device. A reference measurement $\vec{y}$ is simply a fingerprint observation made at a known position $\vec{p}$

$$\vec{y} = \{\vec{p}|x_0, \ldots, x_N\}. \tag{2}$$

A reference map is a set of many reference measurements taken at M different positions

$$\mathbf{Y} = \{y_0, \ldots, y_M\}. \tag{3}$$

The radio map Y then can be thought of as a sparse $M \times N$ matrix of readings and a set of M point locations.

Given an observation and the reference measurements, some measure of similarity (or inversely, distance) needs to be found to gauge which reference points are closer to the observation. For this to work it is important that the assumptions on an unchanged environment and fixed transmitters are fulfilled. Furthermore, in situations with few visible transmitters the similarity estimates will have very large uncertainties that are likely to lead to poor location estimates.

The distance between the sets is commonly expressed as the Minkowski distance

$$d(x, y) = \left(\sum_{i=0}^{n} |x_i - y_i|^\alpha \right)^{1/\alpha}, \tag{4}$$

which for $\alpha = 1$ corresponds to Manhattan (City block or Taxi) distance, and for $\alpha = 2$ to Euclidean distance. To use the distance as a similarity S it is commonly inverted to give a measure that decreases with distance, for example, $S(x, y) = d(x, y)^{-1}$ or $S(x, y) = a - d(x, y)$. A possible improvement is to weight the difference by reference variance σ_i^2, which for $\alpha = 2$ and assuming no off-diagonal covariances corresponds to the Mahalanobis or normalized Euclidean distance

$$d(x, y) = \left(\sum_{i=0}^{n} \frac{|x_i - y_i|^2}{\sigma_i^2} \right)^{1/2}. \tag{5}$$

Both of these measures with varying p and many other variants occur in fingerprinting solutions, and the right choice of similarity measure can reduce location uncertainty by 30% (Retscher and Joksch, 2016; Torres-Sospedra et al., 2015).

A nonlinear transformation of variables can further improve localization by another 10%, for instance, using exponentiation, powers, or z-score (Torres-Sospedra et al., 2015; Burgess et al., 2016). This can be explained by considering the logarithmic distance relation of RSSI. Close to the source, signals have a strong distance dependence, which gets much weaker further away. Thus, one should not expect optimal results when treating all signal strengths the same. In addition to transforming signals, they can also be subject to threshold to reject weak signals. This may reduce the noise sensitivity, at the cost of shorter range for the transmitters. However, this often does not have a large positive impact on localization accuracy (Torres-Sospedra et al., 2015). Furthermore, the calculated distance may be transformed in turn, to better correspond to a distance estimate.

As signal strength is not linear, some methods approach the problem by only considering the RSSI ordering between the two sets using rank-based methods such as Spearman's Footrule, F, and Kendall's τ, K (Kumar and Vassilvitskii, 2010; Machaj et al., 2011). In these approaches the observations $\vec{x}$ and $\vec{y}$ are sorted by RSSI and only the order of the transmitters is retained as $\vec{x}'$, $\vec{y}'$. For instance with an observation $\vec{x} = \{-30, -20, -40\}$ the ranking vector becomes $\vec{x}' = \{2, 1, 3\}$. Spearman's Footrule measures the disarray as the sum of the absolute difference between ranks in a signal.

$$F(\vec{x}', \vec{y}') = \sum_i |\vec{x}_i' - \vec{y}_i'|. \tag{6}$$

Kendall's τ counts the number of pairs of transmitters that are in opposite order in the two observations, as follows

$$K(\vec{x}', \vec{y}') = \frac{2}{n(n-1)} \sum_{i,j \in N} 1 \text{ if } i, j \text{ in opposite order in } \vec{x}' \text{ and } \vec{y}' \text{ else } 0. \tag{7}$$

As ordering is not affected by scaling or offset, these approaches are insensitive to device gain settings. Furthermore, they often produce more robust results than metric similarities. However, some information is lost as the amount of difference leading to rank difference is lost. This is especially at weak signal strengths, where noise can give significant contribution to rank order. Here thresholding can help alleviate the impact of noise. Transformations and rank-based methods can also help to reduce device heterogeneity issues outlined in Section 2.3.

Moreover, it is common that some of the readings in the sparse reference vector lack a common reading for the same transmitter in the observation vector. This is a big problem, as the similarity measures rely on pairwise comparisons between the two sets. Signals can be cut out by the device because they are too weak or because of collisions. The probability that a signal disappears from an observation is inversely correlated to its strength, thus making it a bigger problem for weak signals.

There are several approaches to dealing with missing signals. The simplest approach is to only consider the observations common between the observation and reference point. However, the set of common transmitters is expected to be smaller with increasing distance between observation and reference. For many distance measures this introduces a bias going against the true distance as fewer terms are included in the sums. Thus, this can lead to degraded performance unless compensated for. A more refined approach is to only consider the transmitters visible in the observation, and to fill in some default missing value for missing transmitters in the reference points. Typically, the minimal observable signal is used (e.g., -100 dB) for this. This works and is in common use in practical applications. However, it is a rather blunt approach as the *true* signal could be much stronger or weaker than the default value. In the extreme case where no common signal is seen between the observation and reference, clearly this approach does not produce a meaningful result. The probability that a signal disappears depends on the location, which could be used as an observable in its own right in addition to RSSI to strengthen localization. Optimizing which similarity estimate to use and whether or not to transform RSSI or similarity value is not trivial. Furthermore, the similarity estimation is deeply entangled with the subsequent error estimation. In practice, it is safest to try multiple approaches using well-understood testing data.

2.2 Location and Error Estimation

The most common way to obtain a location from reference points with similarity estimates is the k-nearest neighbors (kNN) (Bahl and Padmanabhan, 2000) and its weighted k-nearest neighbors (WkNN) (Shin et al., 2012) variant. Assuming $k = 1$, the position is assigned to be at the most similar fingerprint point, while the average position is taken for higher k. Like the choice of similarity measure, the optimal value for k is best determined empirically as it depends on the radio environment. Using WkNN, the average points are weighted by similarity, which to some extent reduces the impact of k. While position estimation is straightforward with this method, accurate error estimation is not (Zhuo et al., 2012). Typically some empirical relation to the average distance between reference points and estimate is employed. It should be noted that the kNN methods may be biased by an uneven distribution of reference points. Such unevenness may cause problems in manually measured maps, where it is difficult to ensure perfect reference point distribution. However, unevenly distributed points can also be used to improve results by biasing localization toward much traveled trajectories such as corridors and doorways. To obtain a uniform point density, uneven points can be interpolated to fixed grid, for example, through a Gaussian process, or simple nearest neighbor interpolation.

It is also possible to employ probabilistic methods to try to find the location, for instance, through maximum likelihood methods (Mirowski et al., 2014). However, usually this is orders of magnitude slower than kNN, and still not guaranteed to give better results. Further approaches use the radio map as training data for machine learning approaches such as support vector machines (Wu et al., 2004) or neural networks (NN) (Gogolak et al., 2011). By learning through the map, these approaches tend to avoid the tuning issues with similarity measures and avoid parameters like k. In principle, also localization error can be determined through machine learning when there is enough labeled high-quality training data available. Machine learning methods are less transparent than kNN, and thus less robust to noise and map degradation unless specifically trained for these situations.

Often a preselection phase is applied before location estimation. Preselection can improve calculation speed, reduce memory requirements, and to reduce the risk of large position outliers, *jumps*. In this phase, reference points clearly not compatible with observation are filtered out based on which signals were recently visible. If the preselection can operate with only partial summary information, it can dramatically improve calculation speed and reduce memory footprint, as both directly depend on the number of points to consider (Burgess et al., 2016). Furthermore, the number of reference points available grows with the square of distance to the observation. Clearly incompatible reference points can be identified from having few observation matches and significant difference in RSSI. Jumps occur when there are one or a few observation scans with very few visible transmitters. With fewer transmitters to discriminate with, a larger number of references will give plausible matches. As the similarity gradient is weaker with few readings the final position essentially is a random point inside the matched references. By restricting the range of reference points to consider, the magnitude of these jumps can be reduced.

It is essential that the observation occurs within the range of selected points as most fingerprinting approaches are restricted to producing results in the convex hull of the selected reference points. Preselection can be done by rejecting reference points with less than N matching networks to the observation, through spatial clustering (Ma et al., 2008), or simply by reducing the map by joining nearby fingerprint points. In clustering approaches, nearby points with similar properties are grouped into clusters, for instance, through k-means clustering. The joining approach works similarly but relies on partial hierarchical clustering. Aggregates for the clusters are compared to the observation and references in close clusters are used as preselected points. Further improvements of localization can be attained through the use of a Kalman filter (Yim et al., 2008), which considers the previous estimate and current observation to produce a smooth new estimate.

As with similarity measures, finding the optimal localization strategy is not trivial. Machine learning approaches often are able to show the highest accuracy, at the cost of training data volume and robustness. Trying all combinations of localization algorithms, similarity measures, and tunable parameters also may lead to problems with overtraining (Murphy, 2012). Commonly, an empirical trial and error approach is taken to find a combination of methods good enough for the application at hand.

2.3 Device Heterogeneity

The mobile device market is highly heterogeneous, with a large number of distinct device producers, chipsets, revisions, and firmware updates. Different devices measuring the same transmitter in identical conditions often still report different RSSI, typically by a constant offset or in some cases a linear term (Dong et al., 2017). This is a common problem, as very often the choice of locating terminal (i.e., smart phone make and model) is not influenced by the LBS provided. A difference in RSSI characteristics will bias the similarity estimates. If this difference is large it can severely impact localization accuracy.

Device heterogeneity can be handled by avoiding it altogether or by transforming values to a common scale. The aforementioned rank-based similarity methods from Section 2.1 avoids the problem by ignoring the RSSI value and only considering the sort-order of detected transmitter (Kumar and Vassilvitskii, 2010; Machaj et al., 2011). This works as the strongest network will have the highest rank in both reference and observation, regardless of scaling and offsets difference from gain or bias in the receiving antennas. Another approach is to employ differential fingerprinting where the constant offset is removed by comparing differences between RSSI within the observation to difference within the references (Laoudias et al., 2013). For example, consider an observation reference pair taken at the same location but with a 10-dB offset in the reference: $\vec{x} = \{-30, -50\}$, $\vec{y} = \{-40, -60\}$. Without knowing the offset the points will not be considered similar with Minkowski distances. With differential fingerprinting, instead internal pairwise differences are considered $\vec{x}^{\dagger} = \{-30 - -50\} = \{20\}$, and $\vec{y}^{\dagger} = \{-40 - -60\} = \{20\}$, and the Minkowski distance will be 0. However, these approaches do not work when trying to combine statistics from multiple measurements, for instance, when making a radio map.

If the range of possible RSSI values is exactly known for a transmitter, it is trivial to make a linear transformation to clamp the signals to a predetermined range. However, if the devices have different sensitivity (i.e., lower minimal RSSI threshold) this approach may lead to biased estimates. Furthermore, care has to be taken with special values, that is, $\pm\infty$, not-a-number (NaN), or *magic* values such as 0 or -100 dB. Thus, some sort of predetermined linear fit from simultaneous measurements between devices often is used, either through manual or automatic procedures (Dong et al., 2017). In practice, this entails having two devices measuring the same signal over the full range of relevant signal strengths, and then make a fit to the scatter plot of one device against the other. This has to be done between all potential devices and some standard device to provide universal standardization. It should be noted that maintaining a database with cross-calibration between all common devices is a daunting and error prone task.

2.4 Obtaining and Updating Radio Maps

The radio map is the set of all reference points $Y = \{\vec{y}_1, \ldots, \vec{y}_M\}$. For optimal results, these maps should span the entire navigable area with a uniform point density and high-quality point estimates. The distribution of the reference points through the building affects the localization. If the density of points is too low important features may be missed, resulting in poor accuracy. On the other hand, too high density increases storage, memory, and computational load without improved localization accuracy. Nonuniform point densities may lead to problems when tuning localization algorithms to work optimally with the radio map. Another consideration is that most localization algorithms only produce results interior to the radio map and thus the radio map should span every navigable position. Common point densities range from 1 to 10 m. While only hexagonal grids can give truly uniform point density in two dimensions, often square grids are employed for the sake of simplicity. In practice it may not be possible to measure uniformly, and thus less rigid approximate grids also are common.

Creating accurate radio maps is a resource intensive task (Kaemarungsi and Krishnamurthy, 2004) that additionally requires regular updates due to the evolving radio environment. Manual measurements performed at each reference point is the most straightforward approach to obtaining radio reference maps. It should be noted that the quality of the measurements also affects the localization quality. For instance, low statistics leaves high uncertainty in RSSI estimates, and offsets between actual and reported point locations distort positioning. Measuring conditions such as the user partly shading the signal path, the influence of surrounding crowds, or dynamic environment can further degrade reliability of estimates. Measurements may take up to a minute per reference point, which for a 1000 m^2 venue and 1 m fingerprint density can mean over 15 h of measurements. For typical commercial deployments the radio map needs to be updated every 6–12 months in the author's experience. The amount of measurements and the care needed in making them have been a limiting factor in the commercial adoption of fingerprinting solutions in the indoor navigation market.

Numerous technologies have been developed to lower the cost of radio map creation. For instance, some use collected RSSI and precise transmitter location knowledge (Koo and Cha, 2012), and others propose ray-tracing to calculate fingerprints (Raspopoulos et al., 2012; Renaudin et al., 2017; Tayebi et al., 2009). Several employ Simultaneous Localization And Mapping (SLAM) algorithms (Frese et al., 2005; Ferris et al., 2007; Huang et al., 2011; Murphy, 2012; Koller and Friedman, 2009) to replace costly point-by-point measurements with simple continuous measurements along paths. From the author's practical experience with SLAM approaches they are able to reduce measurement time by a factor of 10 for initial mapping. By crowdsourcing some trajectories from navigating users, it is possible to use SLAM to update radio maps and thus circumventing the need for any additional dedicated measurements beyond initialization (Rai et al., 2012; Laoudias et al., 2013). It should be noted that crowdsourcing poses many privacy concerns, as well as potentially costly amounts of calculations.

3 Summary and Conclusions

In this chapter, the popularity of radio fingerprinting for indoor localization has been explained. The basic assumptions made on radio environment, transmitter, radio fingerprint reference points, and receiving devices have been clarified. After this introductory part follows a discussion on some of the challenges that arise when attempting to satisfy these basic assumptions. Here details are given on how radio map reference point similarity is calculated, how device heterogeneity can be handled, and how radio maps are built. In measuring fingerprint similarity the utility of nonlinear RSSI transformations, the use of rank-based methods to avoid otherwise cumbersome device heterogeneity, and possible dangers when missing signals are discussed. Outlines are given to how location and error estimation can be done using common kNN methods as well as with advanced machine learning alternatives. Furthermore, methods to address device heterogeneity have been elaborated. Finally, an overview of challenges and approaches related to creating maps has been given.

There is no single perfect solution for any one of the problems mentioned above, and certainly not for a combined localization system. Instead, approaches are selected by what fits the intended application best, that is, high accuracy, low latency, low computational complexity, low memory usage, robustness to noise, etc. In general this selection is best done using dedicated evaluation data taken in realistic conditions.

References

ABI Research, 2015. Indoor location in retail: where is the money? Available from: https://www.abiresearch.com/market-research/product/1013925-indoor-location-in-retail-where-is-the-mon/ (Accessed July 2018).

Al Nuaimi, K., Kamel, H., 2011. A survey of indoor positioning systems and algorithms. In: 2011 International Conference on Innovations in Information Technology (IIT). IEEE, pp. 185–190.

Bahl, P., Padmanabhan, V.N., 2000. RADAR: an in-building RF-based user location and tracking system. In: Proceedings IEEE INFOCOM 2000. Nineteenth Annual Joint Conference of the IEEE Computer and Communications Societies, vol. 2, pp. 775–784.

Bekkali, A., Sanson, H., Matsumoto, M., 2007. RFID indoor positioning based on probabilistic RFID map and Kalman filtering. In: Third IEEE International Conference on Wireless and Mobile Computing, Networking and Communications, 2007. WiMOB 2007, p. 21.

Braasch, M.S., 2017. Multipath. In: Springer Handbook of Global Navigation Satellite Systems. Springer, New York, NY, pp. 443–468.

Burgess, T., Dong, B., 2016. WIP: adaptive Kalman filter for indoor navigation. In: 2016 International Conference on Indoor Positioning and Indoor Navigation (IPIN) USB proceedings.

Burgess, T., Müllner, R., Schmitzberger, H., 2016. WIP: FLIP—flexible indoor position estimator. In: 2016 International Conference on Indoor Positioning and Indoor Navigation (IPIN) USB proceedings, pp. 44–53.

Chen, Y., Lymberopoulos, D., Liu, J., Priyantha, B., 2012. FM-based indoor localization. In: Proceedings of the 10th International Conference on Mobile Systems, Applications, and Services, pp. 169–182.

Dong, B., Burgess, T., Neuner, H.B., 2017. indoo. rs CaLibre-Unsupervised database free RSSI calibration indoor navigation. In: 2017 European Navigation Conference (ENC), pp. 198–205.

Ettlinger, A., Neuner, H.B., Burgess, T., 2017. Smartphone sensor-based orientation determination for indoor-navigation. In: Georg, G., Haosheng, H. (Eds.), Progress in Location-Based Services 2016. Springer International Publishing, Cham, pp. 49–68.

Faragher, R., Harle, R., 2015. Location fingerprinting with Bluetooth Low Energy beacons. IEEE J. Sel. Areas Commun. 33 (11), 2418–2428. Available from: http://ieeexplore.ieee.org/xpls/abs_all.jsp?arnumber=7103024.

Ferris, B., Fox, D., Lawrence, N.D., 2007. WiFi-SLAM using Gaussian process latent variable models. In: IJCAI, vol. 7, pp. 2480–2485.

Frese, U., Larsson, P., Duckett, T., 2005. A multilevel relaxation algorithm for simultaneous localization and mapping. IEEE Trans. Robot. 21 (2), 196–207.

Gikas, V., Dimitratos, A., Perakis, H., Retscher, G., Ettlinger, A., 2016. Full-scale testing and performance evaluation of an active RFID system for positioning and personal mobility. In: 2016 International Conference on Indoor Positioning and Indoor Navigation (IPIN), pp. 1–8.

Gogolak, L., Pletl, S., Kukolj, D., 2011. Indoor fingerprint localization in WSN environment based on neural network. In: 2011 IEEE 9th International Symposium on Intelligent Systems and Informatics (SISY), pp. 293–296.

Guo, Y., Chen, Y., Liu, J., 2014. Indoor location estimation based on local magnetic field via hybrid learning. In: Fuchen, S., Kar-Ann, T., Grana, R.M., Kezhi, M. (Eds.), Extreme Learning Machines 2013: Algorithms and Applications. Springer International Publishing, Cham, pp. 189–207. https://doi.org/10.1007/978-3-319-04741-6_14.

Hazas, M., Hopper, A., 2006. Broadband ultrasonic location systems for improved indoor positioning. IEEE Trans. Mob. Comput. 5 (5), 536–547.

Huang, J., Millman, D., Quigley, M., Stavens, D., Thrun, S., Aggarwal, A., 2011. Efficient, generalized indoor WiFi graphslam. In: 2011 IEEE International Conference on Robotics and Automation (ICRA). IEEE, pp. 1038–1043.

Ingram, S.J., Harmer, D., Quinlan, M., 2004. Ultrawideband indoor positioning systems and their use in emergencies. In: Position Location and Navigation Symposium, 2004. PLANS 2004, pp. 706–715.

Jimenez, A.R., Seco, F., Prieto, C., Guevara, J., 2009. A comparison of pedestrian dead-reckoning algorithms using a low-cost MEMS IMU. In: IEEE International Symposium on Intelligent Signal Processing, 2009. WISP 2009, pp. 37–42.

Kaemarungsi, K., Krishnamurthy, P., 2004. Properties of indoor received signal strength for WLAN location fingerprinting. In: The First Annual International Conference on Mobile and Ubiquitous Systems: Networking and Services, 2004. MOBIQUITOUS 2004. IEEE, pp. 14–23.

Kjærgaard, M., 2007. A taxonomy for radio location fingerprinting. In: Location- and Context-Awareness, pp. 139–156.

Koller, D., Friedman, N., 2009. Probabilistic Graphical Models: Principles and Techniques. MIT Press, Cambridge, MA. ISBN 978-0-262-01319-2.

Koo, J., Cha, H., 2012. Unsupervised locating of WiFi access points using smartphones. IEEE Trans. Syst. Man Cybern. Part C Appl. Rev. 42 (6), 1341–1353.

Kumar, R., Vassilvitskii, S., 2010. Generalized distances between rankings. In: Proceedings of the 19th International Conference on World Wide Web. ACM, pp. 571–580. Available from: http://dl.acm.org/citation.cfm?id=1772749.

Kuo, Y.-S., Pannuto, P., Hsiao, K.-J., Dutta, P., 2014. Luxapose: indoor positioning with mobile phones and visible light. In: Proceedings of the Annual International Conference on Mobile Computing and Networking, MOBICOM.

Laoudias, C., Zeinalipour-Yazti, D., Panayiotou, C.G., 2013. Crowdsourced indoor localization for diverse devices through radiomap fusion. In: 2013 International Conference on Indoor Positioning and Indoor Navigation (IPIN), pp. 1–7.

Li, Y., Zhuang, Y., Lan, H., Zhou, Q., Niu, X., El-Sheimy, N., 2016. A hybrid WiFi/magnetic matching/PDR approach for indoor navigation with smartphone sensors. IEEE Commun. Lett. 20 (1), 169–172. Available from: http://ieeexplore.ieee.org/xpls/abs_all.jsp?arnumber=7314907.

Liu, H., Darabi, H., Banerjee, P., Liu, J., 2007. Survey of wireless indoor positioning techniques and systems. IEEE Trans. Syst. Man Cybern. Part C Appl. Rev. 37 (6), 1067–1080.

Ma, J., Li, X., Tao, X., Lu, J., 2008. Cluster filtered KNN: a WLAN-based indoor positioning scheme. In: 2008 International Symposium on a World of Wireless, Mobile and Multimedia Networks, 2008. WoWMoM 2008, pp. 1–8.

Machaj, J., Brida, P., Piché, R., 2011. Rank based fingerprinting algorithm for indoor positioning. In: 2011 International Conference on Indoor Positioning and Indoor Navigation (IPIN), pp. 1–6.

Mautz, R., Tilch, S., 2011. Survey of optical indoor positioning systems. In: 2011 International Conference on Indoor Positioning and Indoor Navigation (IPIN), pp. 1–7.

Mirowski, P., Milioris, D., Whiting, P., Kam Ho, T., 2014. Probabilistic radio-frequency fingerprinting and localization on the run. Bell Labs Tech. J. 18 (4), 111–133.

Murphy, K.P., 2012. Machine Learning: A Probabilistic Perspective. The MIT Press, Cambridge, MA. ISBN 0262018020, 9780262018029.

Otsason, V., Varshavsky, A., LaMarca, A., De Lara, E., 2005. Accurate GSM indoor localization. In: UbiComp 2005: Ubiquitous Computing, p. 903.

Pratama, A.R., Hidayat, R., et al., 2012. Smartphone-based pedestrian dead reckoning as an indoor positioning system. In: 2012 International Conference on System Engineering and Technology (ICSET). IEEE, pp. 1–6.

Rai, A., Chintalapudi, K.K., Padmanabhan, V.N., Sen, R., 2012. Zee: Zero-effort crowdsourcing for indoor localization. In: Proceedings of the 18th Annual International Conference on Mobile Computing and Networking, pp. 293–304.

Raspopoulos, M., Laoudias, C., Kanaris, L., Kokkinis, A., Panayiotou, C.G., Stavrou, S., 2012. 3D ray tracing for device-independent fingerprint-based positioning in WLANs. In: 2012 9th Workshop on Positioning Navigation and Communication (WPNC), pp. 109–113.

Renaudin, O., Burgess, T., Zemen, T., 2017. Ray-tracing based fingerprinting for indoor localization, TD(17)05024. In: Proceedings 5th COST CA15104 IRACON Management Committee Meeting, Graz (Austria), September 2017.

Retscher, G., Joksch, J., 2016. Analysis of nine vector distances for fingerprinting in multiple-SSID Wi-Fi networks. In: Proceedings of the International Conference on Indoor Positioning and Indoor Navigation (IPIN), pp. 4–6.

Rong, P., Sichitiu, M.L., 2006. Angle of arrival localization for wireless sensor networks. In: 2006 3rd Annual IEEE Communications Society on Sensor and Ad Hoc Communications and Networks, 2006. SECON'06, vol. 1, pp. 374–382.

Shin, B., Lee, J.H., Lee, T., Kim, H.S., 2012. Enhanced weighted K-nearest neighbor algorithm for indoor Wi-Fi positioning systems. In: 2012 8th International Conference on Computing Technology and Information Management (ICCM), vol. 2, pp. 574–577.

Tayebi, A., Gomez Perez, J., Saez de Adana, F.M., Gutierrez, O., 2009. The application of ray-tracing to mobile localization using the direction of arrival and received signal strength in multipath indoor environments. Prog. Electromagn. Res. 91, 1–15.

Torres-Solis, J., Falk, T.H., Chau, T., 2010. A review of indoor localization technologies: towards navigational assistance for topographical disorientation. In: Ambient Intelligence. InTech.

Torres-Sospedra, J., Montoliu, R., Trilles, S., Belmonte, Ó., Huerta, J., 2015. Comprehensive analysis of distance and similarity measures for Wi-Fi fingerprinting indoor positioning systems. Expert Syst. Appl. 42 (23), 9263–9278.

Wu, C.L., Fu, L.C., Lian, F.L., 2004. WLAN location determination in e-home via support vector classification. In: 2004 IEEE International Conference on Networking, Sensing and Control, vol. 2, pp. 1026–1031.

Yim, J., Park, C., Joo, J., Jeong, S., 2008. Extended Kalman filter for wireless LAN based indoor positioning. Decis. Support. Syst. 45 (4), 960–971.

Zhao, X., Xiao, Z., Markham, A., Trigoni, N., Ren, Y., 2014. Does BTLE measure up against WiFi? A comparison of indoor location performance. In: Proceedings of European Wireless 2014; 20th European Wireless Conference, pp. 1–6.

Zhuo, W., Zhang, B., Chan, S.H.G., Chang, E.Y., 2012. Error modeling and estimation fusion for indoor localization. In: 2012 IEEE International Conference on Multimedia and Expo (ICME), pp. 741–746.

7

Low-Complexity Offline and Online Strategies for Wi-Fi Fingerprinting Indoor Positioning Systems

Giuseppe Caso, Luca De Nardis, Maria-Gabriella Di Benedetto
DEPARTMENT OF INFORMATION ENGINEERING, ELECTRONICS AND TELECOMMUNICATIONS (DIET), SAPIENZA UNIVERSITY OF ROME, ROME, ITALY

1 Introduction

Indoor localization of wireless mobile devices, also referred to as indoor positioning, is nowadays an intensively investigated research topic, toward the extension of outdoor location-based services to indoor environments. Among the available communication technologies and infrastructures, Wi-Fi appears as an excellent localization support, since it is largely widespread in indoor environments, implying low implementation time and costs (Liu et al., 2007).

Fingerprinting is one of the most investigated techniques for the implementation of Wi-Fi indoor positioning systems (IPSs) (Honkavirta et al., 2009). It relies on a preliminary collection of location-dependent signal propagation data at predefined positions in the area of interest, called reference points (RPs). Received signal strength (RSS) collection is common in implementing Wi-Fi fingerprinting IPSs, and leads to defining the *fingerprint* as the set of RSS values, measured in a given RP, from nearby Wi-Fi access points (APs) (Bahl and Padmanabhan, 2000).

RSS fingerprinting is typically organized in two phases: offline and online. In the offline phase, an RSS fingerprint is collected at each RP, in order to create a discrete radiomap of the area. In the online phase, the unknown position of a target device is estimated as a function of the position of RPs, that best matches the *online reading*, that is the RSS fingerprint measured by the target device.

Accuracy and complexity of fingerprinting mainly depend on two factors: (1) a careful planning of the offline phase, particularly in terms of RP locations and number of measurements, and (2) an optimized definition of the estimation algorithm used in the online phase.

Geographical and Fingerprinting Data for Positioning and Navigation Systems. https://doi.org/10.1016/B978-0-12-813189-3.00007-1

As regards the offline phase, the goal is to achieve a satisfactory trade-off between positioning accuracy, and efforts and time dedicated to the collection of RPs. Previous work highlighted the impact of human body and device orientation on measured RSS values, and the need for multiple measurements at each RP, in order to counteract channel variability and measurement errors (Hossain et al., 2007; Liao and Kao, 2008; Honkavirta et al., 2009; Kessel and Werner, 2011). Furthermore, in order to limit the measurement effort, two main approaches have been proposed:

- *RSS prediction*: Most of the RSS values are predicted, rather than measured. RSS prediction may be used either for the generation of *virtual* RPs, leading to *discrete virtual* fingerprinting, or for the evaluation of continuous RSS distributions, leading to *continuous virtual* fingerprinting. RSS prediction may be obtained by either *indoor propagation modeling*, in which an empirical radio propagation model, trained with a set of initial measurements, is used, or by *interpolation*, in which adjacent real RPs are interpolated (Chintalapudi et al., 2010; Kumar et al., 2016; Hernández et al., 2017).
- *Crowdsourcing*: System users also contribute to the collection of RSS fingerprints (Bolliger, 2008). Application of crowdsourcing entails a further challenge, that is heterogeneity, in terms of RP locations and devices used for collection, as they both depend on users (Laoudias et al., 2013).

Considering the online phase, the goal is to achieve a satisfactory trade-off between positioning accuracy and algorithm complexity, by decreasing the average number of online operations required for position estimation. As regards the online procedure, deterministic (Bahl and Padmanabhan, 2000; Shin et al., 2012; Caso et al., 2015b), and probabilistic (Roos et al., 2002; Youssef and Agrawala, 2004; Le Dortz et al., 2012) weighted k-nearest neighbors (WkNN) schemes are, by far, the most widely investigated. In particular, deterministic WkNN is highly appealing, since it requires the evaluation of a simple deterministic similarity metric between the online reading and each RP fingerprint. Previous work highlighted the impact on the achievable accuracy of the value of k, and the similarity metric used for RPs selection and weighting (Caso et al., 2015b). As regards the reduction of the online operations, the adoption of *two-step* algorithms was proposed (Youssef et al., 2003; Feng et al., 2012; Yu et al., 2014), in which:

- The offline phase is organized in RPs *collection* and *clustering*, during which measurements are collected and then divided into nonoverlapping groups, according to the adopted similarity criterion.
- The online phase is organized in *coarse* and *fine* localization. During coarse localization, the online reading is compared with a fingerprint associated with each cluster, according to the adopted similarity metric; only the RPs within clusters passing a predefined similarity threshold are used in fine localization, where a traditional WkNN estimator can be then applied.

The main goal of two-step algorithms is to reduce the online complexity, by reducing the RP space. While for a generic WkNN *flat* algorithm, in fact, all RPs are compared with the online reading, in order to select the k most relevant ones, in two-step algorithms, only RPs belonging to selected clusters are taken into account. Among others, the use of Affinity Propagation (Frey and Dueck, 2007), that evaluates RPs mutual similarities, has been proposed for the creation of clusters (Feng et al., 2012; Tian et al., 2013). As a result, in Affinity Propagation two-step algorithms, the definition of the similarity metric has an important role at RP clustering, and coarse and fine localization phases.

This chapter analyzes and discusses the trade-off between accuracy and complexity, within offline and online phases of Wi-Fi fingerprinting IPSs. The main goal is to identify and derive low-complexity strategies, leading to a simple system implementation, while preserving the achievable positioning accuracy.

As regards the offline phase, a strategy adopting RSS prediction, in the form of discrete virtual fingerprinting via indoor propagation modeling, is proposed, and compared with traditional, real RPs only, deterministic WkNN (Caso and De Nardis, 2015, 2017; Caso et al., 2016). As for online phase, Affinity Propagation two-step deterministic WkNN is compared with traditional flat deterministic WkNN, in order to highlight factors, that mainly impact on system complexity and accuracy (Caso et al., 2015a).

Experimental results reported in the present work were obtained in the testbed implemented at the first two floors of the Department of Information Engineering, Electronics, and Telecommunications (DIET) of Sapienza University of Rome.

The chapter is organized as follows: Sections 2 and 3 present the proposed low-complexity strategies for offline and online phases, respectively. In both sections, the reference model is first introduced, followed by a description of the testbed, experimental settings, and performance indicators, and a discussion on the obtained results. Section 4 highlights main results, and the advantage of using the proposed low-complexity strategies, and concludes the chapter underlying possible future research lines.

2 Low-Complexity Strategy for Offline Phase

This section focuses on the description of the proposed low-complexity strategy for the offline phase. As introduced in Section 1, a discrete virtual fingerprinting approach via indoor propagation modeling is implemented, in order to decrease the effort in RSS measurements, while maintaining a satisfying positioning accuracy. In particular, the empirical multiwall multifloor (MWMF) indoor propagation model is adopted for the creation of virtual RPs, while a traditional deterministic WkNN estimator is used in the online phase.

2.1 RSS Prediction via MWMF Model

The MWMF model (Damosso, 1999) is an appealing solution for indoor propagation modeling, due to the good trade-off between simplicity and prediction accuracy (Borrelli et al., 2004). MWMF takes into account objects that may obstruct the propagation over an indoor link, leading to the following path loss model (Damosso, 1999):

Table 1 A_{MWMF} Parameters' Description

Parameters	Description
l_c	Constant loss
N_{obj}	Number of different families of 2D objects
I_n	Number of types of 2D objects considered for family n
$N_{n,i}$	Number of 2D obstructing objects of family n and type i
N_f	Number of obstructing floors
$l_{n,i}$	Loss due to 2D objects of family n and type i
l_f	Loss due to obstructing floors
b	Empirical 3D propagation parameter

$$\text{PL}_{\text{MWMF}} = \text{PL}_{\text{OS}} + A_{\text{MWMF}} \quad \text{(dB)}, \tag{1}$$

where PL_{OS} models the path loss over the transmitter-receiver (Tx-Rx) distance d, while A_{MWMF} models the additional loss due to obstructing obstacles, that may be different in nature and type, such as walls, doors, pipelines, and others. Details on both terms of Eq. (1), with a particular focus on the peculiar MWMF term, that is A_{MWMF}, are provided in Caso et al. (2016). As regards A_{MWMF}, a linear combination of different obstructing obstacles is considered (Borrelli et al., 2004), and a generic description of parameters within this term is reported in Table 1.

The use of the MWMF model requires an initial, hopefully small, set of M measurements, and also, for each mth measurement ($m = 1, 2, \ldots, M$), the information regarding number, type, and positions of objects obstructing the Tx-Rx direct path, indicated as the set of topological parameters $\{\mathcal{T}_m\}$. The measurements are used in order to estimate the set of propagation parameters $\{\mathcal{S}\}$, within PL_{OS} and A_{MWMF} terms, that characterizes the model in the area of interest, such as path loss exponent and loss terms due to obstructing objects. An iterative least square fitting procedure can be adopted, so to minimize the difference between RSS initial measurements and predictions in the same M positions, and in turn estimate $\{\mathcal{S}\}_{\text{opt}}$, as follows:

$$\{\mathcal{S}\}_{\text{opt}} = \underset{\{\mathcal{S}\}}{\text{argmin}} \left\{ \sum_{m=1}^{M} |\text{RSS}_m - \widehat{\text{RSS}}_m|^2 \right\}, \tag{2}$$

where, for the mth available measurement, RSS_m and $\widehat{\text{RSS}}_m$ are the actual versus predicted RSS values at the Rx, when considering a Tx emitting a known effective isotropic radiated power (EIRP) $\text{W}_{\text{Tx}}^{\text{EIRP}}$ at distance d_m. At each iteration, $\widehat{\text{RSS}}_m$ is computed as the difference between $\text{W}_{\text{Tx}}^{\text{EIRP}}$ and the path loss evaluated as in Eq. (1), and the iterative procedure stops when the difference with RSS_m is minimized, and thus $\{\mathcal{S}\}_{\text{opt}}$ is found.

The propagation parameters to be optimized, included in $\{\mathcal{S}\}$, may differ depending on the model definition, and may include parameters from both PL_{OS} and A_{MWMF} terms; the set adopted in this work is reported in Caso et al. (2016).

2.2 Offline Phase

Given a set of L Wi-Fi APs at known positions in the area of interest $\mathcal{A}$, initial measurements in a set of N^{r} real RPs are collected, so that an $L \times 1$ RSS fingerprint $\boldsymbol{s}_{n_1}$ is associated with the n_1th RP ($n_1 = 1, 2, \ldots, N^{\mathrm{r}}$). The generic $\boldsymbol{s}_{n_1}$ component, denoted by s_{l,n_1}, contains the RSS measured by a reference Rx, placed at the n_1th RP, from the lth AP. This value is usually obtained by averaging $q > 1$ repeated measurements, in order to counteract propagation channel variability and measurement errors.

The MWMF model is then used for the generation of virtual RPs. The model is first calibrated on the set of N^{r} real RPs, so that the derived propagation parameters are used for the generation of further N^{v} virtual RPs. The component $\hat{s}_{l,n_2}$ of the generic $L \times 1$ fingerprint $\hat{\boldsymbol{s}}_{n_2}$ contains the predicted RSS at the n_2th virtual RP from the lth AP ($n_2 = 1, 2, \ldots, N^{\mathrm{v}}$).

Both the amount and the positions of real RPs are expected to affect the generation of virtual RPs and the achievable positioning accuracy, since the amount, in particular, defines the number of measurements considered in the model fitting procedure of Eq. (2). Given N^{r} real RPs, regularly spaced over a grid in the environment $\mathcal{A}$ of area $|\mathcal{A}|$, their spatial density is given by:

$$d^{\mathrm{r}} = \frac{N^{\mathrm{r}}}{|\mathcal{A}|}, \tag{3}$$

that defines the number of real RPs for each meter square of $\mathcal{A}$.

In Caso and De Nardis (2015, 2017) and Caso et al. (2016), several strategies were proposed for the use of real RPs in Eq. (2), toward the evaluation of the propagation parameters for the MWMF model; in this work, a *Specific AP Fitting* strategy is adopted, where the measurements at all RPs are discriminated considering the AP they refer to, so that a different set of propagation parameters is obtained for each AP.

In analogy to Eq. (3), denoting with N^{v} the number of virtual RPs to be generated, their spatial density is given by:

$$d^{\mathrm{v}} = \frac{N^{\mathrm{v}}}{|\mathcal{A}|}. \tag{4}$$

Position of virtual RPs in $\mathcal{A}$ can be freely defined: in this work a grid placement is adopted, as an extension of the commonly adopted real RPs placement.

2.3 Online Phase

A deterministic WkNN estimation algorithm, using the combination of real versus virtual RPs, is used to infer the target location. Denoting by $N = N^{\mathrm{r}} + N^{\mathrm{v}}$ the total number of RPs in $\mathcal{A}$, $\boldsymbol{s}_n$ the RSS fingerprint of nth RP ($n = 1, 2, \ldots, N$), and $\boldsymbol{s}_i$ the RSS online reading collected during the ith positioning request by a target device in unknown position $\boldsymbol{p}_i = (x_i, y_i, z_i)$, the position estimate relies on the computation of a set of similarity metrics $\mathrm{sim}_{n,i} = \mathrm{sim}(\boldsymbol{s}_n, \boldsymbol{s}_i)$. The W$k$NN algorithm selects the k RPs that present the highest $\mathrm{sim}_{n,i}$ values and provides an estimate of $\boldsymbol{p}_i$ defined as:

$$\hat{\boldsymbol{p}}_i = \frac{\sum_{n=1}^{k} (\mathrm{sim}_{n,i}) \boldsymbol{p}_n}{\sum_{n=1}^{k} \mathrm{sim}_{n,i}}, \tag{5}$$

where $\boldsymbol{p}_n = (x_n, y_n, z_n)$ is the position of the nth selected RP, and $\hat{\boldsymbol{p}}_i = (\hat{x}_i, \hat{y}_i, \hat{z}_i)$ is the estimated position of the target device. The similarity metric can be any deterministic metric computable in the RSS space between vectors $\boldsymbol{s}_n$ and $\boldsymbol{s}_i$. A popular choice is the inverse Minkowski distance of order o, defined as follows:

$$\mathrm{sim}_{n,i} = [\mathcal{D}^o_{n,i}]^{-1} = \left[\left(\sum_{l=1}^{L} |s_{l,i} - s_{l,n}|^o \right)^{1/o} \right]^{-1}, \quad o \geq 1. \tag{6}$$

Typical orders are $o = 1$ (inverse Manhattan distance) and $o = 2$ (inverse Euclidean distance). Similarity metrics based on modified versions of the inner product between RSS vectors have also been proposed (Torres-Sospedra et al., 2015; Caso et al., 2015a). In this work, the inverse Euclidean distance is adopted as similarity metric.

As regards the value of k, it is selected as follows:

$$k = \lceil 0.05(d^{\mathrm{r}} + d^{\mathrm{v}})|\mathcal{A}| \rceil, \tag{7}$$

that is proved to be a reliable estimator of the optimal value of k, defined as the one minimizing the average positioning error (Caso et al., 2016).

2.4 Experimental Setting and Performance Indicators

Experimental analysis of the proposed low-complexity strategy was conducted in the testbed at the DIET of Sapienza University of Rome. The testbed is deployed in an office environment, and covers two floors with an area of approximately 42×12 m^2 each. $L_1 = 6$ Wi-Fi APs working @2.4 and 5 GHz and $L_2 = 7$ Wi-Fi APs working @2.4 GHz, with a beacon transmission period of $T_{\mathrm{b}} = 100$ ms and a transmit power $W_{\mathrm{Tx}}^{\mathrm{EIRP}} = 20$ dBm, are placed at known positions at the first and second floors, respectively. On both floors, APs are placed in the false ceiling of the central corridor, due to deployment constraint.

The second floor was adopted as area of interest $\mathcal{A}$, and the APs on this floor were considered in the fingerprinting measurement campaign, so that $L = L_2$. During the offline phase, $N^{\mathrm{r,tot}} = 72$ RPs were selected for RSS collection on a regular grid within $\mathcal{A}$; fingerprints were also collected in a set of $N^{\mathrm{t}} = 31$ target points (TPs) randomly distributed over $\mathcal{A}$. TPs were used as ground truth, in order to test the positioning accuracy of the proposed scheme. In both cases, fingerprints were obtained as the average of $q = 50$ scans at each location. Furthermore, all measurements were carried out during weekend afternoons, using a MacBook Pro equipped with an AirPort Extreme Network Interface Card, placed on a wooden platform.

Two different analyses were carried out in order to demonstrate the advantage of using the proposed low-complexity strategy:

- *Analysis I*: Reliability of the MWMF model in generating virtual RPs.
- *Analysis II*: Effect of adopting the proposed virtual fingerprinting scheme on the achievable positioning accuracy.

As regards Analysis I, MWMF reliability was evaluated as a function of the parameter ρ, that determines the number N^{r} of RPs used for the model fitting of Eq. (2), out of the total number of measured RPs, $N^{\mathrm{r,tot}}$, so that $N^{\mathrm{r}} = \lceil \rho N^{\mathrm{r,tot}} \rceil$ (Caso et al., 2016). For each considered value of ρ, the MWMF model was fitted, and RSS values in the $N^{\mathrm{r,tot}}$ RP locations were generated and compared with the collected ones, so to evaluate the prediction error, as follows:

$$\delta_{l,n}(\rho) = |s_{l,n} - \hat{s}_{l,n}(\rho)|, \tag{8}$$

where $\hat{s}_{l,n}(\rho)$ is the predicted RSS for the generic (AP_l, RP_n) pair, obtained by using a set of $N^{\mathrm{r}} = \lceil \rho N^{\mathrm{r,tot}} \rceil$ RPs in the model fitting procedure, while $s_{l,n}$ is the measured RSS for the same pair. Assuming the generic $\delta_{l,n}(\rho)$ value as a sample of a random variable $\delta_l(\rho)$ related to the lth AP, the cumulative distribution function (CDF) of $\delta_l(\rho)$, that is $F_{\delta_l(\rho)}(\delta_{l,n}(\rho)) = \Pr\{\delta_l(\rho) \leq \delta_{l,n}(\rho)\}$, and the average error $\bar{\delta}_l(\rho) = \frac{\sum_{n=1}^{N^{\mathrm{r,tot}}} \delta_{l,n}(\rho)}{N^{\mathrm{r,tot}}}$ were also evaluated.

Considering Analysis II, positioning accuracy of the proposed strategy was evaluated as a function of densities d^{r} and d^{v}, and adopting a value of k as in Eq. (7). The analysis was carried out by computing the positioning error $\epsilon_i(d^{\mathrm{r}}, d^{\mathrm{v}})$ for each TP i ($i = 1, 2, \ldots, N^{\mathrm{t}}$) as follows:

$$\epsilon_i(d^{\mathrm{r}}, d^{\mathrm{v}}) = \sqrt{(x_i - \hat{x}_i)^2 + (y_i - \hat{y}_i)^2}, \tag{9}$$

where $(x_i, y_i) = \boldsymbol{p}_i$ and $(\hat{x}_i, \hat{y}_i) = \hat{\boldsymbol{p}}_i$ are the actual and the estimated positions of the ith target device at the DIET second floor, respectively. As in the case of the prediction error $\delta_l(\rho)$, the CDF of positioning error $\epsilon(d^{\mathrm{r}}, d^{\mathrm{v}})$, that is $F_{\epsilon(d^{\mathrm{r}}, d^{\mathrm{v}})}(\epsilon_i(d^{\mathrm{r}}, d^{\mathrm{v}})) = \Pr\{\epsilon(d^{\mathrm{r}}, d^{\mathrm{v}}) \leq \epsilon_i(d^{\mathrm{r}}, d^{\mathrm{v}})\}$, and the average positioning error $\bar{\epsilon}(d^{\mathrm{r}}, d^{\mathrm{v}}) = \frac{\sum_{i=1}^{N^{\mathrm{t}}} \epsilon_i(d^{\mathrm{r}}, d^{\mathrm{v}})}{N^{\mathrm{t}}}$ were evaluated.

2.5 Results and Discussions

Fig. 1A shows the CDF of the prediction error $\delta_l(\rho)$ for the AP fitting strategy, a reference AP, and four different values of ρ (note that $\rho = \{0.1, 0.2, 0.5, 1\}$ lead to $N^{\mathrm{r}} = \{8, 15, 35, 72\}$ uniformly distributed RPs used in the model fitting procedure). Results show that slightly different $\delta_l(\rho)$ errors are obtained as ρ increases from 0.1 to 1, suggesting that a relatively small value of ρ (and in turn of N^{r} and d^{r}) is sufficient to obtain a reliable estimation of the propagation parameters for the generation of virtual RPs, although a few measurements are still required, due to the empirical nature of the MWMF model. Fig. 1B shows the CDF of $\delta_l(\rho)$ for the AP fitting strategy ($\rho = 1$) against a baseline *No Fit* strategy with no fitting, that is no use of measurements, and by obtaining RSS predictions using propagation parameters estimated for a different area, and reported in Borrelli et al. (2004). Results show that the prediction error significantly increases when no site-specific model fitting is carried out.

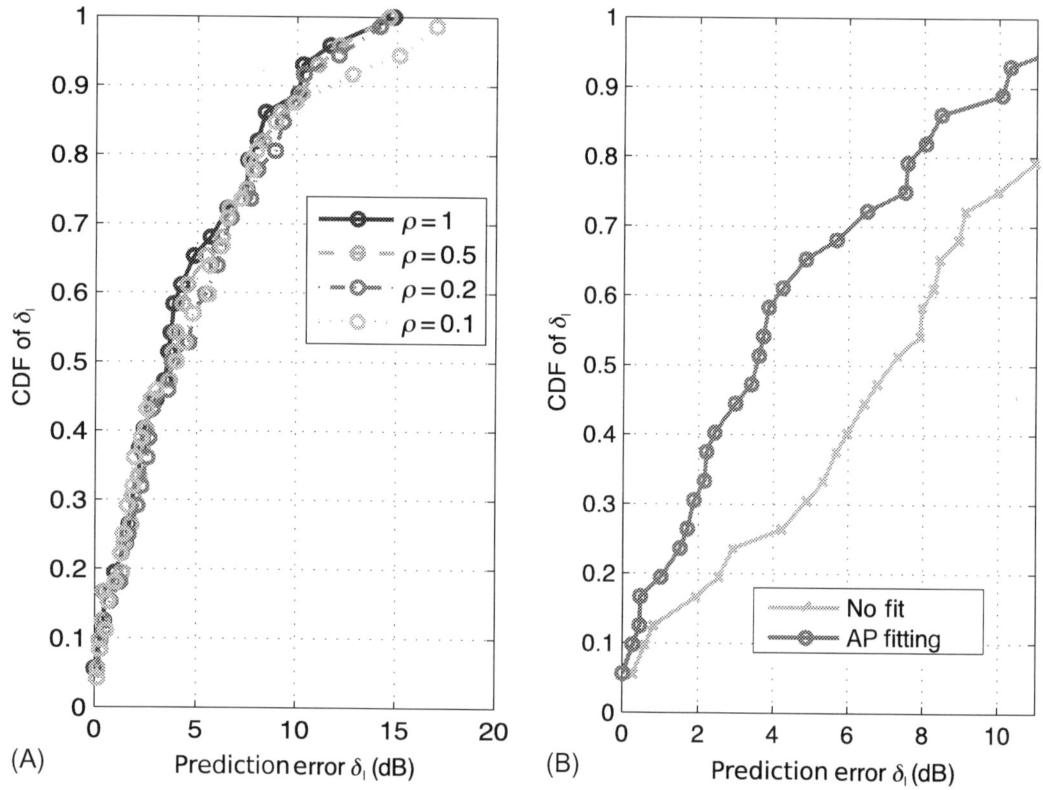

FIG. 1 CDFs of the prediction error $\delta_l(\rho)$ for a reference AP: AP fitting strategy (real RPs selection parameter $\rho = \{0.1, 0.2, 0.5, 1\}$) (A), no fit versus AP fitting strategy ($\rho = 1$) (B).

Fig. 2 shows the average positioning error $\bar{\epsilon}(d^r, d^v)$ of virtual fingerprinting, as a function of d^v, and four different values of d^r, corresponding to the values of ρ and N^r adopted in the previous analysis (see Eq. 3). The value of k is fixed as in Eq. (7). Fig. 2 also reports the average error obtained by adopting real RPs only fingerprinting, with a number of real RPs corresponding to minimum and maximum values of d^r. Results highlight a few important aspects: on one hand, the introduction of virtual MWMF fingerprints does not lead to performance improvement when a large enough amount of real ones is collected (see, in particular, $d^r = 0.14$ as a function of d^v curve vs. $d^r = 0.14$ with $d^v = 0$ one). On the other hand, a significant accuracy improvement is obtained, when a large amount of virtual RPs ($d^v = 10$) was predicted from a limited set of initial real measurements in the offline phase, and used in the online phase (see $d^r < 0.14$ as a function of d^v curves vs. $d^r = 0.02$ and $d^r = 0.14$ with $d^v = 0$ ones). Furthermore, note that the error obtained with virtual fingerprinting almost accommodates on the error of the optimal real RPs only scheme, where a large amount of real RPs is used. The accuracy decrease is of about 60 cm in the extreme case ($d^r = 0.02$ as a function of d^v).

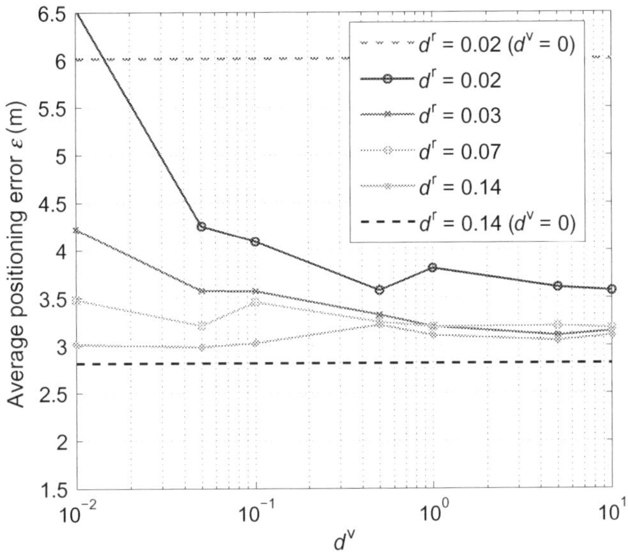

FIG. 2 Average positioning error $\bar{\epsilon}(d^r, d^v)$ for virtual fingerprinting, as a function of virtual RPs density d^v, and four different values of real RPs density d^r; upper and lower bounds of the error are also shown, for real RPs only fingerprinting with $d^r = 0.02$ and $d^r = 0.14$.

Results clearly show that creation and adoption of a large enough amount of virtual RPs can significantly reduce the offline phase complexity, in terms of number of needed initial measurements, while preserving achievable accuracy.

3 Low-Complexity Strategy for Online Phase

This section focuses on the description of the proposed low-complexity strategy for the online phase. As introduced in Section 1, an Affinity Propagation two-step algorithm is implemented, in order to decrease the number of online operations required for a position estimate, while maintaining a satisfying positioning accuracy. In particular, the offline phase foresees an RP clustering step via Affinity Propagation, while the online phase is divided into coarse (cluster selection) and fine (RPs selection and weighting) localization.

3.1 RP Clustering via Affinity Propagation

Affinity Propagation is a clustering algorithm, that divides a set of elements in clusters, and elects for each cluster a representative clusterhead, also dubbed as exemplar (Frey and Dueck, 2007). The algorithm usually follows a distributed and iterative approach: elements are seen as network nodes which exchange messages containing computed values, that measure the *affinity* of one element with another element, until it converges to a stable set of exemplars and corresponding clusters.

In the context of indoor positioning, since elements correspond to RPs, the algorithm is centralized and, for each iteration, the requested values are evaluated by a central processing unit based on an initial measure of similarity $\text{sim}^{\text{CL}}_{n_1,n_2} = \text{sim}^{\text{CL}}(\mathbf{s}_{n_1}, \mathbf{s}_{n_2})$ (with $n_1, n_2 = 1, 2, \ldots, N$, $n_1 \neq n_2$) between each RP pair, where the superscript CL indicates that such similarity values are evaluated during the clustering step, measuring how well the RP_{n_2} is suited to be the exemplar for RP_{n_1}. The self-similarity value $\text{sim}^{\text{CL}}(\mathbf{s}_n, \mathbf{s}_n)$ (with $n = 1, 2, \ldots, N$), that is also dubbed as *preference*, indicates the possibility that RP_n may become an exemplar. In order to give all RPs the same chance to become an exemplar, their preferences are initially set to a common finite value, typically defined as:

$$\text{pref}(\mathbf{s}_n) = \text{sim}^{\text{CL}}(\mathbf{s}_n, \mathbf{s}_n) = \gamma \cdot \text{median}\{\text{sim}^{\text{CL}}(\mathbf{s}_{n_1}, \mathbf{s}_{n_2})\}, \quad \forall n_1, n_2 \in \{1, 2, \ldots, N\}, \quad n_1 \neq n_2, \tag{10}$$

where γ is a tunable parameter (Feng et al., 2012; Frey and Dueck, 2007), equal to 1 in the present work.

The definition of exemplars relies on the iterative evaluation of two values between each RP pair:

- *Responsibility* $\text{resp}(\mathbf{s}_{n_1}, \mathbf{s}_{n_2})$: It reflects the accumulated evidence for how well-suited RP_{n_2} is to serve as the exemplar for RP_{n_1}, taking into account other potential exemplars for RP_{n_1}.
- *Availability* $\text{avail}(\mathbf{s}_{n_1}, \mathbf{s}_{n_2})$: It reflects the accumulated evidence for how appropriate it would be for RP_{n_1} to choose RP_{n_2} as its exemplar, taking into account the support from other RPs that RP_{n_2} should be an exemplar.

These values are updated according to the following equations:

$$\text{resp}(\mathbf{s}_{n_1}, \mathbf{s}_{n_2}) = \text{sim}^{\text{CL}}(\mathbf{s}_{n_1}, \mathbf{s}_{n_2}) - \max_{n_3}\left\{\text{avail}(\mathbf{s}_{n_1}, \mathbf{s}_{n_3}) + \text{sim}^{\text{CL}}(\mathbf{s}_{n_1}, \mathbf{s}_{n_3})\right\}, \tag{11}$$

$$\text{avail}(\mathbf{s}_{n_1}, \mathbf{s}_{n_2}) = \min\left\{0, \text{resp}(\mathbf{s}_{n_2}, \mathbf{s}_{n_2}) + \sum_{n_3} \max\{0, \text{resp}(\mathbf{s}_{n_3}, \mathbf{s}_{n_2})\}\right\}, \tag{12}$$

$\forall n_1, n_2, n_3 \in \{1, 2, \ldots, N\}$, $n_1 \neq n_2$, $n_3 \neq n_2$ in Eq. (11), $n_3 \neq n_1, n_2$ in Eq. (12).

In order to facilitate convergence of the iterative procedure and avoid ringing oscillations, a *damping* factor $\text{DF} \in [0.5, 1)$ is typically introduced leading to the following expressions for the new values of responsibility and availability:

$$\begin{aligned} \text{resp}_{\text{new}}(\mathbf{s}_{n_1}, \mathbf{s}_{n_2}) &= \text{DF} \cdot \text{resp}_{\text{old}}(\mathbf{s}_{n_1}, \mathbf{s}_{n_2}) + (1 - \text{DF}) \cdot \text{resp}'_{\text{new}}(\mathbf{s}_{n_1}, \mathbf{s}_{n_2}), \\ \text{avail}_{\text{new}}(\mathbf{s}_{n_1}, \mathbf{s}_{n_2}) &= \text{DF} \cdot \text{avail}_{\text{old}}(\mathbf{s}_{n_1}, \mathbf{s}_{n_2}) + (1 - \text{DF}) \cdot \text{avail}'_{\text{new}}(\mathbf{s}_{n_1}, \mathbf{s}_{n_2}), \end{aligned} \tag{13}$$

$\forall n_1, n_2 \in \{1, 2, \ldots, N\}$, $n_1 \neq n_2$, with $\text{resp}'_{\text{new}}(\mathbf{s}_{n_1}, \mathbf{s}_{n_2})$ and $\text{avail}'_{\text{new}}(\mathbf{s}_{n_1}, \mathbf{s}_{n_2})$ evaluated by using Eqs. (11), (12), respectively. $\text{DF} = 0.6$ is generally adopted, and thus also used in the present work.

Two main issues are identified in the application of Affinity Propagation:

- *Degeneracies*: Degeneracies can arise if, for example, the similarity metric is commutative and two elements (RPs) are isolated from all the others. In this case

oscillations in deciding which of the two elements should be the exemplar might appear. The solution proposed in Frey and Dueck (2007), and also adopted in the present work, is to add a small amount of random noise to similarities values to avoid such deadlock situations.
- *Outliers*: When applied to RP clustering, the algorithm might occasionally lead to an RP belonging to a cluster, but being physically far away from the cluster exemplar. In Feng et al. (2012), taking advantage of the knowledge of each RP position, each outlier is forced to join the cluster characterized by the exemplar at minimum distance from the outlier itself. This solution is also adopted in the present work.

3.2 Offline Phase

Affinity Propagation is used for grouping the RPs collected in the offline phase. Given a set of L Wi-Fi APs that can be detected in $\mathcal{A}$ (differently from the assumption given in Section 2.2, position of APs may be unknown and it is not required for the application of the present low-complexity online phase strategy), initial measurements in a set of N RPs are collected, so that an $L \times 1$ RSS fingerprint $\boldsymbol{s}_n$ is associated with the nth RP ($n = 1, 2, \ldots, N$).

After the RSS collection, Affinity Propagation clustering takes place, and the RPs are divided into $N_c < N$ clusters. The definition of similarity, used during the iterative process, may be inherited from Frey and Dueck (2007); in this case, given a pair of RPs, $\text{sim}^{\text{CL}}(\boldsymbol{s}_{n_1}, \boldsymbol{s}_{n_2})$ is as follows:

$$\text{sim}^{\text{CL}}(\boldsymbol{s}_{n_1}, \boldsymbol{s}_{n_2}) = -[\mathcal{D}^2(\boldsymbol{s}_{n_1}, \boldsymbol{s}_{n_2})]^2 \quad \forall n_1, n_2 \in \{1, 2, \ldots, N\}, \quad n_1 \neq n_2, \tag{14}$$

where $\mathcal{D}^2(\boldsymbol{s}_{n_1}, \boldsymbol{s}_{n_2})$ expresses the Euclidean distance between two RP fingerprints. In order to use the Affinity Propagation algorithm in its traditional settings, this definition is also adopted in this work. Detailed analysis on the impact of using different definitions can be found in Caso et al. (2015a).

3.3 Online Phase

Once the offline phase is complete, the position estimate is obtained through coarse and fine localization steps. In coarse localization, $N_{c,i} \leq N_c$ clusters that best match the $\boldsymbol{s}_i$ online reading are selected, through the computation of N_c similarity values $\text{sim}^{\text{C}}_{n_c,i} = \text{sim}^{\text{C}}(\boldsymbol{s}_{n_c}, \boldsymbol{s}_i)$ (with $n_c = 1, 2, \ldots, N_c$) between the online reading and a fingerprint selected as the n_cth cluster's representative, denoted as $\boldsymbol{s}_{n_c}$. In the present work, $\boldsymbol{s}_{n_c}$ is a synthetic fingerprint, generated by averaging the fingerprints of the RPs within a cluster. The selection of the clusters is performed by comparing each $\text{sim}^{\text{C}}_{n_c,i}$ with a threshold α, defined as follows (Feng et al., 2012):

$$\alpha = \alpha_1 \cdot \max_{\boldsymbol{e} \in \mathcal{E}} \left\{\text{sim}^{\text{C}}(\boldsymbol{s}_i, \boldsymbol{e})\right\} + \alpha_2 \cdot \min_{\boldsymbol{e} \in \mathcal{E}} \left\{\text{sim}^{\text{C}}(\boldsymbol{s}_i, \boldsymbol{e})\right\}. \tag{15}$$

Clusters with similarity values above α are selected. In Eq. (15), $\mathcal{E}$ denotes the set of cluster fingerprints, and $\alpha_1 + \alpha_2 = 1$. The values of α_1 and α_2 allow to adjust the number of selected clusters: as an example, the smaller the number of desired selected clusters, the higher should be the value of α_1 (and conversely, the lower the value of α_2). $\alpha_1 = 0.95$ and $\alpha_2 = 0.05$ are the values adopted in the present work.

Denoting as N_i the total number of RPs within the selected $N_{c,i}$ clusters, k out of N_i RPs are then selected during the fine localization, by computing N_i similarity values $\text{sim}^{\text{F}}_{n,i} = \text{sim}^{\text{F}}(\boldsymbol{s}_{n_i}, \boldsymbol{s}_i)$ (with $n_i = 1, 2, \ldots, N_i$) between the online reading and the RP fingerprints. Similarly to Section 2.3, final position estimate is then obtained via WkNN, so that:

$$\hat{\boldsymbol{p}}_i = \frac{\sum_{n=1}^{k} \left(\text{sim}^{\text{F}}_{n,i}\right) \boldsymbol{p}_n}{\sum_{n=1}^{k} \text{sim}^{\text{F}}_{n,i}}. \tag{16}$$

It is worth noting that $\text{sim}^{\text{C}}_{n_c,i}$ and $\text{sim}^{\text{F}}_{n,i}$ are properly defined similarity metrics, and superscripts C and F indicate that such similarities are evaluated during the coarse and fine localization, respectively. Similarly to the discussion in Section 2.3, several definitions may be adopted. In this work, the inverse Euclidean distance is considered at both steps, so that, overall, the proposed Affinity Propagation two-step algorithm adopts the similarity definition of Eq. (14) for RP clustering ($\text{sim}^{\text{CL}}_{n_1,n_2}$), while the one of Eq. (6) for coarse and fine localization ($\text{sim}^{\text{C}}_{n_c,i}$ and $\text{sim}^{\text{F}}_{n,i}$), respectively.

Detailed analysis on the impact of using different metric combinations at different steps can be found in Caso et al. (2015a).

As regards parameter k in the fine localization WkNN, a *dynamic* k selection scheme is used, in which the value of k is adjusted at each positioning request, as also proposed in Shin et al. (2012), Marcus et al. (2013), and Caso et al. (2015b). In general, this scheme relies on the definition of a threshold λ taking values in the same domain of the similarity metric, and on the selection of the RPs that show a value of the metric above the threshold. In this case, given the ith positioning request, λ is evaluated as a function of the average on the RPs similarity values $\text{sim}^{\text{F}}_{i,n}$, that is:

$$\lambda_i\left(\text{sim}^{\text{F}}_{n,i}\right) = c \cdot \overline{\text{sim}}^{\text{F}}_{n,i} = c \cdot \frac{\sum_{n=1}^{N} \text{sim}^{\text{F}}_{n,i}}{N}, \tag{17}$$

where c is a tuning parameter, ranging from 0.1 to 2 in the present work.

3.4 Experimental Setting and Performance Indicators

Experimental analysis of the proposed low-complexity strategy was conducted in the testbed implemented at the DIET of Sapienza University of Rome, described in Section 2.4.

In this case $N_1 = 65$ and $N_2 = 69$ RPs were collected, in a grid fashion, at the first and second floors, respectively, for a total number of $N = 134$ RPs. For each RP, RSS values received from all detected APs, were collected, by averaging $q = 5$ measurements. Once the offline stage was completed, the total number of detected APs was $L = 133$, including physical and virtual APs as well as temporary and mobile connection points. In particular L also contained $L_1 = 6$ and $L_2 = 7$ testbed-dedicated Wi-Fi APs at DIET first and second floors, as described in Section 2.4. Fingerprints were also collected in a set of $N^{\text{t}} = 70$ TPs, randomly distributed at both floors. TPs were used as ground truth, in order to test the positioning accuracy of the proposed scheme. All measurements were carried out during weekdays, using an Android Samsung Tablet held on by a human surveyor.

Two different analyses were carried out in order to demonstrate the advantage of using the proposed low-complexity strategy:

- *Analysis I*: Effect of adopting the proposed two-step scheme on the achievable positioning accuracy.
- *Analysis II*: Effect of adopting the proposed two-step scheme on the complexity of the online phase, in terms of average number of operations required for obtaining a position estimate.

As regards Analysis I, positioning accuracy of two-step versus flat algorithms was evaluated as a function of the threshold parameter c defined in Eq. (17). The flat algorithm corresponds to a traditional WkNN scheme, with no RP clustering, and thus no coarse localization. Only fine localization is used, by comparing the online fingerprint with each RP fingerprint. The analysis was carried out by computing the 3D positioning error $\epsilon_i(c)$ for each TP i ($i = 1, 2, \ldots, N^{\mathrm{t}}$), as follows:

$$\epsilon_i(c) = \sqrt{(x_i - \hat{x}_i)^2 + (y_i - \hat{y}_i)^2 + (z_i - \hat{z}_i)^2}, \tag{18}$$

where $(x_i, y_i, z_i) = \boldsymbol{p}_i$ and $(\hat{x}_i, \hat{y}_i, \hat{z}_i) = \hat{\boldsymbol{p}}_i$ are the actual and the estimated position of the ith target device in the coordinate system including DIET first and second floors, respectively. CDF of positioning error $\epsilon(c)$, that is $F_{\epsilon(c)}(\epsilon_i(c)) = \Pr\{\epsilon(c) \leq \epsilon_i(c)\}$, and the average positioning error $\bar{\epsilon}(c) = \frac{\sum_{i=1}^{N^{\mathrm{t}}} \epsilon_i(c)}{N^{\mathrm{t}}}$ were also evaluated as a function of c.

Regarding the computational complexity, the selected performance indicator was the number of similarity values N_{sim} to be computed for obtaining a position estimate. In the case of the two-step algorithm, N_{sim} for the generic ith online reading can be expressed as follows:

$$N_{\mathrm{sim}} = N_{\mathrm{c}} + N_i, \tag{19}$$

where N_{c} is the number of RP clusters and N_i is the number of RPs passing the coarse localization. Noting that in case of the flat algorithm $N_{\mathrm{sim}} = N$ for each positioning request, one can observe that, on average, the adoption of the two-step algorithm will lead to a reduction of computational complexity if $\overline{N}_{\mathrm{sim}} = N_{\mathrm{c}} + \overline{N}_i < N$, where $\overline{N}_{\mathrm{sim}}$ is the average number of similarity computations, depending in turn on the average number of selected RPs $\overline{N}_i = \frac{\sum_{i=1}^{N^{\mathrm{t}}} N_i}{N^{\mathrm{t}}}$.

3.5 Results and Discussion

Before discussing the results of the analysis described in Section 3.4, a brief mention to the results of the RP clustering is reported. By adopting Affinity Propagation on 100 iterations, with $\mathrm{sim}^{\mathrm{CL}}(\boldsymbol{s}_{n_1}, \boldsymbol{s}_{n_2})$ as in Eq. (14), and tuning parameters as reported in Section 3.1, a total number of 13 clusters were obtained in the area of interest, 6 on the first floor, and 7 on the second one, respectively. Fig. 3 reports the clusters obtained at DIET first floor, with corresponding exemplars. An average of 10 RPs is included in each cluster, with maximum cluster amount of 15 and minimum of 5 RPs.

Two-step and flat algorithms were then compared in terms of positioning accuracy. Fig. 4 shows the average positioning error $\bar{\epsilon}(c)$ for both schemes, as a function of the threshold parameter c, adopted in fine localization WkNN (in case of two-step, cluster selection in the coarse localization was performed by using the threshold α of Eq. 15). Results show that the two-step algorithm always leads to comparable or better results than

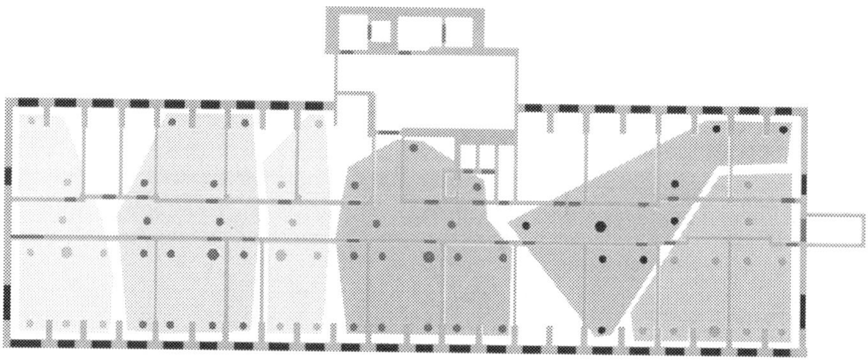

FIG. 3 RP clustering at DIET first floor (areas represented in *different colors* indicate different clusters; *larger dots* indicate exemplars).

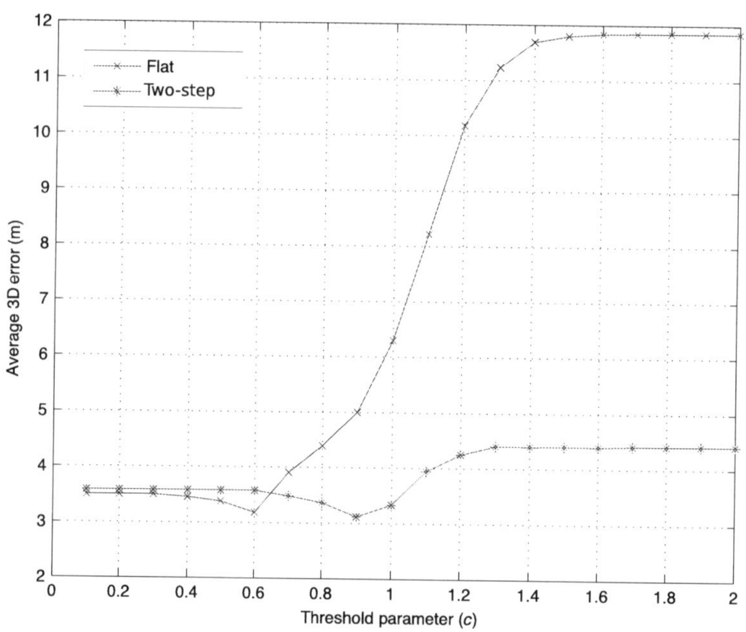

FIG. 4 Flat vs. two-step: 3D average positioning error $\bar{\epsilon}$ as a function of the threshold parameter c.

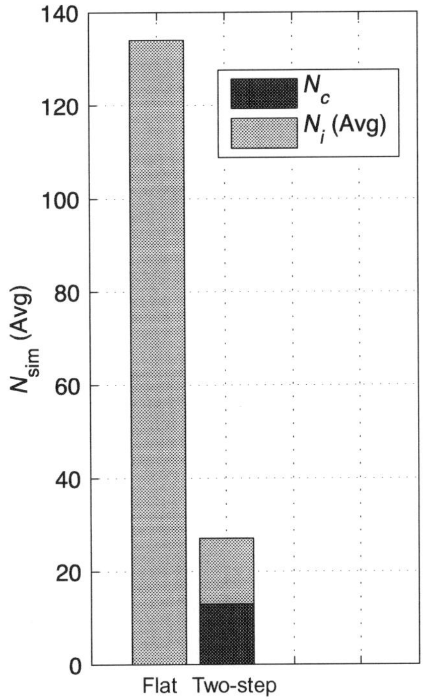

FIG. 5 Flat vs. two-step: Average number of similarity values $\overline{N}_{sim}$ to be computed for obtaining a position estimate.

the flat algorithm. Performance improvement is particularly significant when the adopted value of c leads to thresholds, that allow a large number of RPs to be selected. Under these conditions, the RP space reduction provided by two-step schemes leads to significant reduction in $\bar{\epsilon}$, and in turn improved positioning accuracy.

Fig. 5 shows the average number of similarity values N_{sim} to be computed for obtaining a position estimate, and confirms the main expected advantage of the use of two-step algorithms, that is the reduction of online complexity. While, in fact, the number of computed similarity values is always equal to N for the flat algorithm, a significantly lower average value, of about 27, is obtained for the two-step scheme.

4 Conclusion and Future Work

In this work, a theoretical and experimental analysis of Wi-Fi RSS fingerprinting IPSs was presented, focusing on the trade-off between performance and complexity of offline and online fingerprinting phases. Two low-complexity system implementation strategies were proposed and analyzed in the experimental testbed at the DIET of Sapienza University of Rome.

Considering the offline phase, in order to limit the efforts related to the RSS collection, the use of the MWMF indoor propagation model was proposed for the generation of virtual fingerprints. The MWMF accuracy in predicting RSS values in the area of interest and the impact of using such values in the online phase were analyzed. Experimental results showed that a significant reduction of measurement collection is possible, since positioning accuracy is preserved thanks to the use of virtual RPs.

As regards the online phase, an Affinity Propagation two-step WkNN algorithm was analyzed. Experimental results showed a significant decrease of required operations for obtaining a position estimate, with a preserved or even improved positioning accuracy, when the two-step algorithm was compared with a traditional flat scheme.

Moving from this work, several research lines can be identified. The joint application of the proposed solutions is under investigation, and is being tested in different environments, in order to generalize the obtained results. Moreover, the analysis may be extended to different positioning approaches, such as continuous virtual fingerprinting in the offline phase, and probabilistic estimation in the online one. Finally, considering recent research trends, the definition of low-complexity strategies for hybrid IPSs, that exploit different wireless technologies in order to provide accurate localization, is expected be extremely important.

References

Bahl, P., Padmanabhan, V.N., 2000. RADAR: an in-building RF-based user location and tracking system. In: Proceedings IEEE INFOCOM 2000. Nineteenth Annual Joint Conference of the IEEE Computer and Communications Societies, vol. 2, pp. 775–784.

Bolliger, P., 2008. Redpin-adaptive, zero-configuration indoor localization through user collaboration. In: Proceedings of the First ACM International Workshop on Mobile Entity Localization and Tracking in GPS-Less Environments, pp. 55–60.

Borrelli, A., Monti, C., Vari, M., Mazzenga, F., 2004. Channel models for IEEE 802.11 b indoor system design. In: 2004 IEEE International Conference on Communications, vol. 6, pp. 3701–3705.

Caso, G., De Nardis, L., 2015. On the applicability of multi-wall multi-floor propagation models to WiFi fingerprinting indoor positioning. In: Future Access Enablers of Ubiquitous and Intelligent Infrastructures, pp. 166–172.

Caso, G., De Nardis, L., 2017. Virtual and oriented WiFi fingerprinting indoor positioning based on multi-wall multi-floor propagation models. Mob. Netw. Appl. 22 (5), 825–833.

Caso, G., De Nardis, L., Di Benedetto, M.G., 2015a. A mixed approach to similarity metric selection in affinity propagation-based WiFi fingerprinting indoor positioning. Sensors 15 (11), 27692–27720.

Caso, G., De Nardis, L., Di Benedetto, M.G., 2015b. Frequentist inference for WiFi fingerprinting 3D indoor positioning. In: 2015 IEEE International Conference on Communication Workshop (ICCW), pp. 809–814.

Caso, G., De Nardis, L., Lemic, F., Handziski, V., Wolisz, A., Di Benedetto, M.G., 2016. ViFi: virtual fingerprinting WiFi-based indoor positioning via multi-wall multi-floor propagation model. ArXiv preprint arXiv:1611.09335.

Chintalapudi, K., Padmanabha Iyer, A., Padmanabhan, V.N., 2010. Indoor localization without the pain. In: Proceedings of the Sixteenth Annual International Conference on Mobile Computing and Networking, pp. 173–184.

Damosso, E., 1999. COST ACTION 231: digital mobile radio towards future generation systems. Final Report, Tech. Rep., European Communities, EUR 18957.

Feng, C., Au, W.S.A., Valaee, S., Tan, Z., 2012. Received-signal-strength-based indoor positioning using compressive sensing. IEEE Trans. Mobile Comput. 11 (12), 1983–1993.

Frey, B.J., Dueck, D., 2007. Clustering by passing messages between data points. Science 315 (5814), 972–976.

Hernández, N., Ocaña, M., Alonso, J.M., Kim, E., 2017. Continuous space estimation: increasing WiFi-based indoor localization resolution without increasing the site-survey effort. Sensors 17 (1), 147.

Honkavirta, V., Perala, T., Ali-Loytty, S., Piché, R., 2009. A comparative survey of WLAN location fingerprinting methods. In: 6th Workshop on Positioning, Navigation and Communication, 2009. WPNC 2009, pp. 243–251.

Hossain, A.K.M.M., Van, H.N., Jin, Y., Soh, W.S., 2007. Indoor localization using multiple wireless technologies. In: IEEE International Conference on Mobile Adhoc and Sensor Systems, 2007. MASS 2007, pp. 1–8.

Kessel, M., Werner, M., 2011. SMARTPOS: accurate and precise indoor positioning on mobile phones. In: Proceedings of the First International Conference on Mobile Services, Resources, and Users, MOBILITY, pp. 158–163.

Kumar, S., Hegde, R.M., Trigoni, N., 2016. Gaussian process regression for fingerprinting based localization. Ad Hoc Netw. 51, 1–10.

Laoudias, C., Zeinalipour-Yazti, D., Panayiotou, C.G., 2013. Crowdsourced indoor localization for diverse devices through radiomap fusion. In: 2013 International Conference on Indoor Positioning and Indoor Navigation (IPIN), pp. 1–7.

Le Dortz, N., Gain, F., Zetterberg, P., 2012. WiFi fingerprint indoor positioning system using probability distribution comparison. In: 2012 IEEE International Conference on Acoustics, Speech and Signal Processing (ICASSP), pp. 2301–2304.

Liao, I.-E., Kao, K.-F., 2008. Enhancing the accuracy of WLAN-based location determination systems using predicted orientation information. Inform. Sci. 178 (4), 1049–1068.

Liu, H., Darabi, H., Banerjee, P., Liu, J., 2007. Survey of wireless indoor positioning techniques and systems. IEEE Trans. Syst. Man Cybern. C (Appl. Rev.) 37 (6), 1067–1080.

Marcus, P., Kessel, M., Werner, M., 2013. Dynamic nearest neighbors and online error estimation for SMARTPOS. Int. J. Adv. Internet Technol. 6 (1–2), 1–11.

Roos, T., Myllymäki, P., Tirri, H., Misikangas, P., Sievänen, J., 2002. A probabilistic approach to WLAN user location estimation. Int. J. Wirel. Inform. Netw. 9 (3), 155–164.

Shin, B., Lee, J.H., Lee, T., Kim, H.S., 2012. Enhanced weighted K-nearest neighbor algorithm for indoor Wi-Fi positioning systems. In: 2012 8th International Conference on Computing Technology and Information Management (ICCM), vol. 2, pp. 574–577.

Tian, Z., Tang, X., Zhou, M., Tan, Z., 2013. Fingerprint indoor positioning algorithm based on affinity propagation clustering. EURASIP J. Wirel. Commun. Netw. 2013 (1), 272.

Torres-Sospedra, J., Montoliu, R., Trilles, S., Belmonte, Ó., Huerta, J., 2015. Comprehensive analysis of distance and similarity measures for Wi-Fi fingerprinting indoor positioning systems. Expert Syst. Appl. 42 (23), 9263–9278.

Youssef, M., Agrawala, A., 2004. Handling samples correlation in the Horus system. In: INFOCOM 2004. Twenty-third AnnualJoint Conference of the IEEE Computer and Communications Societies, vol. 2, pp. 1023–1031.

Youssef, M.A., Agrawala, A., Shankar, A.U., 2003. WLAN location determination via clustering and probability distributions. In: Proceedings of the First IEEE International Conference on Pervasive Computing and Communications, 2003 (PerCom 2003), pp. 143–150.

Yu, F., Jiang, M., Liang, J., Qin, X., Hu, M., Peng, T., Hu, X., 2014. 5 G WiFi signal-based indoor localization system using cluster k-nearest neighbor algorithm. Int. J. Distrib. Sens. Netw. 10 (12), 247525.

8

Study and Evaluation of Selected RSSI-Based Positioning Algorithms

Stefan Knauth

HFT STUTTGART, UNIVERSITY OF APPLIED SCIENCES, STUTTGART, GERMANY

1 Introduction

This chapter reports on the authors' work on radio signal strength indicator (RSSI) Smartphone indoor positioning and discusses ideas and theory behind it. It documents and extends the authors' presentation at the First International Workshop on Challenges of Fingerprinting in Indoor Positioning and Navigation, 2016 held at Barcelona and the participation at the IPIN EvAAL competition 2015 which was collocated with the IPIN 2015 International Conference on Indoor Positioning and Indoor Navigation.

Smartphone indoor positioning is a substitution for GNSS as typically GNSS signals are to weak to be used inside buildings. A wide span of technologies and schemes are investigated to provide location estimations and tracking inside buildings. Back to 2000, Microsoft RADAR (Bahl and Padmanabhan, 2000) was among the first Wi-Fi RSSI-based positioning experiments. Since then, countless approaches have been proposed and developed for the Wi-Fi RSSI-based positioning problem. Upon the more prominent solutions there are proximity, centroid, fingerprinting, multilateration, and radio tomography. Unfortunately propagation properties in indoor scenarios are quite individual and results of a certain algorithm may vary considerably between different deployments and locations or laboratories.

RSSI is of course not the only possible solution for indoor positioning. There are other technologies suitable for smartphones like light modulation (Philips Lighting Holding B.V., 2017), sound and ultrasound (Smith et al., 2004; Ward et al., 1997; Knauth et al., 2009, 2015a), magnetic fingerprinting, and dead reckoning (Kang and Han, 2015; Willemsen et al., 2015; Knauth and Koukofikis, 2016) using the built-in sensors of smartphones like barometer, compass, acceleration, gyroscope, etc. These methods may also use RSSI as supporting technology.

RSSI-based indoor positioning using standard smartphones is particularly interesting because users may apply this technology in existing environments where Wi-Fi access points (APs) are deployed. Smartphone-based localization may be performed even if the

Geographical and Fingerprinting Data for Positioning and Navigation Systems. https://doi.org/10.1016/B978-0-12-813189-3.00006-X

Wi-Fi installation has not been specifically prepared for that use case and, for example, characteristics and positions of the Wi-Fi nodes are unknown.

In Section 2 some important features of the indoor radio propagation are recalled. Section 3 is on proximity and centroid methods which most of them do not need effortful recording of reference maps, at the cost of limited accuracy. Wi-Fi fingerprinting principles and the authors' algorithms are presented in Section 4. Section 5 focuses on fingerprint calibrated weighted centroid, which is a special method combining fingerprinting and centroid. In Section 6 running algorithms of the presented methods are discussed and the results are compared against each other and "third-party" solutions. Section 7 outlines some more exotic methods, and Section 8 closes the chapter with a short summary.

2 Indoor Radio Propagation

2.1 The Free Space Model

For the smartphone signal strength-based positioning problem, it is interesting how the signal at a receiver depends on the environment and the distance to a transmitter, for example, a Wi-Fi AP. In free space, radio waves expand more or less undisturbed on straight trajectories. For frequencies below 20 GHz and ranges up to some kilometers, the atmospheric absorption can be neglected. So at typical Wi-Fi frequencies, the received power P_r is given by geometric considerations: A part of the transmitted radio power P_t reaches the receiver antenna, the rest of it beams in other directions. In the isotropic case, the received energy is given by the effective antenna cross-section (which may differ from the physical one) with the sphere with radius d around the transmitter. Here d denotes the distance between transmitter and receiver. The ratio $P_\mathrm{r}/P_\mathrm{t}$ is named "path loss." One should keep in mind that path loss does not automatically mean that the power is absorbed by some means, in the described case the antenna captures less power because the radio beams diverge.

The effective antenna area of typical antennas is proportional to the square of the wavelength λ. With the sphere surface being $4\pi d^2$ this leads to the Friis equation (Friis, 1946):

$$P_\mathrm{r}/P_\mathrm{t} = G_\mathrm{r} G_\mathrm{t} \frac{\lambda^2}{4\pi d^2} \tag{1}$$

The Friis equation describes the relation between distance and path loss in free space, where P_r is the received power and P_t expresses the transmitted power, in nonlogarithmic units. Further variables are the antenna gains G_r and G_t. The formula can be written also as

$$P_\mathrm{r}/P_0 = d^{-2} \tag{2}$$

with P_0 being the received power at a distance of 1 m and d given in meters.

2.2 Indoor Propagation

Unfortunately, the free space model performs quite poor in indoor scenarios. Strong deviations from the model arise mostly due the following effects (Rappaport, 2002):

- Radio waves are heavily reflected in the inside of modern buildings. Iron and steel, copper, concrete walls, and a lot more of modern building materials strongly reflect radio waves, making the inner of a building like a mirror cabinet for the latter. Long corridors act as wave guides and considerably reduce the path loss.
- Multipath fading is caused by interference of signals from different path, for example, by reflection. Interference may be constructive or destructive, the destructive interference leading to strong fading up to vanishing of a signal, even when close to the source. Multipath fading caused by reflections is a major cause for the difficulties in reliably predict path losses in buildings with the Friis formula.
- Absorption of radio power is caused, for example, by the bodies of people and by certain building materials, mainly those who can keep moisture. Absorption and reflection leads to attenuation of the signals when traversing through walls and ceilings.
- The orientation of the mobile antenna depends highly on the users pose and the position of the smartphone. Real existing antennas may be isotropic in two dimensions, but not in all three dimensions, and will therefore always have some directional behaviors.

The effects can to some extend be approximated by using an exponent n different from the free space value of 2 for the distance d in Eq. (2) (see e.g., Hashemi, 1993), leading to

$$P_r/P_0 = d^{-n} \tag{3}$$

$$d = \left(\frac{P_0}{P_r}\right)^{1/n}, \tag{4}$$

or, with logarithmic units

$$d = d_{\mathrm{RSSI}} = 10^{-S[\mathrm{dBm}]/(10\cdot n)} \cdot C, \tag{5}$$

where S is the received RSSI value in dBm and C is a constant expressing P_0. Measured in linear units like watt, the path loss is the quotient of transmitted power and received power. Going to logarithmic units, for example, dBm, the path loss L formula changes to $L = P_t\,[\mathrm{dBm}] - RSSI + c$.

In Fig. 1 curves for several values of n are shown. In buildings, typical values are between 2.5. Depending on antennas and transmission power, there may be a different offset C and the curves may have to be shifted up or down, but the characteristics remain the same. Considering a noise level of, for example, 10 dBm it is obvious that a range estimation is most useful for distances of a few meters only.

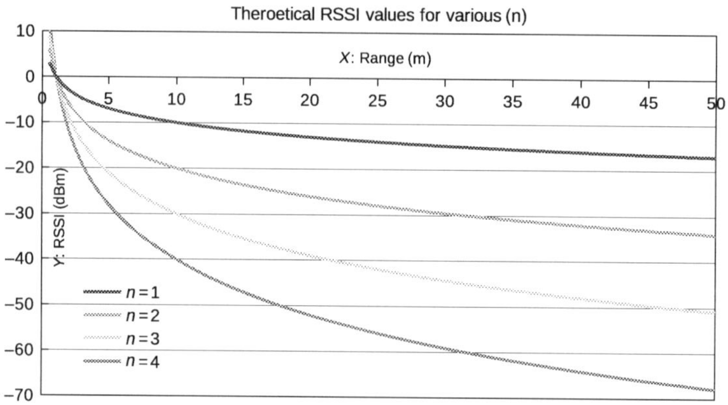

FIG. 1 Received RSSI value for different propagation exponent *n*. Depending on transmitter power, all curves may have to be shift up or down.

2.3 The RSSI Measure

In the following chapters mostly the RSSI will be used. The RSSI represents the path loss in logarithmic units. From experience, the authors assume that typical values reported from smartphones lie in the interval 0,…,−95 dBm. 0 dBm would be regarded as a very high RSSI value, it is likely that a smartphone detecting this value is very close to an Wi-Fi AP. −50 dBm could indicate a 5.25-m distance or a signal which passed some walls. Values of −95 dBm indicate a possible high distance, or strong attenuation, for example, by waves crossing several concrete ceilings or walls. Already these estimates show that application of the modified Friis path loss model is of limited use for a single position estimate and must be seen as a statistical approach.

3 Wi-Fi Positioning by Centroid Methods

3.1 The Centroid Method

The problem of RSSI positioning typically introduces a "rover," also referred to as "mobile node," whose position R shall be determined using path loss measurements between this point and a set of fixed nodes, typically Wi-Fi APs at a known or unknown position R_q. It is assumed that the fixed nodes emit radio packets and the rover determines the received signal strength (RSSI), but operating vice versa would not change the path loss. As we are discussing mainly smartphone Wi-Fi RSSI positioning, the rover can be identified with the smartphone, and the reference points are the AP.

A quite simple approach for estimating the position is to report the position of the AP, which is received with the highest RSSI value, as the rovers position. This is often referred to as the proximity method. The accuracy of this method is obviously related to the density

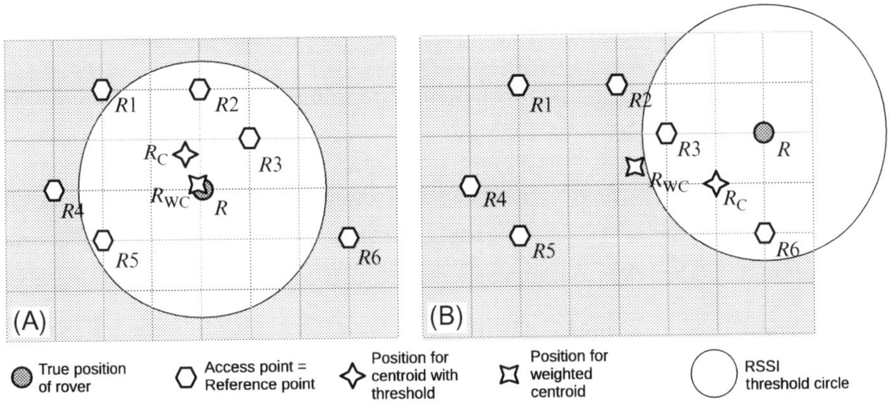

FIG. 2 Two sample scenarios, for the centroid and weighted centroid methods each. *Threshold circle* marks the area over which summation of access point positions is performed for the unweighted centroid. (A) A scenario where a rover is surrounded by a set of access points. Obtained positions more or less match the rover position, for both methods. (B) An anisotropic scenario where the rover position lies outside of the convex hull of the access point set. For the threshold-based centroid, the result is still reasonable, while the weighted centroid is "clamped" by the access point positions.

of APs. Looking at Fig. 2, for case (A) this would be R_3, for case (B) this would be either R_3 or R_6, whichever would be received with a slightly higher RSSI value.

Better results are obtained when calculating the "average" position of all fixed points R_q for which the path losses to the rover are below a certain threshold. Mathematically this is performed by averaging the coordinates of the respective points (the centroid):

$$R_C = \frac{\sum_q R_q}{\sum_q 1}, \tag{6}$$

where q runs over all fixed points fulfilling the threshold criterion.

3.2 Weighted Centroid Method

A more sophisticated approach is the weighted centroid:

$$R_{WC} = \frac{\sum_q R_q \cdot w(q)}{\sum_q w(q)} \tag{7}$$

$$w(i) \propto |R_q - R_{WC}|^{-1} \tag{8}$$

Each reference point R_q is given a weight $w(q)$ with respect to the current measurement at R. Eq. (8) defines the "ideal" weight, that is, a weight which is inverse proportional to the distance d_q between the rover R and a fixed point R_q. Unfortunately, the rover position and the distances are not known in advance. A real weight function shall estimate a range based on the observed path loss or RSSI value.

In Fig. 2 two sample scenarios for the centroid and weighted centroid methods are sketched. APs cover a certain area. The real position of a rover is indicated. For the threshold-based nonweighted centroid, the "threshold circle" marks the area over which the summation of AP positions is performed. For case (A), k would be in {1, 2, 3, 5} while in case (B) this would be {3, 6}. The positions obtained using threshold-based centroid (R_C) according to Eq. (6) and by weighted centroid (R_{WC}) (Eq. 7) are indicated for the two different scenarios. In Fig. 2A the rover is surrounded by APs. In this case, both methods deliver a more or less reasonable result. Obviously, the centroid method always delivers a result, which lies within the convex hull of the set of APs, as long as no negative weights are considered. So the true position R should also be within the convex hull, otherwise reasonable errors will be introduced.

For the situation outlined in Fig. 2B, the convex hull condition is not fulfilled. For the threshold-based centroid, the result will at least lie within the threshold circle. For the weighted centroid the situation is different. The spatial distribution of the APs is quite anisotropic, with respect to the position of the rover. The obtained result is then "compressed," that is, shifted more to the center of gravity of the APs. As it can be already estimated from Fig. 2, centroid is useful especially under conditions where an area is regularly covered by APs, and the rover position lies within that area.

If the convex hull condition for the centroid method is not given, for example, positioning shall also occur in areas outside of an AP "cloud," centroid results may not be satisfying. A possible solution could be the usage of multilateration algorithms operating on the ranges obtained by Eq. (5). However, classic multilateration does not take into account weights. This is not feasible for RSSI, since low RSSI values are typically quite noisy, and the obtained range is therefore quite error prone. Nevertheless, optimized multilateration has been applied (Gau et al., 2008; Mautz et al., 2007) to wireless sensor networks. An alternative approach is probability map-based algorithms. An example is given in Section 7.

4 Wi-Fi Fingerprinting

4.1 The Radio Map

The last sections described methods for position estimation based on the Friis model or else model. While the positions of the APs have to be known, no site surveying, etc., is used here. The expected accuracy will be, for example, limited by the density of APs. As a rule of thumb, the expected accuracy will not exceed 0.5,...,1 times the average spacing of the APs.

Higher accuracy may be obtained by surveying the area under interest: Before the system can determine the position of a rover, a database of "fingerprints" needs to be created. A single fingerprint is created by recording RSSI values for all receivable APs at a certain position R_i, where i is an index for the reference positions, and the number of reference positions is n. The received RSSI values at position R_i form a row vector

$$S_i = \{s_{i,1}, s_{i,2}, \ldots, s_{i,q}, \ldots, s_{n,m}\} \tag{9}$$

with q being an index for a certain AP, m being the total number of APs, and $s_{i,q}$ being the RSSI value received at R_i from AP with index q. The APs index is an enumeration. The actual identification of the AP is performed via the APs' MAC address. Special handling is needed if a certain AP is not received at a certain position. Since a vector needs values for each coordinate, some special value may be used to indicate that no signal has been received. For example, if measured values lie in the interval 0,..., −100 dBm, a "null" value of −101 could indicate that no value has been recorded for that AP. The row vector consisting of all surveyed reference point vectors S_i forms the radio map which can also be seen as a matrix S

$$
\begin{array}{cc} & \begin{array}{cccc} AP_1 & AP_2 & \ldots & AP_m \end{array} \\ S = \begin{array}{c} S_{i=1} \\ S_2 \\ \vdots \\ S_n \end{array} & \left(\begin{array}{cccc} s_{1,1} & s_{1,2} & \cdots & s_{1,4} \\ s_{2,1} & s_{2,2} & \cdots & s_{2,4} \\ \vdots & \vdots & s_{i,q} & \vdots \\ s_{n,1} & s_{n,2} & \cdots & s_{n,4} \end{array} \right), \end{array} \tag{10}
$$

where the row vectors S_i are the reference vectors measured at position R_i (Eq. 9), and the column vectors AP_q are the measured RSSI values for the qth AP. When a radio map is available, position estimates can be calculated by performing the k-Nearest Neighbor (kNN) method with the following steps:

- Observe RSSI values of APs at the unknown position R_x and create a measurement vector S_x.
- For each reference vector S_i calculate the similarity $\sigma_{x,i}$ between S_x and S_i.
- Take the set of the k reference vectors S_i which are most similar to S_x.
- Calculate a final estimate based on the positions R_i of the k most likely S_i.

The final estimate will typically be again a centroid method. Weights could be based on the similarity measure σ or could just be fixed. The authors typically use $k = 3$ and use fixed weights of 2–1–1 in descending similarity order.

4.2 RSSI Vector Similarity Measures

Popular algorithms like the above-discussed kNN fingerprinting compare measured RSSI vectors with the RSSI vectors in a radio map database. A crucial step in this algorithm is of course the comparison itself. There is quite a choice for similarity measures. The most common are

$$
\text{Euclidian distance: } \sigma_{x,i} = \sqrt{\sum_q (s_{x,q} - s_{i,q})^2}, \tag{11}
$$

$$
\text{Manhattan distance: } \sigma_{x,i} = \sum_q (s_{x,q} - s_{i,q}), \tag{12}
$$

$$\text{Euclidian or cosine norm: } \sigma_{x,i} = \frac{\sqrt{\sum_q (s_{x,q} \cdot s_{i,q})^2}}{|S_x| \cdot |S_i|}. \tag{13}$$

While the Euclidian distance is widely used as comparison norm, there are quite some more choices like, for example, the Manhattan distance, the cosine norm and others. Focusing on the three mentioned, they have advantages and disadvantages. First, it must be said, that for kNN, the similarity measure is used mainly for ranking of the RSSI vectors. Since the measures are monotone, if there is a reasonable difference in the top-similar fingerprints and a small is k used, either norm will produce the same ranking. In particular when using fixed weighting and a small k, for the three mentioned norms it is likely that the same result will be obtained for the estimated position on either norm. However, this is not generally guaranteed. A comprehensive analysis of the Wi-Fi similarity problem is given, for example, in Torres-Sospedra et al. (2015).

There is a principal problem that different phones even of the same model may have a different offset in the RSSI values. This may be caused, for example, by different Wi-Fi chipsets, protection cases around the phone and different pose of the phone during measurement. Such an offset in the RSSI values may, in particular for the distance-based norms, lead to failures in the ranking of the vectors. Avoidance is typically performed either by include already different phones and poses in the mapping phase or by normalizing procedures, for example, normalizing of the map vectors and the measured vectors. The normalizing is self-included when using the cosine norm. Actually the cosine norm is more or less performing a correlation of the vectors more than a difference calculation.

There are other peculiarities when dealing with the similarity:

- In practical measurements, depending on the scale of the deployment, a majority of APs will not be received at all. The formulas have to be modified to cope with such "null" values. One solution is to replace "null" values with the lowest overall obtained RSSI value. But in situations, where an AP is taken out of operation, or during the scan period has not been captured, this would lead to reasonable error in similarity estimation. Another approach is to compare only those RSSI values, which are not null in both, in the radio map data and in the online received data. Of course, this would have to be taken into account also into the normalization such that the normalization should only count values, which are available on both sides.
- The RSSI-based similarity does not take into account the "credibility" of the RSSI value. As stated in Section 2, RSSI values are typically quite noisy. A difference of ± 10 dBm relates to a small range difference when being at a level of -20 dBm, but gives a high spatial range difference at -70 dBm. This can be addressed by performing a weighting of the RSSI values before comparison. A natural choice for the weighting function is the inverse distance calculated from Eq. (5):

$$d^{-1} = 10^{S|\text{dBm}|/(10 \cdot n)} \tag{14}$$

with the propagation exponent n. The constant C expressing the transmitter power and antenna gain has been omitted since it will vanish on the normalization step in the similarity measure, visible, for example, in Eq. (13).

5 Fingerprint Calibrated Weighted Centroid

Wi-Fi fingerprinting algorithms do not need to know the position of the APs, since all necessary information is captured by the radio map generation. In some situations, however, the size of the radiomap may be a burden for a specific implementation. The radio map contains n reference vectors, each of it containing an array of q values for the APs. The data set of the 2015 IPIN EvAAL competition had about 500 APs and more than 800 reference points leading to about 500,000 values, leading to a database size of about 1 MB, in compressed format. Also, each position estimate has a complexity of about 400,000 calculations. While this is not a serious task for a smartphone, embedded devices based on small microcontrollers reach their resource limits with these numbers.

The fingerprint calibrated weighted centroid (FCWC) algorithm (Knauth et al., 2015b) processes the radio map data to estimate the positions of the APs. A rover may then use these positions to calculate a position estimate, for example, via weighted centroid. The size of the database is thereby collapsed to the positions of the APs, which will be not more than 500 coordinates in the described case, thus reducing the size of the database down to some kilobyte. As the estimates are performed by weighted centroid (Eq. 7), the computational effort reduces to a number proportional to the AP count, in the given case to 500.

The method works by first estimating the positions of the APs offline by "reverse positioning": By transposing of the RSSI matrix (10), the qth row vector contains now a set of RSSI values for the qth AP. Each RSSI value corresponds to a certain known reference position R_i. The path loss is independent of the propagation direction, that is, the path loss from a transmitter to a receiver is the same if the roles of transmitter and receiver are exchanged. So the unknown position R_q of the AP can be estimated by the weighted centroid method using Eqs. (7), (8).

Once the positions of the APs have been estimated, they may be used for online position detection. A position estimate can now be based on the RSSI report S_i of an unknown location and the AP positions R_q using again Eqs. (7), (8), but substituting R_i with R_q. It should be noted that the estimated AP positions not necessarily have to be close to their real positions. The estimated positions are called "virtual positions" and might be interpreted as just being parameters for a Wi-Fi-based parametric approximation, based on the radio map data.

A comparable approach has been described for ultrasound positioning in Knauth et al. (2013).

For both, the fingerprinting and the FCWC algorithm, in multifloor scenarios also the floor has to be identified. Therefore, each reference point is assigned a z-coordinate based on the floor ID given in the dataset. The centroid calculation also processes the z coordinates such that the virtual APs and the position estimates contain a z-coordinate. This coordinate is then used to assign a floor to the result.

6 Validation of the Described Fingerprint and FCWC Schemes

6.1 Validation Data

The validation is performed using publicly available measurement data provided by the UJIIndoorLoc database (UJIIndoorLoc Data Set at UCI Machine Learning Repository, 2014; Torres-Sospedra et al., 2014). This data formed the base for the EvAAL 2015 competition on indoor localization (EvAAL: Evaluating AAL Systems through Competitive Benchmarking, 2015) track on "Wi-Fi fingerprinting in large environments (off-site)." The presented algorithms participated at the competition as one of four competitors, allowing to compare the results of the two algorithms (fingerprinting and FCWC) between each and also to other schemes and implementations (Torres-Sospedra et al., 2017).

The provided data comprises a dataset for training and adjusting of the algorithms, which is enriched with ground truth information, and a competition dataset for which the ground truth is not known to the competitors. The UJIIndoorLoc training table comprises about 20,000 measurements (rows) collected with a variety of mobile equipment at 933 different positions, distributed within 3 huge university buildings each having 4 to 5 floors. The area is equipped with 520 Wi-Fi APs. Each data row consists of 520 RSSI values, one for each AP, and additional information like the position of the measurement, the used equipment, a code for the person performing the measurement, a building ID and a floor ID.

6.2 Algorithm Implementations

As the described algorithms run on local metric coordinates, latitude/longitude position data are converted into a local x–y coordinate with respect to a reference point in the middle of the spawned area and a reference direction, which is aligned to the main structures of the given data. A z-coordinate is also introduced, which is set to floor ID times 5 m. Note that, due to the linearity of Eq. (7), the kNN calculation used in both algorithms is independent of the used coordinate system and could actually be performed with latitude/longitude values instead of local coordinates. The local coordinates are convenient for debugging purposes and calculation of deltas between given true and estimated positions. Also the floor spacing value does not need to match the true floor spacing, as it is only used to determine the floor ID, not the real z position of an item.

A fingerprint-based algorithm "SPCF" (scalar product correlation fingerprinting) has been set up as described in Section 4. The algorithm employs the cosine norm (Eq. 13) on weighted RSSI values (Eq. 14) as similarity measurement and uses *k*NN with fixed weights for final position estimation.

The parameters used were 2–1–1 as fixed weights, $k = 3$, and n in [2.5,..., 4], depending on the building. These values have been determined manually by iteratively optimizing the algorithm output against the provided true ground positions in the validation data. The optimization criteria were the average position error not regarding floor and building penalties.

A further optimizing has been reached by removal of "bad" APs: For each AP, the algorithm was run one time with a particular AP included, one time with the specific AP excluded. If the result was better when the specific AP was not included, it was considered as "bad AP" and its data were ignored in the processing. This led to some decimeters of overall accuracy gain.

Also an FCWC algorithm was implemented, as described in Section 5: A reverse positioning of the APs is performed. The training table with the calibration data is transposed such that for each AP, the RSSI values measured at different positions R_i are available. The yet unknown position R of an AP is now calculated using Eq. (7) with the weight function (14). The parameters used were mostly same as for the SPCF algorithm. Also, the described optimizing was applied.

However, some parameter settings were altered for FCWC: the different parameter settings evaluated were the exponent n and the lower RSSI threshold $RSSI_{\min}$. Obtained values were $RSSI_{\min} = -85\,\text{dBm}$ and $n = 1.0$. This value is surprisingly low as typical indoor propagation exponents are above 2.0. In the given scenario the low exponent favors high RSSI values, because weak readings will be assigned a particularly low weight. This seemed to be important in the FCWC case, where the weighted centroid is applied twice.

6.3 Results SPCF and FCWC

The overall mean delta of FCWC against the provided validation data file is about 9.7 m (see also Fig. 3). This accuracy is x–y based and does not include the height displacement. Using the calculated positions the floor and building ID is estimated. It is found that about 94% of floor IDs are reported correctly, and the building ID has been assigned correctly to all measurements. For all measurements, a position has been estimated. The discussed and implemented SPCF (scalar product correlation fingerprinting algorithm) delivered an average delta of 7.7 m and a floor ID detection rate of about 96% against the "validationData.csv" file and thus performed reasonably better on the validation data. As it will be seen in the next section, for the private test set of the competition, the difference in performance between the two algorithms was much lesser.

6.4 Validation Against Competing Algorithms

Four teams have participated at the offsite track of the 5th EvAAL competition. Detailed results have been published at the competition website (EvAAL: Evaluating AAL Systems through Competitive Benchmarking, 2015). Table 1 lists the results of the four teams.

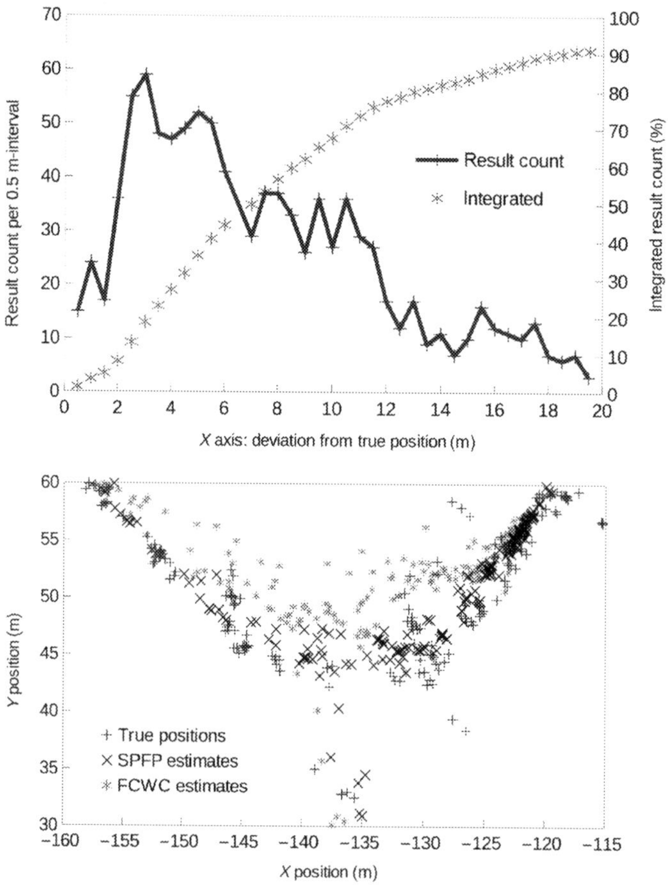

FIG. 3 *Top*: Delta between calculated positions and true positions, for the public test set, for FCWC. The *solid line* displays the number of measurements obeying the indicated (*x*-axis) delta. The interval for the value count is 0.5 m. The total measurement count is 1110, of which about 10% have a delta of 20 m and above and lie outside of the figure. The *asterisk line* indicates the integrated number of measurements, in % of the total number. The 50th percentile delta value is about 7 m, the 75th percentile about 11.5 m. *Bottom*: FCWC and SPCF results. Results of the centroid algorithm always lie within the convex hull of the reference points. Therefore, they more accumulate in the centers of covered areas.

The authors team "HFTLoc" is listed in the second and the third columns. The ranking of the teams has been performed according to the row "third-quartile error," with better results listed more right. For the "HFTLoc" team, the "official" results have been obtained using the SPCF algorithm. The extra column lists the results obtained with the FCWC algorithm and is included to allow comparison of FCWC and SPCF competition performance. A topological overview of the FCWC and SPCF results is given in Fig. 3.

Team "MOSAIC" uses a probabilistic Wi-Fi sensor model and calculates the mean mutual information. Maximum likelihood estimation and *k*NN are then applied as localization scheme (Berkvens and Weyn, 2015). Team "ICSL" (Choi et al., 2015) uses machine

Table 1 EvAAL 2015 Offsite Track Results

Team Name	MOSAIC	HFTLoc SPCF	HFTLoc FCWC	ICSL	RTLSUM
Building success rate (%)	98.65	100.00	100.00	100.00	100.00
Floor success rate (%)	93.86	96.25	93.71	86.93	93.74
Floor error rate (%)	6.14	3.75	6.29	13.07	6.26
Floor ±1 rate (%)	98.88	99.98	100.00	100.00	99.79
B&F success rate (%)	92.59	96.25	93.71	86.93	93.74
Mean error if B&F success	7.97	8.40	8.55	6.80	5.71
Mean error	11.64	8.49	8.69	7.67	6.20
First quartile error	3.26	3.69	4.43	3.10	2.51
Second quartile error	6.72	6.99	7.41	5.88	4.57
Third quartile error	12.12	11.60	11.79	10.87	8.34

learning techniques and a deep learning approach. Team "RTLSUM" (Moreira et al., 2015) employed besides fingerprinting also temporal filtering. Details can be found in Torres-Sospedra et al. (2017).

An important finding is that, for the three left columns, the results are quite close together, while the approaches taken are quite different. This may be an indication that the information content of the dataset does not allow to come to a more accurate result. For the convincing results in the rightmost column, the team among others applied temporal filtering.

The presented SPCF approach turned out to be particularly successful in estimating the floor. This may be attributed to the usage of 3D coordinates, that is, using x, y, and z, when averaging positions for calculating of a result. The FCWC algorithm scored quite comparable, which is an amazing result since it means that in large deployments it is possible to use only one vector describing the virtual positions of the APs instead of the whole fingerprint database, and still get reasonable results. The described algorithms can well compete with algorithms published by other groups and may therefore be helpful for implementing fingerprint smartphone indoor location systems.

7 Wi-Fi Probability-Based Positioning and BLE

7.1 The Probability Density Function

A drawback of the centroid and the *k*NN step in fingerprinting is that obtained positions are always reported within the convex hull of the APs, as averaging can not go "out" of the convex hull. This restriction is not present for Multilateration. However, the latter is heavily disturbed by noise, offsets, etc. The parametric approach presented in this section merges to some extend the concepts of proximity and multilateration to obtain accuracies more close to those of fingerprinting algorithms. There is only minor mapping effort needed, in comparison to fingerprint-based approaches.

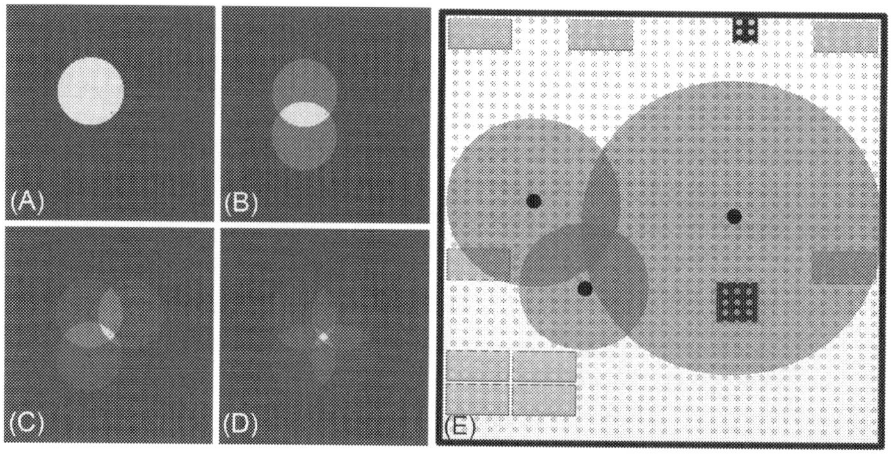

FIG. 4 (A) Probability map (representation of a probability density function) of a single receiver, for an RSSI value corresponding to the diameter of the circle. (B)–(D) Merged maps for several receivers. (E) Merged map showing also the underlying calculation raster, and some orientation marks in the room. Overall size of (E) is about 6 × 6 m. *Circle areas with small radius* correspond to high RSSI values/low path losses.

The probability-based positioning algorithm makes use of a parametric probability density function $p(d, s)$. The function describes, for a received RSSI value s, the probability p for the rover to be at a distance d from the reference position. In Fig. 6B, a 2D gray scale map representation of such a function is displayed where d is the distance between a map pixel and the position of a receiver R_i. R_i is the dark point in the center of the rings.

Fig. 4 sketches a simple variant of $p(d, s)$: here the probability is 1 for points R within a certain range $d = |R - R_i| < d_{\text{RSSI}}$ around the fixed node position R_i, and 0.3 for other positions. The figure represents this function as a gray scale image. d_{RSSI} is given by the observed RSSI value according to Eq. (5). Already this easy probability density function respects some important properties of the indoor path loss, namely that the distance between fixed node and mobile node can be small, even if only low RSSI values are reported (high attenuation), but for a high RSSI value, it is unlikely that there is a large distance between transmitter and receiver. The figure sketches the principle of operation: for each AP, there is a probability density function $p(d, s)$ describing the probability of the mobile node to be at a certain position, based on the RSSI information of that AP. Fig. 4B–D represents subsequent introduction of further APs. The probability maps of the APs are merged by multiplying thus creating an overall merged map. The merged map indicates the overall probability of presence of the mobile device for a given coordinate. This map could, for example, be used to seed new particles in a particle filter. In the presented work we estimate the position of the mobile node by finding the position of highest probability. Fig. 4E displays again the merging of three probability functions, and indicates the discrete points for which probability values are calculated. In practice, more complex probability density functions are used. For example, the function may model the directional behavior of the fixed node antenna. A more smooth function will typically result in smoother absolute maximum regions in the merged probability map.

The choice of a suitable $p(d, s)$ is important for the success of the method. For the example shown in Fig. 6, the function was defined such that at distances related to d_{RSSI} (Eq. 5) the function has the highest probability. A typical function will look like

$$p(d, s) = \frac{1}{(d - d_{\mathrm{RSSI}}(s))^2 + c}, \tag{15}$$

where $d_{\mathrm{RSSI}}(s)$ is the RSSI-based distance estimation for an RSSI value s, and c is a constant defining the "sharpness" of the function. An example for $d(s) = 2$ m and $c = 0.5$ is plotted in Fig. 6C.

Eq. (15) gives the probability from one AP. In order to get a probability map which considers all APs, the probabilities are multiplied. The residual probability including m APs is

$$p(R, S) = \prod_{q=1,\ldots,m} p(|R_q - R|, s_q) \tag{16}$$

where R is the position of the mobile node, R_q is the position of the qth AP, and S is the measured RSSI vector at position R. The above result delivers the probability distribution for all APs.

7.2 A Probability-Based Setup and Algorithm

A simple approach to get a position estimate is to use the position of the maximum of Eq. (16). For the validation of the presented probability-based positioning algorithm, the maximum is found by evaluating $p(R, S)$ on a grid of positions with a certain spacing. The spacing should be smaller than the expected accuracy and also smaller than the features of the probability density function $p(d, s)$. A typical value is 25 cm. For large deployments performance might be an issue. In that case, a two-phase approach will be useful: First, the maximum on a coarse grid is searched using a smooth $p_{\mathrm{coarse}}(d, s)$. Then, around the obtained coarse position, a fine search is performed using a sharp $p(d, s)$. The algorithm may be used to calculate probability distributions or "maps." These maps can be used as one of the input data for probabilistic positioning schemes using Markov chains or for seeding in particle filter approaches.

The algorithm has been implemented in a small scale lab setup (Knauth et al., 2014). Instead of using Wi-Fi, a different setup consisting of smartwatches and microcontroller-based receivers was used. The principles are applicable also to Wi-Fi setups. A general operation schema is shown in Fig. 5. Fig. 6B sketches the actual setup in the Lab: Fixed receiver nodes were installed in a 6 × 6 m sized room. Six CC1110DK MINI nodes were positioned on tables in the lab. Eight ground positions were defined and marked on the floor. For these positions, measurements were recorded. The mobile node consisted of a student wearing an EZ430-Chronos smart watch. The student visited the eight marked positions. A second person recorded accurate timing information, that is, at what time

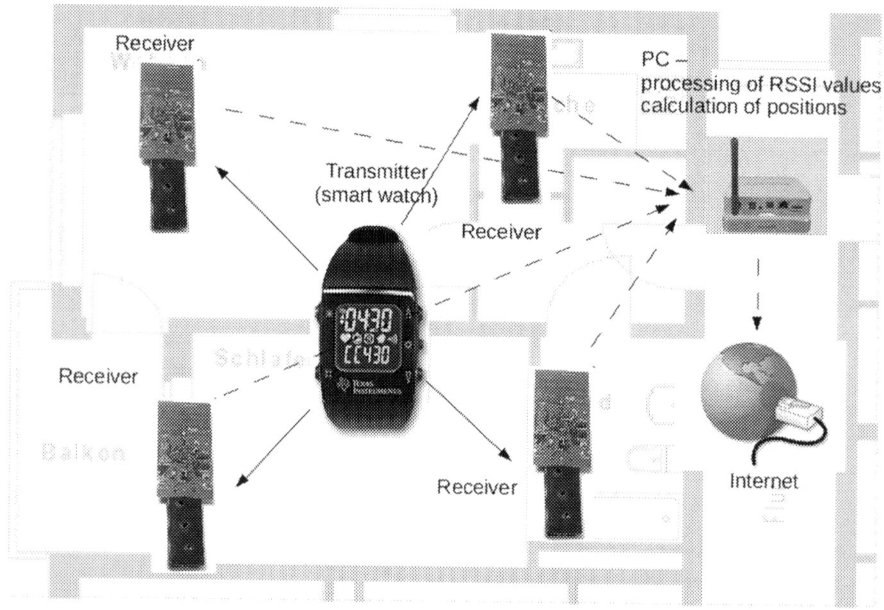

FIG. 5 Schematic setup used for validation of the probability-based parametric method: An EZ430 Smartwatch emits 868 MHz radio packets, which are received by CC1110DK MINI nodes (*solid arrows*). These fixed nodes transmit their obtained RSSI values to a central PC (*dashed arrows*), which logs the values and runs the algorithm. Note that in the experiment, six receiver nodes were used.

which position was visited. The smart watch emits 868 MHz RF Packets of about 3 ms duration each, at a rate of 30 Hz. The six receivers collect their observed RSSI values and transmit them to a PC equipped with an 868 MHz USB dongle receiver. A TDMA scheme is used to avoid collisions.

For each test position, the probability density is calculated according to Eq. (16) using Eq. (15) as probability density function. The calculation is performed over a grid of R values, each grid point position related to a pixel of a bitmap graphic, representing the probability map. The resulting distribution is shown in Fig. 6, inset 1–8, for each of the reference points. The maps indicate the probability for the mobile node to be at a certain position.

7.3 Probability-Based Results

Table 2 lists the positioning results of the algorithm. Obtained errors lie in the range of some centimeters up to about 2.5 m, with an average error of about 1.2 m. As it can be seen, high errors were obtained for positions 6, 7, and 8 (see also Fig. 6). This might be attributed to the fact that these points lie more or less outside of the area spanned by the reference nodes, which generally leads to less good algorithmic conditions for range-related positioning methods. As far as general statements can be deducted from such a

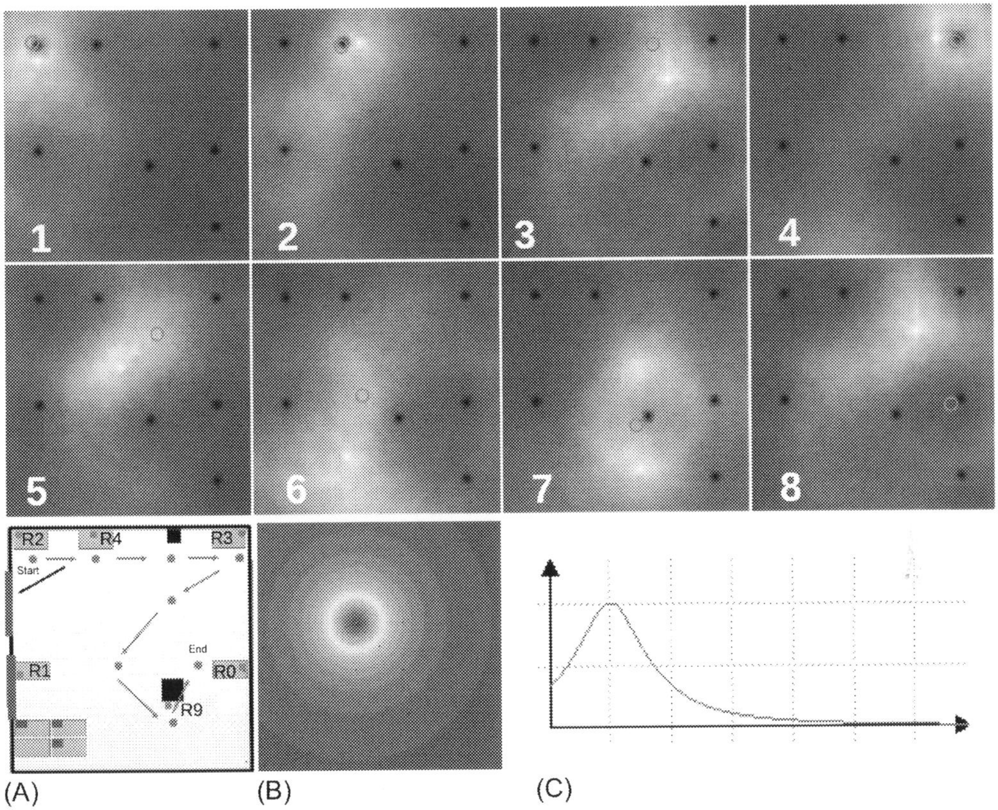

FIG. 6 *Top 1–8*: Obtained probability maps for eight measurement points. The map numbering corresponds to the sequence of the points in the path indicated in the *left lower inset*. The positions of the fixed nodes are indicated as *dark points* in the maps. *Bottom* (A): Situation in the lab. *Gray rectangles* indicate tables, receiver positions are marked with *asterisks*. The test path is indicated with *arrows*, measurement positions are indicated by *filled circles*. (B) Typical probability map as used in the experiments, in a gray scale representation. *Dark parts* indicate low probability, *bright parts* indicate high probability. The size of the map is 10 × 10 m, the d_{RSSI} value is 2 m. For lower RSSI values, the radius of highest probability (*ring with bright gray* values) would be larger, and vice versa. (C) Plot of a probability function $P(s, d)$ for an RSSI value s related to 2 m distance.

Table 2 Probability-Based Results

	Real Pos.	Estimated	Error
1	0.00/5.50	0.00/5.38	0.12
2	2.00/5.50	2.44/5.94	0.62
3	4.00/5.50	4.70/5.00	0.86
4	6.00/5.50	5.35/6.00	0.82
5	3.70/4.00	2.80/3.70	0.95
6	2.30/2.45	2.11/0.60	1.86
7	3.70/2.00	3.50/0.25	1.76
8	5.00/2.45	4.51/4.83	2.43

minimal proof-of-concept setup, obtained results indicate that the accuracy of the method could be comparable to weak-mapped fingerprinting, but besides a one-time calibration to resemble the internal path loss and antenna pattern of the reference node model (the constant C in Eq. (5)), no calibration has to be performed.

The method employs an RSSI-dependent probability density function, which is applied for each fixed node. The results are combined by multiplying to get the overall mobile nodes spatial probability distribution. Values were calculated for a grid of positions. The method has the advantage that no fingerprint maps have to be generated, certainly at the cost of a somewhat lower accuracy but allowing for less engineering effort during deployment. While in this experiment the "brightest point," that is, the most likely position has been chosen as the reported position estimate, the generated maps allow also for more sophisticated position estimators, especially when using them as seeding information in particle filters or other probabilistic methods.

7.4 BLE Beacon RSSI Weighted Centroid

Weighted centroid can be a good solution for Bluetooth Low Energy (BLE) beacon-based positioning. The following laboratory scale experiment was setup: A number of 14 BLE transmitters ("iBeacons") were installed at defined positions in a hall. BLE uses the same 2.4 GHz band as Wi-Fi, so no principal differences in radio propagation are to be expected between the both. A Samsung Galaxy S4 Mini smartphone was used for detection. Sixty-five test points were marked on the floor. A test person visited the test points and recorded measurements for about 10 s, at each test point. For each beacon and test position, on average eight RSSI values were recorded in that time.

In Fig. 7 the geometry and results for two test scenarios are shown: 14 transmitters are arranged around the about 300 m^2 test area. The weighted centroid algorithm is applied and in the left case, the average deviation was about 4 m. Note, that in this case, the orientation of the test person and smartphone was random.

In the experiment run demonstrated at the right-hand side of Fig. 7, the test person always kept herself and the smartphone in a certain pose, that is, facing the top border with respect to the image. The obtained average deviation for this case is about 4.5 m, but as it can be seen, this is not equally distributed over the test area but more pronounced at the lower parts. A quite likely explanation of that behavior is that the directional antenna pattern and the RF shading of the test persons body reduces weights of the "lower" transmitters, mainly IDs 1, 2, 7, 10, and 11.

The displayed results were actually not obtained with a pure weighted centroid approach, but by combining weighted centroid and introducing a lower RSSI threshold of −85 dBm such that only APs (here iBeacons) which have an RSSI value above this lower RSSI threshold are considered in the weighted centroid algorithm. So it could be called "proximity weighted centroid." The threshold value was determined by optimizing the mean error.

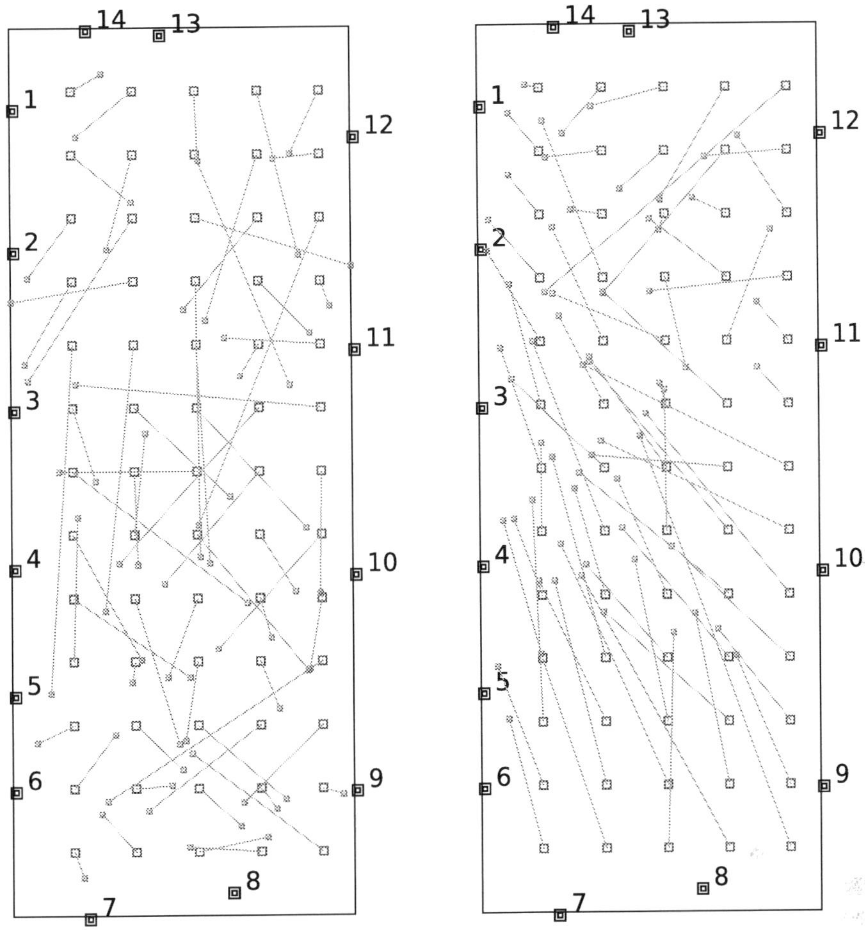

FIG. 7 Results of two weighted centroid measurements recorded in a laboratory test setup. *Larger double squares on the borders* indicate positions of radio transmitters (iBeacons). The regular grid marked with *squares* indicates the real positions of measurements. The spacing of these pints is 2 m. The size of the marked frame is about 28 × 11 m. The vectors starting at the test points indicate the calculated position. Both shown graphs differ in pose of the test person: (*right*) the test person rotated around herself while performing a measurement, (*left*) the test person kept herself and the phone in a certain orientation, that is, facing the top border.

8 Summary

Selected RSSI-based positioning algorithms have been studied. The backgrounds of radio propagation and the Friis model and modifications have been recalled, and aspects of indoor propagation have been discussed. Based thereon, centroid and weighted centroid methods have been explained. They form the base for more sophisticated approaches like fingerprinting. The fingerprinting method, the radio map concept and mathematic backgrounds like common similarity measures have been described.

In Chapter 5 the FCWC algorithm is described. It uses a radiomap for calculation of AP positions, which are then used for weighted centroid. This algorithm and the SPCF algorithm, which is a typical radiomap-based fingerprinting algorithm, have been described in detail and the performance of the algorithms has been assessed by comparing against each other. Furthermore, the algorithms participated at the IPIN EvAAL 2015 competition on indoor localization (offline track) (EvAAL: Evaluating AAL Systems through Competitive Benchmarking, 2015), which allowed to compare also against different approaches of other competitors, based on a very large dataset spawning four university buildings. Analysis showed that both algorithms well competed with other approaches, details have been outlined in Section 6.4.

Further examples of RSSI-based methods described are, for example, a BLE setup employing weighted centroid, and a probability map-based setup using proprietary 868 MHz communications of smart watches. Both algorithms are described in detail and can be regarded as alternative approaches, mostly useful in small-scale deployments where additional infrastructure is acceptable.

References

Bahl, P., Padmanabhan, V.N., 2000. RADAR: an in-building RF-based user location and tracking system. In: INFOCOM, vol. 2, pp. 775–784.

Berkvens, R., Weyn, M., 2015. Mean mutual information of probabilistic Wi-Fi localization. In: 2015 International Conference on Indoor Positioning and Indoor Navigation (IPIN).

Choi, S., Jaehyun, Y.O.O., Kim, H.J., 2015. Machine learning for indoor localization: deep learning and semi-supervised learning. In: 2015 International Conference on Indoor Positioning and Indoor Navigation (IPIN).

EvAAL: Evaluating AAL Systems through Competitive Benchmarking, 2015. EvAAL-ETRI on-site and off-site indoor localization competition in conjunction with IPIN 2015. Available from: http://evaal.aaloa.org/ (Accessed 20 June 2018).

Friis, H.T., 1946. A note on a simple transmission formula. Proc. IRE 34 (5), 254–256. ISSN 0096-8390. https://doi.org/10.1109/JRPROC.1946.234568.

Gau, Y.H., Chu, H.C., Jan, R.H., 2008. A weighted multilateration positioning method for wireless sensor networks. Int. J. Pervasive Comput. Commun. 3 (3), 289–303. https://doi.org/10.1108/17427370710856246.

Hashemi, H., 1993. The indoor radio propagation channel. Proc. IEEE 81 (7), 943–968. ISSN 0018-9219. https://doi.org/10.1109/5.231342.

Kang, W., Han, Y., 2015. SmartPDR: smartphone-based pedestrian dead reckoning for indoor localization. IEEE Sensors J. 15 (5), 2906–2916. ISSN 1530-437X. https://doi.org/10.1109/JSEN.2014.2382568.

Knauth, S., Koukofikis, A., 2016. Smartphone positioning in large environments by sensor data fusion, particle filter and FCWC. In: 2016 International Conference on Indoor Positioning and Indoor Navigation (IPIN), pp. 1–5.

Knauth, S., Jost, C., Klapproth, A., 2009. iLoc: a localisation system for visitor tracking and guidance. In: Proc. 7th IEEE Int. Conf. on Industrial Informatics INDIN2009, Cardiff, UK.

Knauth, S., Andrushevich, A., Kaufmann, L., Kistler, R., Klapproth, A., 2013. The iLoc+ ultrasound indoor localization system for AAL applications at EvAAL 2012. In: Chessa, S., Knauth, S. (Eds.), Evaluating

AAL Systems Through Competitive Benchmarking, vol. 362, Communications in Computer and Information Science. Springer, Berlin, Heidelberg, pp. 83–94.

Knauth, S., Griese, T., Tran, Y., Ortega, A.B., 2014. Towards smart watch position estimation employing RSSI based probability maps. Proc. First BW-CAR Baden-Württemberg CAR Symposium on Information and Communication Systems (SInCom 2014), Furtwangen, Germany, ISBN 978-3-00-048182-6, pp. 75–78.

Knauth, S., Kaufmann, L., Andrushevich, A., Kistler, R., Klapproth, A., 2015a. Evaluating the iLoc indoor localization system: competition outcomes and lessons learned. J. Ambient Intell. Smart Environ. 7, 287–300.

Knauth, S., Storz, M., Dastageeri, H., Koukofikis, A., Mäher-Hipp, N.A., 2015b. Fingerprint calibrated centroid and scalar product correlation RSSI positioning in large environments. In: 2015 International Conference on Indoor Positioning and Indoor Navigation (IPIN), pp. 1–6.

Mautz, R., Ochieng, W.Y., Brodin, G., Kemp, A., 2007. 3D wireless network localization from inconsistent distance observations. Ad Hoc Sensor Wirel. Netw. 3 (2–3), 140–170.

Moreira, A., Nicolau, M.J., Meneses, F., Costa, A.D., 2015. RTLS@UM—WiFi fingerprinting competition at IPIN 2015. In: 2015 International Conference on Indoor Positioning and Indoor Navigation (IPIN).

Philips Lighting Holding B.V., 2017. Available from: http://www.lighting.philips.com.au/systems/themes/led-based-indoor-positioning (Accessed 20 June 2018).

Rappaport, T.S., 2002. Wireless Communications: Principles and Practice, Prentice Hall Communications Engineering and Emerging Technologies Series. Prentice Hall PTR. ISBN 9780130422323. Available from: https://books.google.de/books?id=TbgQAQAAMAAJ.

Smith, A., Balakrishnan, H., Goraczko, M., Priyantha, N.B., 2004. Tracking moving devices with the cricket location system. In: 2nd International Conference on Mobile Systems, Applications and Services (Mobisys 2004), Boston, MA.

Torres-Sospedra, J., Montoliu, R., Martínez-Usó, A., Arnau, T.J., Avariento, J.P., Benedito-Bordonau, M., Huerta, J., 2014. UJIIndoorLoc: a new multi-building and multi-floor database for WLAN fingerprint-based indoor localization problems. In: Proceedings of the Fifth International Conference on Indoor Positioning and Indoor Navigation, Busan, Korea.

Torres-Sospedra, J., Montoliu, R., Trilles, S., Belmonte, Ó., Huerta, J., 2015. Comprehensive analysis of distance and similarity measures for Wi-Fi fingerprinting indoor positioning systems. Expert Syst. Appl. 42 (23), 9263–9278. ISSN 0957-4174. https://doi.org/10.1016/j.eswa.2015.08.013. Available from: http://www.sciencedirect.com/science/article/pii/S0957417415005527.

Torres-Sospedra, J., Moreira, A., Knauth, S., Berkvens, R., Montoliu, R., Belmonte, O., Trilles, S., Nicolau, M.J., Meneses, F., Costa, A., Koukofikis, A., Weyn, M., Peremans, H., 2017. A realistic evaluation of indoor positioning systems based on Wi-Fi fingerprinting: the 2015 EVAAL-ETRI competition. J. Ambient Intell. Smart Environ. 9 (2), 263–279. https://doi.org/10.3233/AIS-170421.

UJIIndoorLoc Data Set at UCI Machine Learning Repository, 2014. Available from: https://archive.ics.uci.edu/ml/datasets/UJIIndoorLoc (Accessed 20 June 2018).

Ward, A., Jones, A., Hopper, A., 1997. A new location technique for the active office. IEEE Pers. Commun. 4 (5), 42–47.

Willemsen, T., Keller, F., Sternberg, H., 2015. A topological approach with MEMS in smartphones based on routing-graph. In: 2015 International Conference on Indoor Positioning and Indoor Navigation (IPIN), pp. 1–6.

9

Mapping Indoor Environments: Challenges Related to the Cartographic Representation and Routes

Luciene S. Delazari*, Leonardo Ercolin Filho†, Rhaissa Viana Sarot*, Pedro Paulo Farias‡, Amanda Antunes*, Scarlet Barbosa dos Santos‡

*GEODETIC SCIENCES GRADUATE PROGRAM, FEDERAL UNIVERSITY OF PARANÁ, CURITIBA, BRAZIL
†DEPARTMENT OF GEOMATICS, FEDERAL UNIVERSITY OF PARANÁ, CURITIBA, BRAZIL
‡CARTOGRAPHIC ENGINEERING UNDERGRADUATE COURSE, FEDERAL UNIVERSITY OF PARANÁ, CURITIBA, BRAZIL

1 Introduction

The growth in the size and complexity of public buildings, universities, airports, and shopping centers has led to the need for efficient indoor navigation. However, indoor maps have not received a lot of attention in Cartography and most of the indoor maps are very similar to a floor plan.

Nevertheless, users are claiming for location-based services for indoor environments. Big companies such as Apple, Nokia, Microsoft, Google, and Motorola have been investing in this market. Some market researches pointed out the indoor mapping market will rise 37% between 2014 and 2019 (Markets, 2014). A great part of the research on indoor navigation has been done by studies concentrating on the positioning technology and its feasibility, which have been tested in several technical implementations (Basiri et al., 2017; Correa et al., 2017; Gunduz et al., 2016; Xia et al., 2017). Moreover, it is also necessary to consider that these environments became bigger and more complex, and in some cases, the navigation task can be a challenge, since indoor environments have some characteristics that make them different from outdoor environments: orientation and navigation are different and landmarks can change frequently. For this reason, the efficient management of indoor spatial information is a crucial demand in large buildings and a few commercial services have been already provided, such as Google Indoor, to meet such demand.

Geographical and Fingerprinting Data for Positioning and Navigation Systems. https://doi.org/10.1016/B978-0-12-813189-3.00009-5

An example of an indoor map is the "You are here" map, which is a reference map that is typically large scale and placed within the surrounding area that it depicts. This map can include an arrow or some other symbol representing the location and perhaps the heading of a person viewing the map. There is generally a symbol indicating the location of the person viewing the map. The main objective of a YAH map is to aid in the navigation process, but there are some issues concerning its use, such as those related to misalignment (Montello, 2010), object rotation, and self-location (Lobben, 2004).

In addition to the YAH maps, indoor environments are represented using floor plans, mainly for emergency maps, which have a high level of detail and can be difficult to use in terms of understanding the whole building. Three-dimensional models (depending on the application) can also be used to represent these environments; however, complex interfaces are needed to enable users to manipulate the model. Nossum (2011) proposed the indoor tube map, inspired by Beck's metro map design of the London Underground.

Problems can arise when the buildings are complex and the navigating tasks from one point to another becomes a real challenge. However, there is not a consensus on which of these different ways of representing indoor environments is the best for the user, or which is most helpful for navigation tasks.

At the same time, schematic maps are helpful in spatial problem-solving tasks such as way-finding in outdoor environments or for representing underground railways, surface railways, tram and bus routes. Except in the case of buses, the routes do not change frequently, which makes schematic maps very suitable for their representation. In fact, schematic maps are not used to represent dynamic routes, not only because it is difficult to design a schematic map automatically, but also because the process by which such maps are produced has not been totally codified in cartography (Avelar and Hurni, 2006).

In order to obtain the answers to these questions, we have proposed a research that is being conducted at Federal University of Paraná (Brazil). We have developed a database with different buildings of the University, and, for some of these buildings, all interior spaces were stored and categorized by its use (classroom, staff rooms, administrative spaces, commercial areas, and bathrooms).

We have proposed the use of two different maps: a schematic and the traditional floor plan. The first one will help the users in the wayfinding and navigation since corridors are mapped as lines and rooms as point features. The second (floor plan) shows the interior division of the space with its characteristics. We also developed an adaptation of a PgRouting extension to perform the indoor routes. The obtained results can be seen at www.campusmap.ufpr.br (only in Portuguese).

This chapter is organized as follows: Section 2 presents some researches related to indoor mapping, with focus on the schematic maps and routes tasks; Section 3 describes the study area; Section 4 is dedicated to the database construction and Section 5 to the indoor routing description; Section 6 presents the development environment; and Section 7 presents some results. The chapter finishes with the conclusions and future developments.

2 Related Work

In this research, we deal with two different types of map: schematic and indoor maps. The study of schematic maps is mainly used to transport networks, and there has been significant research on methods for obtaining a schematic representation from a topological structure (Avelar and Hurni, 2006; Ware et al., 2006; Weihua et al., 2008a,b).

Schematic maps are diagrammatic representations based on highly generalized lines and are generally used for showing routes of transportation systems, such as subways, trams, and buses (Avelar, 2002). Also Avelar (2002) was developed a schematic map on demand for a transport network. The approach used a network that was simplified using the Douglas-Peucker algorithm and developed a method for preserving topological relations among the linear features. The schematization process modifies the original road network based on common-sense manual displacement constraints used in many existing schematic maps.

Other relevant research in schematization was developed by Ware et al. (2006) who proposed an algorithm that automates the production of schematic maps for mobile GIS applications. The algorithm uses the simulated annealing optimization technique. Authors described a prototype software and the experimental results showed that the algorithm successfully produced schematic maps that meet user-defined constraints within a reasonable time.

In Hurter et al. (2010) the existing metro map design was adapted for use by air traffic controllers. The authors defined specific mathematical cost functions that measure the quality of schematic map of flight routes to assist air traffic controllers. The simulated annealing algorithm, with these adapted cost functions and optimizations, produces visualizations that fulfill the defined constraints. A method was also proposed for generating colors for representing the different flight routes which consider their semantics and perceptual distances with respect to other colors.

In traditional cartography, the development of new representation methods mostly considers outdoor environments. The focus has only recently changed to indoors, because of the growth of such environments. Indoor environments have some characteristics that make them different from outdoors environments, specifically, orientation and navigation. The main challenge, however, according to Nossum (2013a), is the added dimension introduced by multistory buildings. This remains a problem, with no convergent solution, and in recent studies, different approaches have been taken to this question (Giudice and Li, 2011; Goetz, 2012; Henry and Polys, 2010; Nossum, 2013a; Nossum et al., 2013).

A review of several types of indoor spaces representation is made in Nossum (2013a) and three available solutions are highlighted: architectural style floor plans, abstract floor plans, and augmented reality systems. Architectural floor plans are rich in detail and are available for most buildings for the purposes of emergency plans. However, the level of detail can be too high, making such maps esthetically unsuitable as consumer products. Abstract floor plans are normally less detailed, and the use of colors and symbols is aimed at the consumer market. Augmented reality systems are relatively new in the consumer market (Nossum, 2013a) and are still mostly found in research projects.

In Nossum (2013a), a new design for an indoor map called an IndoorTube Map, the design which is inspired by Beck's metro map of the London Underground is proposed. In this proposed design, corridors correspond to metro lines, rooms to stations, and elevators/stairs to connected stations where metro lines cross. This design was applied to a hospital and some user tests were conducted to verify if this map is better than floor plans for way-finding tasks. The preliminary results do not confirm that IndoorTube maps are significantly better at supporting way-finding than floor plan maps (Nossum, 2013b). At this stage, the design follows the traditional map design approach and is completely manual.

According to Casakin et al. (2000), schematic maps are helpful in spatial problem-solving tasks such as way-finding. This author states that one of the challenges in constructing schematic maps is establishing clear relationships between the detailed information found in the environment and the abstract/conceptual structures contained in the map. Also, an important aim of any schematic way-finding map is to support information for finding a destination efficiently.

Recently, the Open Geospatial Consortium (OGC) started a discussion to specify a standard called IndoorGML, which specifies an open data model and XML schema of indoor spatial information. According to the OGC (OGC, 2015), IndoorGML intentionally focuses on modeling indoor spaces for navigation purposes. According (Kang and Li, 2017) since its publication, several studies on the basic concepts and applications have been done, such as geo-tagging in indoor space by IndoorGML, indoor navigation map for visually impaired people as an extension of IndoorGML and comparison between IndoorGML and CityGML LoD 4.

Considering the importance of developing methodologies for indoor mapping and the possibilities for schematic maps in supporting way-finding tasks, we propose a semiautomatic method for generating a schematic map from a floor plan to support indoor navigation tasks.

3 Context and Study Area

The Federal University of Paraná (UFPR) has 26 different *Campi* in several cities in the Parana State, Brazil; in total, 11 million m^2 of area, with 500 thousand m^2 of constructed area and 316 buildings. UFPR has more than 6000 employees—staff and administrative—about 50,000 undergraduate students and 10,000 graduate students. A great part of this academic community does not know completely the space where they work and study. If we consider the external public who has access to the UFPR these figures are even bigger.

The unfamiliarity with space and its characteristics has direct impacts in several issues, such as management of resources (humans and materials), *Campi* infrastructure management, not only of the exterior but also of the interior of the buildings, security, and

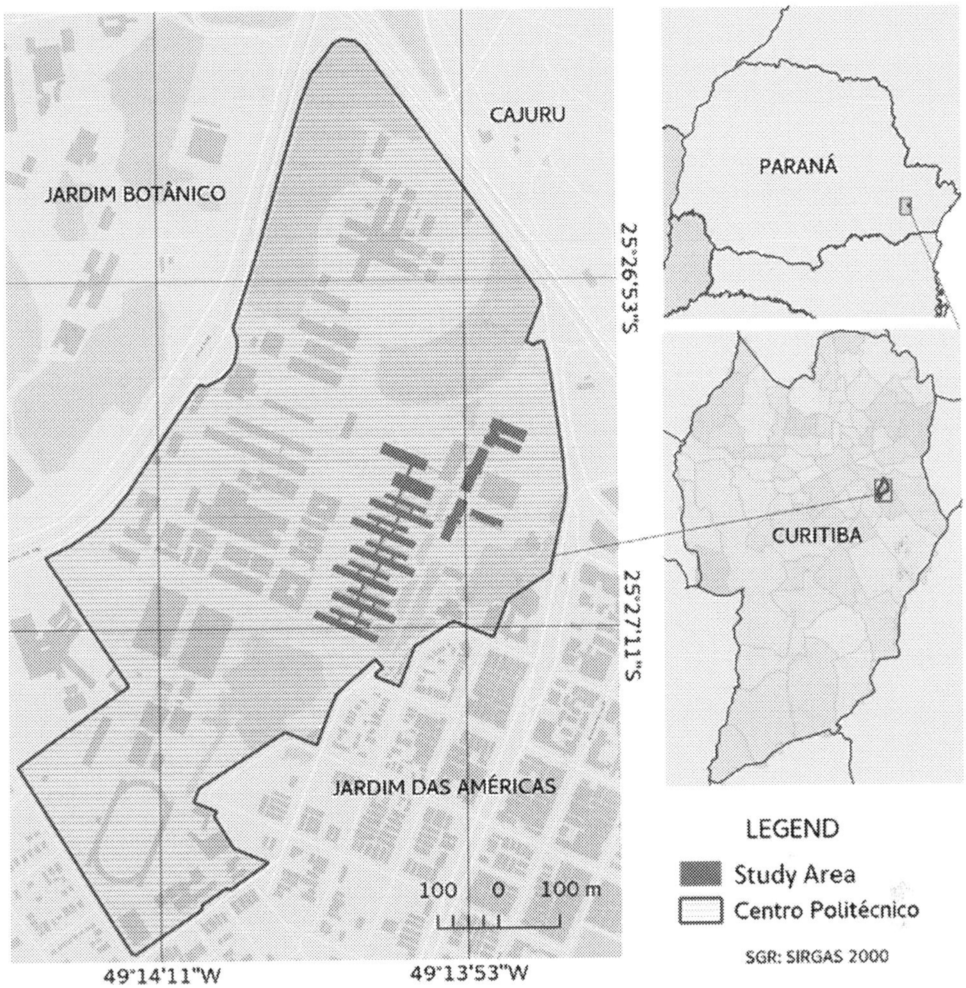

FIG. 1 Study area.

other issues that can be supported by the use of geoinformation. From this perspective, we have started a Project named UFPR CampusMap (UCM) whose main goal is to implement a Geographic Information System with information from the indoor and outdoor environments.

One of the UFPR's *Campi* is Centro Politécnico (Fig. 1). The study area is a set of buildings with symmetric design. There are building with two, three and five floors and it is common users get lost when walking into these buildings. There are different types of information that must be considered: classrooms, bathrooms, coffee shops, laboratories, administrative areas, and offices.

4 Database Construction

The database construction comprises three different activities described in the sequence. First, it is necessary to design the database from the features that are present in the buildings. The second step is to build the database and, finally, the data stored in the database needs to be edited to create the cartographic representations using both the floor plan and the schematic map.

4.1 Database Conceptual Model

Although the project has started at one *Campi* of UFPR we intend to expand it to all other *Campi*. So, we have developed a model which consider all the possible features in the database. This model contains nongeometric and geometric features, presented in Table 1.

From these features we use Unified Modeling Language (UML) for design the class diagram, presented in Fig. 2.

4.2 Database Implementation

Initially, the nonspatial classes were created in Postgres 9.4.3/PostGIS 2.3. These classes are presented in grey in Fig. 2 and are described as follows:

- Institute and department: These classes are important because they are administrative units within the University.
- Type: This class is related to the function of each space, for example, classroom, laboratories, administrative.
- Access: This class has a direct relationship with the navigation task. It links the path with the restrictions, so free access areas do not present any restrictions. Otherwise, there is a restriction, for example, as doors with specific opening hours. Furthermore, it has the information about the access to people with disabilities.

Table 1 Database Features

Features	Attribute
Campus	Name
Building	Name
Room	Classroom/laboratory/office/service/coffee shop/other
Corridor	Free access/restricted access
Bathroom	Ladies room/men's room/handicap bathroom
Transition point	Stairs, way out, lift
Institute	Name
Department	Name
Block	Name

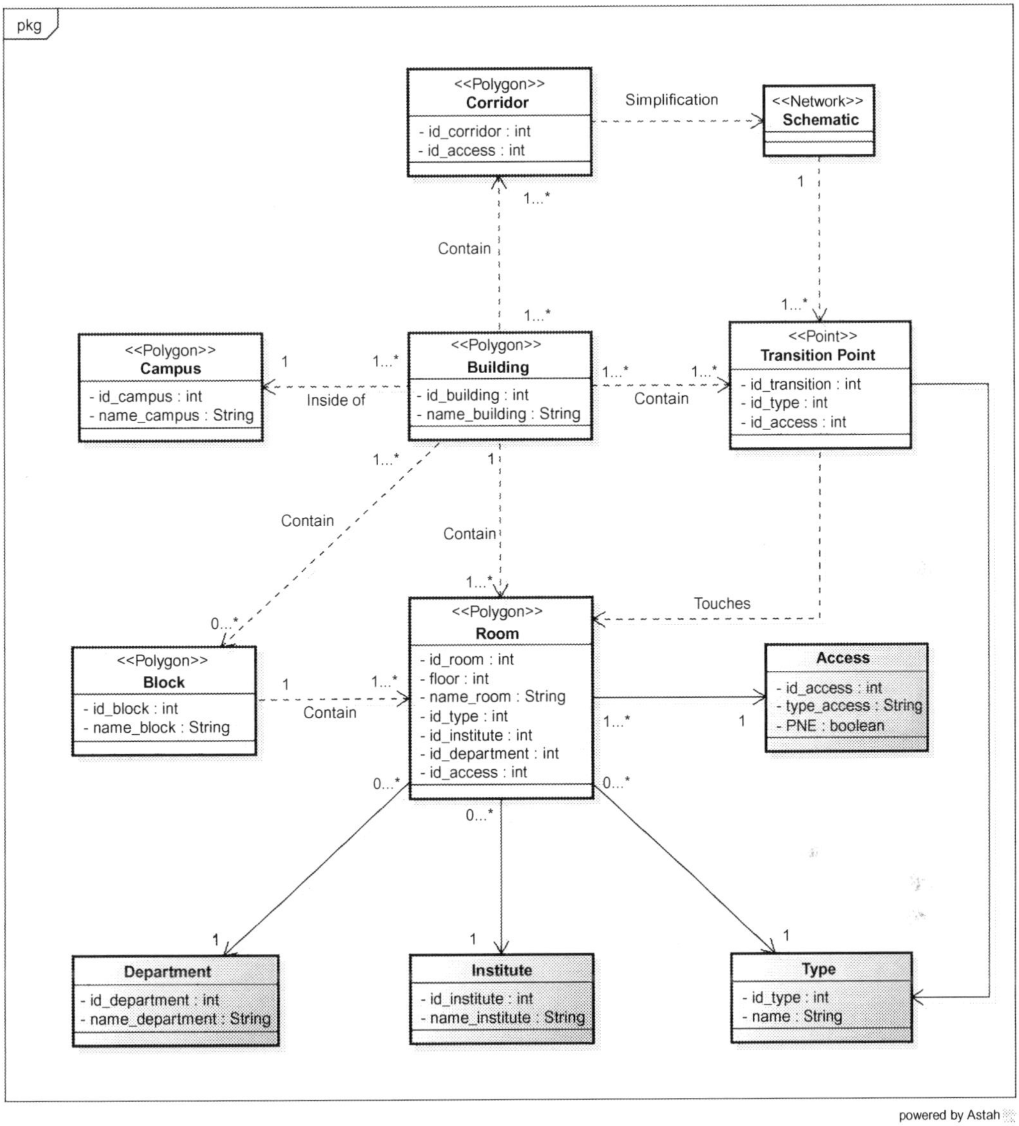

FIG. 2 Classes diagram for UCM.

The spatial classes are represented in white in the diagram:

- Campus: UFPR has 26 different *Campi*, which are spread over the city of Curitiba and in other cities of the Parana State.
- Building: Represents the geometry of the buildings.
- Block: Each building can or cannot belong to a block. One building can have more than one block.

- Room: This class contains the geometries of the interior of the buildings. This geometry is linked to the classes "type" and "access."
- Corridors: This class contains the geometry of the corridors, which are the areas used for navigation and is directly linked to the class "Access."
- Transition points: This class contains the geometry and the classification of the transition points inside the building. Stairs and elevators make the transition between different floors; doors allow the transition between different spaces (rooms and corridors, for example), and way-out allows the transition between indoor and outdoor spaces. This class is important to create routes between points.

4.3 Cartographic Database Construction

In this research we decided to use the cartographic representation in two different aspects: the floor plan that presents the building and rooms and a schematic map, when the visualization scale is bigger, to present the position of a room and the corridors. The schematic representation is also used as a basis for the routing algorithm.

Both cartographic representations were derived from a database obtained in a vector file, which details such as doors, windows, stairs, and text. We used QGIS 2.18 to edit and create the floor plan that has the geometry of the building and rooms. The process to create this database was semiautomatic, since some tasks were performed using specific functions of QGIS. Fig. 3 presents the original database and Fig. 4 presents the final floor plan, after the editing process.

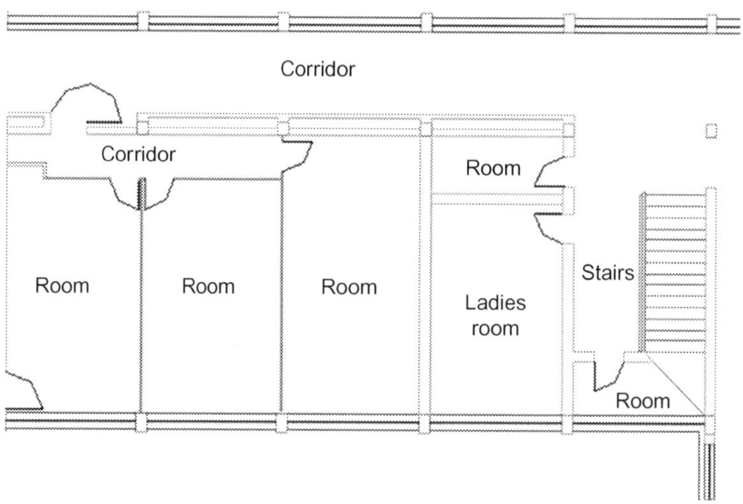

FIG. 3 Part of the architectural plan of the study area.

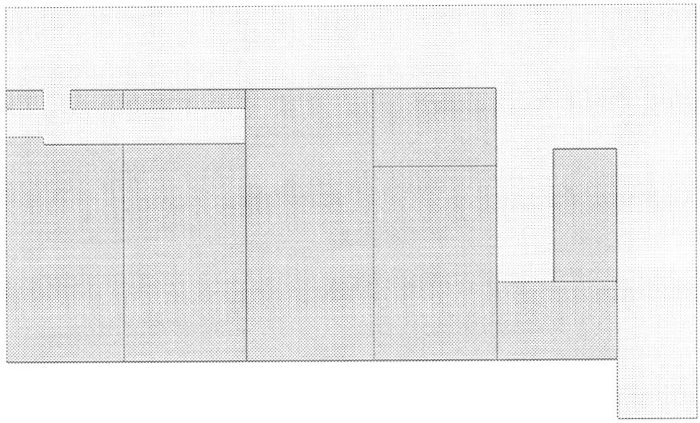

FIG. 4 Part of the floor plan after editing.

After editing the floor plan, we have started the design of the schematic map. Schematization involves reducing the complexity of map details, while at the same time preserving the important characteristics (especially topology). In this research, the final goal is to produce a schematic map from floor plans, using a semiautomatic process. In this process, we have used both computational algorithms and manual processes.

The schematic map is formed by the transition points—which represent the transition between corridors and rooms, stairs, and bathrooms, for example—and the lines connecting them. The position of the transition points is defined by the following rule: if the transition is a door, the point is placed in the middle of its length, and if the transition is a corridor, for example, a turning point, the point is placed at the end of one line and the beginning of the other. The lines representing the corridors were created manually, using QGIS software, by drawing the central line of the corridor's polygon. In Fig. 5A the transition points are represented in white dots and in Fig. 5B the corridor lines were created. In this step it was used a manual process since the schematization algorithm is under development.

In the sequence it was used the PostGIS function *ST-ShortestLine* to create the link between the transition points and the central lines of the corridors. This function creates the smaller line segment connecting two geometries. These lines are represented in dashed lines in Fig. 5C. As a result, we obtained the schematic representation in Fig. 5D.

5 Indoor Routing

Routing is the process of selecting a path for traffic in a network, or between or across multiple networks. One of the most commonly used routing algorithms is Dijkstra's algorithm (Dijkstra, 1959). Dijkstra's algorithm finds the shortest path between two nodes by building a shortest-path tree, and stopping once the destination node has been reached.

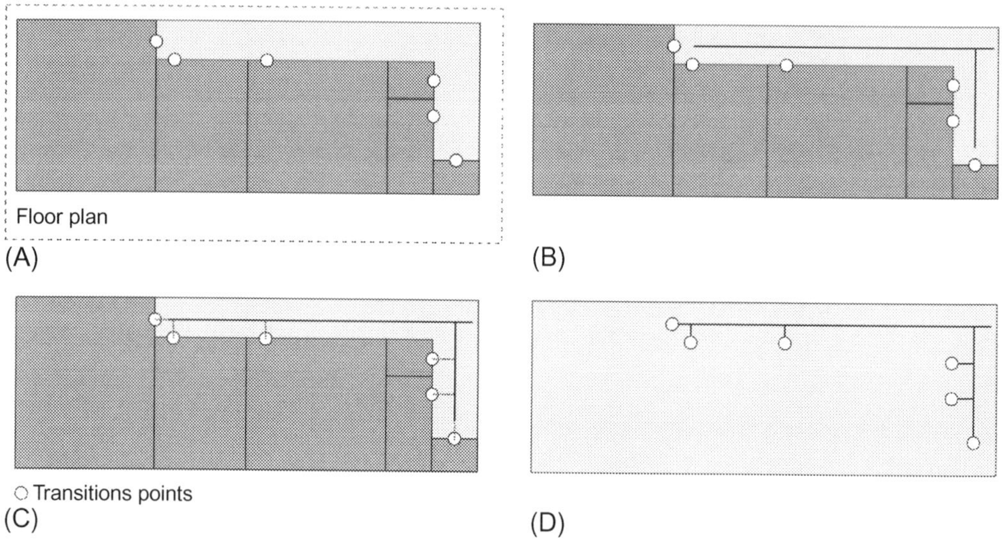

FIG. 5 Schematic map creation process. (A) Representation of transitions points. (B) Creation of corridor lines. (C) Connection between transition points and corridor lines. (D) Final representation.

Normally in routing applications, Dijkstra's algorithm is used to find the shortest route between two locations.

In Fig. 6A it is represented a graph "G" with nodes $V = \{\text{I, II, III, IV, V}\}$ and a set of edges $E = \{8, 7, 9, 4, 5, 3, 6, 7\}$, where the elements represent the weight (cost) of the route. Then, the graph is given by $G = (V, E)$. Assuming we want to go from node I to node V, in Fig. 1B there are a set "S" of paths (II, III, and IV) to use for calculation. The algorithm searches the node with a smaller cost, in this case, the node III (Fig. 6C). Then the search is repeated until we have the set S with the final path, in this case, I, III, and V (Fig. 6D).

In this research, the cost is the distance between the nodes. The graph is composed of the schematic map and streets of the campus, so it is possible to have not only indoor routes but also routes indoor/outdoor. The streets were obtained from OpenStreetMap data.

After the schematic map is finished it is necessary to create a graph that allows the route calculation. To do this, the first step is to build the topology to have the connectivity between the elements. It was used the PostGIS function *topology.CreateTopology* and as a result, the points at the beginning and at the end of the lines become the nodes. If the line segments are not automatically adjusted to the nodes, we can define a tolerance (in this case 10 cm), so the segments will be connected. After this, the line segments become the edges of the graph. When the topology is created, each edge must have one node in the starting and ending points.

The database is composed of buildings with a different number of floors (from two to five floors). Furthermore, the floors have the same corridors structure, which means the geometries of the schematic maps for the corridors in different floors are equal. If we simply apply the function to build the topology, the algorithm would create a graph

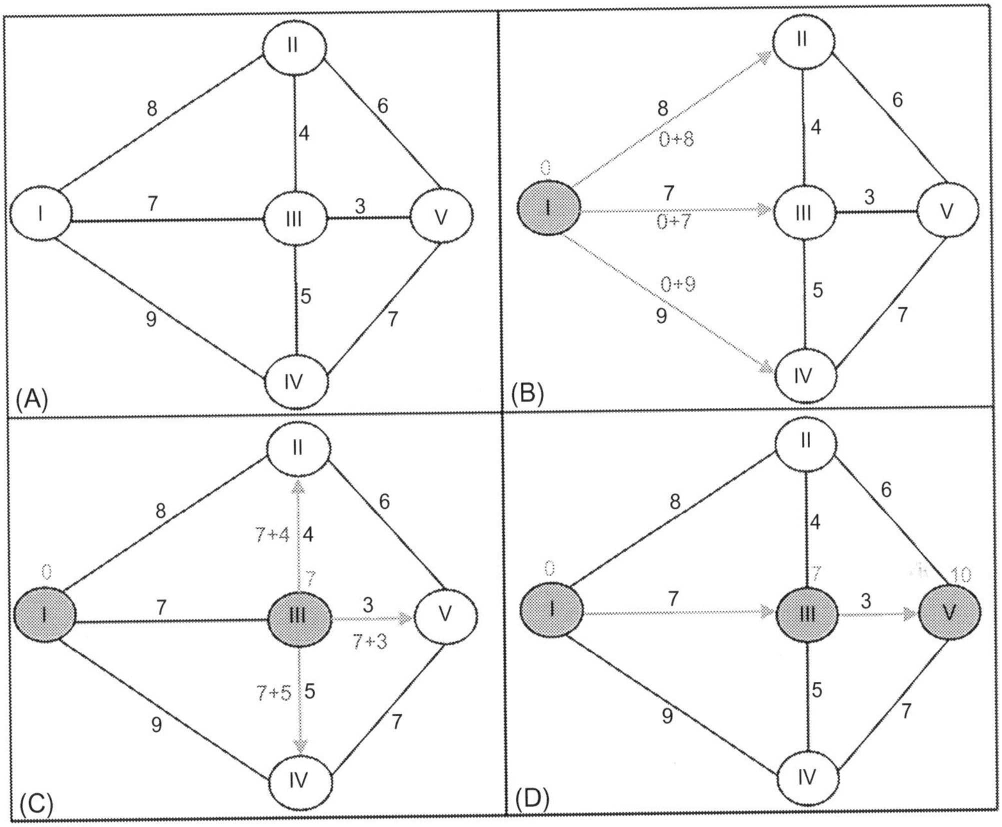

FIG. 6 Graphic demonstration of Dijkstra's algorithm.

with data of different floors. To overcome this issue, subgraphs for each floor were created, applying the function previously described. The only exception is the ground floor, which is the union of data of OpenStreetMap and schematic map.

If the points of origin and destiny are on different floors, the subgraphs must be joined. This join is accomplished creating edges that link the transition points with the end/start of the subgraphs. It was adopted the weight equal to the Euclidian distance between the starting/ending of the route and the transition points (elevators or stairs). In practice, this means a maximum distance of 2 m, which can be considered a small distance when compared to the total distance between the origin and destiny.

The routes are calculated using the PgRouting function *pgr_dijkstra*.

6 Development Environment

The UCM works in a server-client architecture: the server stores data and performs the GIS functions, and the client is the environment where the user can interact with the map. The programming languages used are HTML, CSS, Javascript, PHP, and PLSQL.

The programming languages in the client and in the server are different and are described in the sequence:

(a) Server side:

- PostgreSQL 9.4.3—Database management
- PostGIS 2.3—GIS functions
- PgRouting 2.3—routing
- ThorMap 0.1—program developed for map symbolization
- Apache Web Server 2.5
- PHP 7 programming language

(b) Client side:

- Web browser
- Leaflet.js—for map manipulation and exhibition
- Leaflet extensions: Thormap.js and MarkerCluster.js
- Phonon Framework and JQuery—to design the interface

The Thormap is an application developed in PHP which prepares the stored data and using PostGIS queries, recovers the data, and applies a style stored in a qml file (Qt Modeling Language). Using JSON (Javascript Object Notation) is created a file with geometries, attributes, and symbology for the map. This file is then loaded in the Leaflet.js using the Thormap.js plugin. An example of the code is presented below:

```
{
        "fonte": {
                "tipo": "postgresql",
                "nomeBanco": "ufpr_indoor",
                "senha": "123456",
                "usuario": "postgres",
                "maquina": "200.17.225.171",
                "SRID": 4326,
                "saida": "ucm.tm"
        },
        "mapas": [
              {
                        "nomeMapa": "Planta Baixa",
                        "colunaGrupo": "andar",
                        "valoresGrupo": [0],
                        "tabelas": [{
                                "nomeTabela": "public.sala",
                                "colunas": [
                                        "geom",
                                        "nome_sala",
```

```
                                        "sigla_sala",
                                        "cod_sala",
                                        "andar"],
                                 "estilo": "sala.qml"
                                 }, {
                             "nomeTabela": "public.corredor",
                              "colunas": ["geom", "andar"],
                             "estilo": "corredor.qml"
                                  }]
            },{
                    "nomeMapa": "Esquematico",
                    "colunaGrupo": "andar",
                    "valoresGrupo": [0],
                    "tabelas": [{
                          "nomeTabela": "public.transicao",
                           "colunas": ["geom", "andar"],
                          "estilo": "transicao.qml"
                       },{
                          "nomeTabela": "public.esquematico",
                          "colunas": ["geom"],
                          "estilo": "esquematico.qml"
                       }]
            }]
}
```

7 Results

Results will be presented in two parts: first, the results concerning the representation of the indoor maps, and the second part presents the results related to the indoor routes.

An interface was designed to present the interactive map. The functions implemented are:

- search for campus name
- search for rooms name
- route between two points

7.1 Indoor Cartographic Representation

Two different map designs were proposed: the first considers the floor plan and uses different colors to depict the rooms, based on its classification; the second one is the schematic map where all rooms are represented as point symbols. Some symbols were used for representing bathrooms, stairs, elevators, and other particular classes of the features.

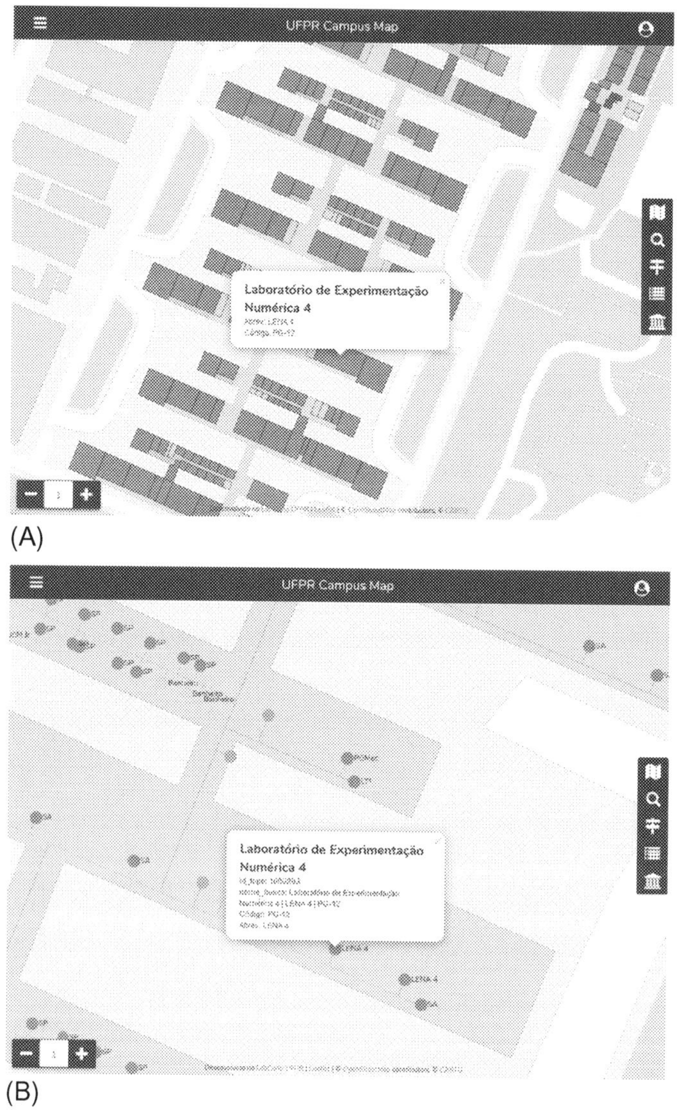

FIG. 7 (A) Floor plan. (B) Schematic map.

After the applying the symbology, the result is presented in Fig. 7. Fig. 7A presents the floor plan where it was selected a specific room to show its attributes. Fig. 7B presents the schematic map for the same area. According with the visualization scale, the system displays the appropriate map. When the scale is bigger the schematic map is presented.

7.2 Indoor Routes

Regarding indoor routes, first the user inserts the points of origin and destiny, by using the specific field (typing the name of the room) or clicking and dragging the markers to the respective start and end points of the routes. In the example of Fig. 8, several points on different floors were selected and the result is shown using different colors. The dashed gray line represents the path on the first floor and the blue solid line represents the path on the ground floor. In Fig. 9, it can be seen a detail of the transition point used to change from the first to the ground floor.

8 Conclusion and Future Developments

This research presented a method to produce a schematic map of an indoor area and the route calculation between points in this environment. The results achieved to date comprise the semiautomatic development of a schematic map and the design of two different maps for the indoor environment.

The cartographic representation of the indoor environments is an issue open for research since there are few studies related to it. Some preliminary studies have shown

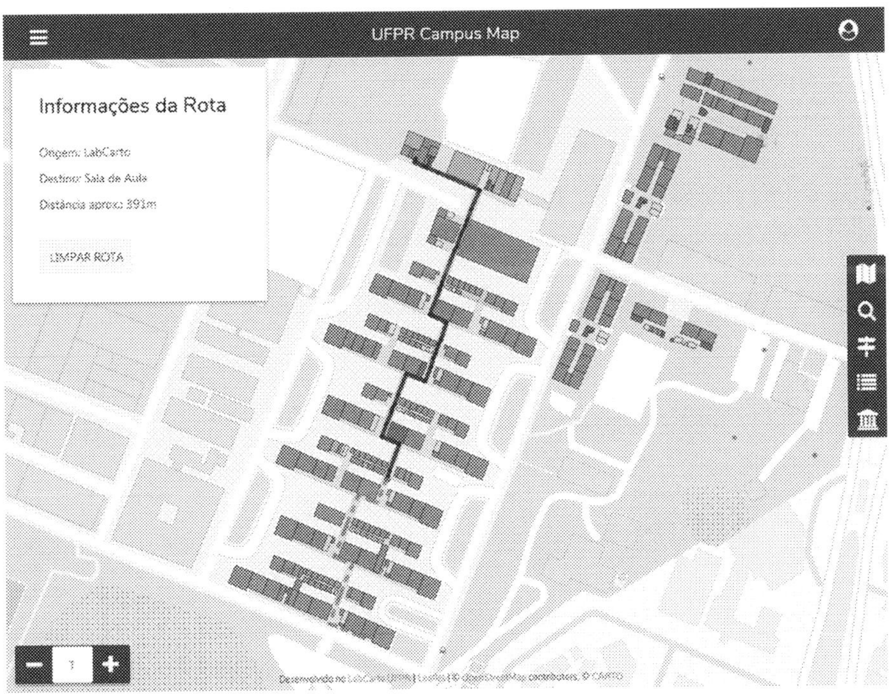

FIG. 8 Route between two points.

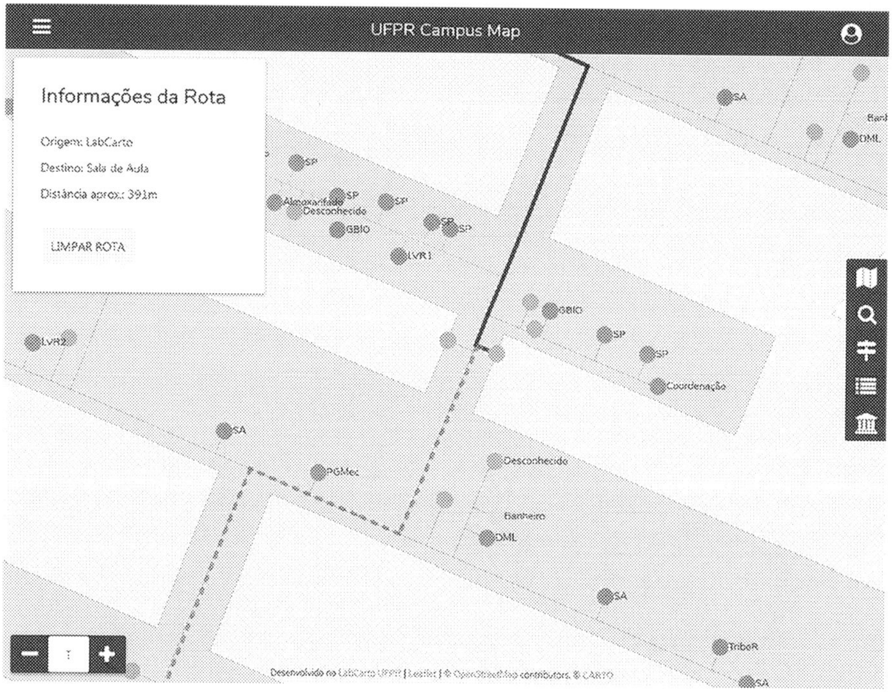

FIG. 9 Detail of the route showing the transition point.

users have a better comprehension of the indoor space when using a schematic map. Therefore, it is necessary further investigations to confirm this.

The routing algorithm implemented in this research allows the determination of the shortest path between two points in an indoor environment. The algorithm also allows paths between indoor/outdoor points. Future implementation should consider routes using different ways of transport, such as bicycle, for example.

Concerning the schematic map creation, we are working on a fully automatic method which will be capable of getting the dwg file and derive not only the points and lines of the map but also the edges and nodes for the graph. This is being developed using QGIS and Python.

Acknowledgments

The authors would like to thank the Conselho Nacional de Desenvolvimento Científico e Tecnológico (CNPq) and Coordenação de Aperfeiçoamento de Pessoal de Nível Superior (CAPES) for their financial support (Process: 459300/2014-8—Edital Universal; Process: 301980/2014-4—Produtividade em Pesquisa)

References

Avelar, S., 2002. Schematic Maps on Demand: Design, Modeling and Visualization. Zurich: Swiss Federal Institute of Technology.

Avelar, S., Hurni, L., 2006. On the design of schematic transport maps. Cartogr. Int. J. Geogr. Inf. Geovis. 41, 217–228. https://doi.org/10.3138/A477-3202-7876-N514.

Basiri, A., Simona Lohan, E., Moore, T., Winstanley, A., Peltola, P., Hill, C., Amirian, P., Figueiredo Silva, P., 2017. Indoor location based services challenges, requirements and usability of current solutions. Comput. Sci. Rev. 24, 1–12. https://doi.org/10.1016/j.cosrev.2017.03.002.

Casakin, H., Barkowsky, T., Klippel, A., Freksa, C., 2000. Schematic maps as wayfinding aids. In: Freksa, C., Brauer, W., Habel, C., Wender, K. (Eds.), Spatial Cognition II. Springer-Verlag, New York, NY, pp. 54–71.

Correa, A., Barcelo, M., Morell, A., Vicario, J.L., 2017. A review of pedestrian indoor positioning systems for mass market applications. Sensors (Switzerland) 17. https://doi.org/10.3390/s17081927.

Dijkstra, E.W., 1959. A note on two problems in connection with graphs. Numer. Math. 1, 269–271.

Giudice, N.A., Li, H., 2011. The effects of visual granularity on indoor spatial learning assisted by mobile 3D information displays. In: Stachniss, C., Schill, K., Uttal, D. (Eds.), Spatial Cognition VIII. Springer-Verlag, New York, NY, pp. 163–172. https://doi.org/10.1007/978-3-642-32732-2.

Goetz, M., 2012. Using crowdsourced indoor geodata for the creation of a three-dimensional indoor routing web application. Futur. Internet 4, 575–591. https://doi.org/10.3390/fi4020575.

Gunduz, M., Isikdag, U., Basaraner, M., 2016. A review of recent research in indoor modelling & mapping. In: International Archives of the Photogrammetry, Remote Sensing and Spatial Information Sciences—ISPRS Archives. https://doi.org/10.5194/isprsarchives-XLI-B4-289-2016.

Henry, J.A.G., Polys, N.F., 2010. The effects of immersion and navigation on the acquisition of spatial knowledge of abstract data networks. Proc. Comput. Sci. 1, 1737–1746. https://doi.org/10.1016/j.procs.2010.04.195.

Hurter, C., Serrurier, M., Alonso, R., Tabart, G., Vinot, J., 2010. An automatic generation of schematic maps to display flight routes for air traffic controllers: structure and color optimization. In: AVI'10, pp. 233–240.

Kang, H.-K., Li, K.-J., 2017. A standard indoor spatial data model—OGC IndoorGML and implementation approaches. ISPRS Int. J. Geoinform. 6, 116. https://doi.org/10.3390/ijgi6040116.

Lobben, A.K., 2004. Tasks, strategies, and cognitive processes associated with navigational map reading: a review perspective. Prof. Geogr. 56, 270–281.

Markets, M., 2014. Indoor Location Market, by Component, Deployment Mode, Application and Region Global Forecast to 2018. Report. Markets and Markets.

Montello, D.R., 2010. You are where? The function and frustration of you-are-here (YAH) maps. Spat. Cogn. Comput. 10, 94–104. https://doi.org/10.1080/13875860903585323.

Nossum, A.S., 2011. Indoortubes a Novel Design for Indoor Maps. Cartogr. Geogr. Inf. Sci. 38, 192–200.

Nossum, A.S., 2013a. Developing a framework for describing and comparing indoor maps. Cartogr. J. 50, 218–224. https://doi.org/10.1179/1743277413Y.0000000055.

Nossum, A.S., 2013b. Exploring New Visualization Methods for Multi-Storey Indoor Environments and Dynamic Spatial Phenomena. PhD Thesis. Trondheim: Norwegian University of Science and Technology.

Nossum, A.S., Li, H., Giudice, N.A., 2013. Vertical colour maps—a data-independent alternative to floor-plan maps. Cartogr. Int. J. Geogr. Inf. Geovis. 48, 225–236. https://doi.org/10.3138/carto.48.3.1641.

OGC, 2015. OGC® IndoorGML, Open Geospatial Consortium. Available from: http://www.opengeospatial.org/. Accessed in July, 2018.

Ware, J.M., Taylor, G.E., Anand, S., Thomas, N., 2006. Automated production of schematic maps for mobile applications. Trans. GIS 10, 25–42. https://doi.org/10.1111/j.1467-9671.2006.00242.x.

Weihua, D., Jiping, L., Qingsheng, G., 2008a. Visualizing schematic maps through generalization based on adaptive regular square grid model. In: The International Archives of the Photogrammetry, Remote Sensing and Spatial Information Sciences, pp. 379–384.

Weihua, D., Qingsheng, G., Jiping, L., 2008b. Schematic road network map progressive generalization based on multiple constraints. Geospatial Inf. Sci. 11, 215–220. https://doi.org/10.1007/s11806-008-0090-z.

Xia, S., Liu, Y., Yuan, G., Zhu, M., Wang, Z., 2017. Indoor fingerprint positioning based on Wi-Fi: an overview. ISPRS Int. J. Geoinf. 6, 135. https://doi.org/10.3390/ijgi6050135.

10

OGC IndoorGML: A Standard Approach for Indoor Maps

Ki-Joune Li*, Giuseppe Conti†, Evdokimos Konstantinidis†, Sisi Zlatanova‡, Panagiotis Bamidis§

*PUSAN NATIONAL UNIVERSITY, BUSAN, SOUTH KOREA †NIVELY, NICE, FRANCE ‡UNIVERSITY OF NEW SOUTH WALES, SYDNEY, AUSTRALIA §MEDICAL SCHOOL OF THE ARISTOTLE UNIVERSITY OF THESSALONIKI, THESSALONIKI, GREECE

1 Introduction

Many recent technological progress in the field of information and communication technologies are not achieved by a single dominant technology but a combination of diverse technologies. This diversity provides a fundamental of technology progress and even accelerates it. The interoperability is a crucial condition to bind diverse technologies, and the standardization is one of the most promising approaches to get the interoperability. It offers a linking mechanism to integrate several technology building blocks to setup an ecosystem. The ecosystem achieved by integrating diverse technologies through standards is expected more flexible than by a single dominant technology.

Indoor location and spatial information services and technologies that mostly started from 2000s have different backgrounds. It includes indoor positioning technologies mostly developed by communication communities, building construction technologies by architectural engineering communities and architecture, Simultaneous Locationing And Mapping (SLAM) by robotics communities, and indoor spatial information technologies from geospatial information communities. And all these background technologies comprise building blocks of indoor location and spatial information services and systems (Afyouni et al., 2012). It is therefore crucial to provide a mechanism to integrate different technologies into an ecosystem.

A very important component of the ecosystem for indoor spatial information is the indoor map. An indoor map may contain a lot of information, just like an outdoor map. But the requirements and specifications of indoor maps are differently defined depending on divers perspectives as listed earlier. For example, indoor maps for indoor positioning are much different from those for simple visualization. We therefore need common specifications of indoor maps to integrate different indoor spatial technologies

Geographical and Fingerprinting Data for Positioning and Navigation Systems. https://doi.org/10.1016/B978-0-12-813189-3.00010-1

to an ecosystem. The OGC IndoorGML (OGC, 2012b) has been developed to respond to this challenge.

The IndoorGML serves as a standard data model and exchange encoding rule in XML for interfacing different components in an ecosystem of indoor spatial services. And an extension has been also defined to meet the requirements for indoor navigation. We will discuss the details of these concepts in the following sections. First, we will discuss the requirements of indoor maps in Section 2 and explain the basic concepts of IndoorGML in Section 3. The data models of IndoorGML composed of the *Core Module* and the *Navigation Module* are to be explained in Section 4. Important technical issues for implementation will be discussed in Section 5 and use-cases for real applications will be presented in Section 6. We will summarize this chapter in Section 7.

2 Requirements for Indoor Maps

In this section, we investigate the requirements of indoor maps, which served as the starting points of OGC IndoorGML design. More detail discussion on the requirements is found at Kang and Li (2017). Note that the simple visualization of indoor maps is not within the scope of our study and we rather focus on the requirements for application services of indoor spatial information.

2.1 Complex Structures of Indoor and Connectivity

Let us assume an example of map application; distance estimation between two points. Computing distances in indoor space is, however, a complicated process not only because of the constraints such as walls, doors, and other obstacles, but also due to vertical structures between multiple floors and lifts and moving walks. It is in fact one of the most important differences between indoor and outdoor spaces. Even indoor space is a type of constraint space, it differs from other types of constraint spaces in outdoor space such as road network spaces due to the vertical connections.

Therefore, a key requirement of indoor maps is how to properly represent the constraints and structures in indoor space. The structure of indoor space is mainly determined by architectural structures that have unique properties. First, the indoor space consists of a number of cells surrounded by architectural components, such as walls, ceilings, and floors, where each cell is separated from the others. Cells in indoor space are horizontally or vertically connected in sophisticated ways via specific types of architectural components like doors and stairs. Furthermore, indoor spatial properties, such as cell geometry and the connectivity structures between cells, differ depending on the type of buildings. For example, subway stations are normally composed of long hallways and platforms on different levels, while office buildings have normally a number of small office rooms connected via corridors.

Second, indoor spaces of complex buildings are often composed of areas with different purposes. For example, a shopping mall has a number of stores, warehouses, control

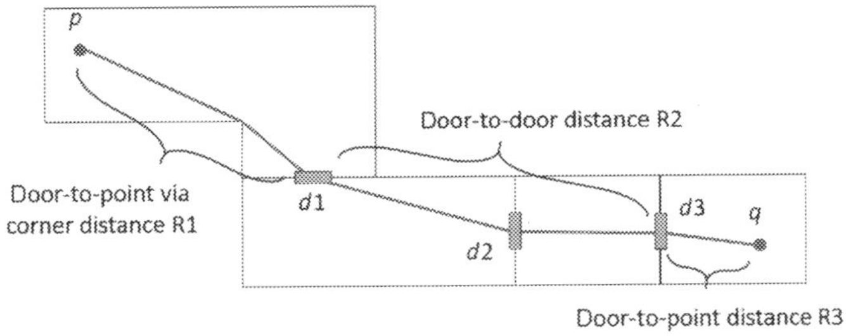

FIG. 1 Indoor distance.

rooms, cinemas, sports centers, subway stations, etc., each of which has unique requirements and functions. This complex buildings make the indoor space very complicated. Third, a single indoor space may be interpreted in different viewpoints. For example, an indoor space is partitioned into rooms and corridors, while it is also partitioned into public areas and private areas regarding its security levels. Since each interpretation forms a space layer with a proper partitioning criteria, an indoor space may have multiple space layers.

We therefore conclude that indoor maps should contain the information about indoor structures, particularly the connectivity. First, the indoor connectivity, called the indoor accessibility graph (Li and Lee, 2008; Lu et al., 2012), has to be prepared, as we need a road network graph to compute distance on the road network. An indoor accessibility graph is represented as $G = (V, E)$, where a node $n \in V$ is a room or a space unit in indoor space and an edge $e \in E$ represents the connectivity between two adjacent space units, for example, via the door. The edge may contain any additional attributes such as the length of the connection.

However, a simple edge connecting two nodes does not fully reflect the distance information particularly when the room connected to the edge has a big area or complicated geometry (Xie et al., 2015). In order to address this issue, we need the geometry information of the room or space unit surrounded by architectural components, such as walls and doors. Once indoor geometry information is provided, the indoor distance is computed by dividing the total path into point-to-door and door-to-door distances as shown in Fig. 1. In this figure, the path from point p to point q is divided into several subpaths; the first subpath from p to door d_1, the second from d_1 to door d_3, and the third path from d_3 to q. While the distances from p to d_1 and from d_3 to q are called point-to-door distance, the distance from d_1 to door d_3 is called door-to-door distance. The point-to-door distance may be computed by line-of-sight (Yuan and Schneider, 2010) or the Minkowski sum (Xie et al., 2015), while the door-to-door distance can be easily computed by the shortest path algorithm with precomputed door-to-door graph (Yuan and Schneider, 2010).

2.2 Cell-Based Context Awareness

In order to provide proper indoor spatial services, the contextual information of user is a fundamental component. As claimed by an early work on ubiquitous and context-aware computing (Schilit et al., 1994), the context has three important aspects; where you are, with whom you are, and what resources are nearby. The first aspect is related with the location of the user, which is normally represented as (x, y, z) coordinates in outdoor space. However, it is more relevant to represent the location of the user in indoor space with the room number than (x, y, z) coordinates, since the context is mainly determined by the type or function of the room. For example, staying in a classroom for an hour has a totally different context from staying in a washing room for an hour. This concept of location is also useful for the second and third aspects. For example, it is possible to implement an interesting service of exchanging digital name cards between the users in the same room. In this example, the concept of location based on room number is more useful than (x, y, z) coordinates.

One of the most basic functions of indoor context awareness services is therefore to identify the room or space unit where the user is currently located. In general, we call the unit of indoor space the *cell*, and rooms, corridors, and staircases are examples of cells. Therefore, the indoor spatial data model should contain the notion of cell to support the *indoor cell awareness*. The indoor cell-awareness is defined as being aware of the cell where the user is currently located. In order to implement it, indoor maps must meet the following requirements. First, the geometry of cell must be clearly defined either in 2D or 3D to facilitate the point-in-polygon or point-in-polyhedra operations. Second, each cell must contain semantic information such as its classification, usage, and other relevant attributes.

2.3 Integrating Multiple Data Sets

Like maps for outdoor space, indoor maps contain multiple layers, each of which has its own data source and interpretation. Overlaying and integration of multiple data sets is a fundamental requirement of indoor maps. First, the integration of indoor and outdoor spatial data sets is crucial for seamless services between indoor and outdoor spaces, for example, indoor parking services. Second, several standards for indoor spatial information have been developed, such as IFC (BuildingSmart, 2009), CityGML (OGC, 2012a), and IndoorGML (OGC, 2012b), each of which has its strengths and weakness. Third, it is often necessary to interpret and configure a single indoor space from multiple viewpoints. For example, the layout of an indoor space is given as a topographic map, while another layer for CCTV coverage is also useful for security purpose.

In general, there are two approaches for the integration. The first approach is a physical integration of multiple data sets of different standards into a single data set. For example, CityGML provides a mechanism called Application Domain Extension (ADE) (OGC, 2012a), which extends CityGML to include additional information. A spatial data model in another standard may be redefined as an ADE of CityGML, and a conversion process from a data set to CityGML ADE is required. The second approach is to link multiple data sets in different standards via external references without physical integration. For example, each feature in a data set D_A of a standard data model has an external reference or foreign key to a feature in another data set D_B of a different standard data model and vice versa. This approach is simple and practical when the correspondence between features in D_A and D_B is one-to-one. Thus, the standard data model should support the integration via the extension mechanism or an external reference.

3 Basic Concepts of OGC IndoorGML

In order to respond to the requirements discussed at the previous section, a common standard data model and encoding schema has been published by Open Geospatial Consortium (OGC) (OGC, 2012b). In addition to the requirements for indoor maps, several requirements for a standard data model have been also considered as below:

- Reflecting the properties of indoor space
- Cellular space model
- Minimal set of specifications
- Interoperability with other standards
- Extensibility

As discussed in Kang and Li (2017) and in the previous section, the indoor cell awareness is a basic requirement of the indoor spatial data model. For this reason, the key concept of IndoorGML is based on the *cellular space model.* A cellular space is defined as a set of nonoverlapping cells, where each cell has an identifier and the union of cells is a subset of the entire indoor space. A cell in this model means a unit space in indoor space such as a room and a corridor. Note that no overlapping between cells is allowed, and the union of all cells is a subset of the given indoor space. This means that there may be shadow areas, which are not covered by any cell.

Based on the cellular space model, IndoorGML introduces four main concepts: cell geometry, topology between cells, cell semantics, and multilayered space model. We study details of each concept in the subsequent sections.

CELL GEOMETRY

The geometry of a cell is defined as a 2D surface or 3D solid of ISO 19107 Standard (ISO TC211, 2003), which provides a basic set of geometry types used in spatial information systems. There are three options to define the cell geometry. The first option is to exclude any geometric description from a data set in IndoorGML and only to include topological relationships between cells, which will be explained in the next section. The second option is to include its geometry within IndoorGML data. For example, the geometry of a cell is defined as a 3D solid of ISO 19107. The third option is to include references to objects in external data sets that may contain geometric data. For example, a cell in IndoorGML data has a pointer to an object in CityGML via GML identifier (OGC, 2007), where the object in CityGML has geometric property. These options are not exclusive and may be combined together. For example, while no geometry is included in an IndoorGML data (Option 1), it contains external references to objects in other data set (Option 2). And it does not always require 3D representation of indoor map but multilevel 2D floor plans, since 2D geometry is also allowed in IndoorGML.

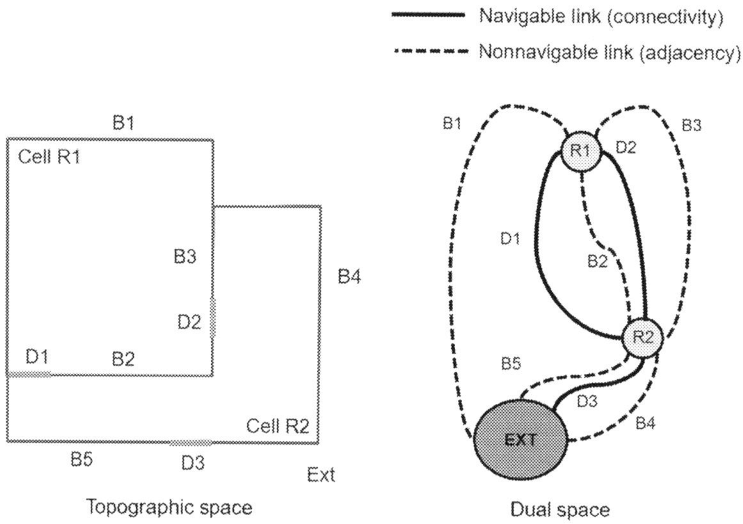

FIG. 2 Derivation of the adjacency graph from cell geometry (OGC, 2012b).

TOPOLOGY BETWEEN CELLS

Once cells are determined with their identifiers and geometric properties, topological relationships between cells have to be determined, which are essential to most of the indoor navigation applications. The topology between cells in IndoorGML can be derived from cell geometries by Poincaré duality (Lee, 2004). A k-dimensional object in the N-dimensional topographic space is mapped to a $(N - k)$-dimensional object in the dual space. The 3D geometry of a cell is, for example, transformed to 0D node of the corresponding graph in dual space and a 2D boundary surface shared by two cells is transformed to a 1D edge of the graph in dual space. This transformation with Poincaré duality results in a topological graph connecting adjacent cells in indoor space as shown in Fig. 2. Furthermore, several application graphs can be derived from the adjacency graph considering edge properties or constraints. With the edges representing doors, we may, for example, derive the connectivity graph, where each edge represents a connectivity between two cells. By using more attributes such as distances, directions, and types of doors on the edge, it may be possible to derive more divers graphs. Note that the edge may contain a line string geometry to depict the path between two rooms via a door. The geometry property of edge is useful particularly when we compute distance in indoor space.

CELL SEMANTICS

Since every cell in indoor space has its proper function and usage, we need to specify the semantics of cells. In the current version of IndoorGML, we classify the types of cells in terms of indoor navigation and expect that other classifications would be necessary for different applications, such as indoor facility management. A detail classification of cells is given as code list with hierarchical classification in the building by OmniClass (ISO/TC59/SC13, 2015). In addition to cells, the semantics of cell boundary are also useful to describe the types and properties of boundary.

MULTILAYERED SPACE MODEL

A same indoor space can be interpreted and represented in different ways. A mechanism, called the multilayered space model is offered in IndoorGML to represent an indoor space by multiple interpretations (OGC, 2012b; Becker et al., 2009). Each interpretation corresponds to a cellular space layer with its own geometric and topological properties. For example in Fig. 3, there are two different layer configurations of an indoor space due to a step in Room 3; the walkable layer and wheelchair layer. Since each layer forms a cellular space, it includes the geometries of cells, topologies between cells given as a graph and cell semantics. In addition to simple aggregation of cellular space layers, a special type of edge, called interlayer connection, is also offered in IndoorGML to represent the relationships between nodes in different layers. In Fig. 3, we have Room 3 in the walkable space layer corresponding with Room 3a and Room 3b in the wheelchair layer because Room 3 is partitioned into Room 3a and Room 3b due to a step. The multilayered space model is also useful for many applications, such as in describing the hierarchical structure of indoor space or in tracking moving objects from sensor data. Further detailed discussion is found in Becker et al. (2009).

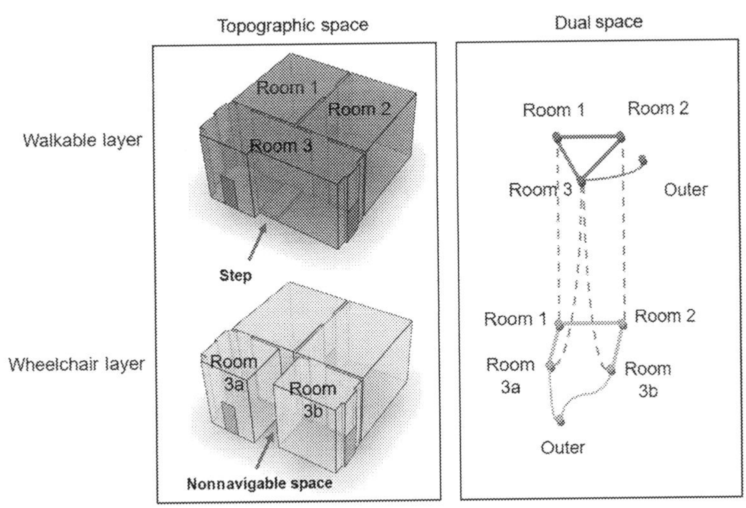

FIG. 3 Example of the multilayered space model (Becker et al., 2009).

4 Modular Structure of IndoorGML

For the sake of extensibility, IndoorGML has a modular structure as shown in Fig. 4. The core module of IndoorGML contains the data model for cell geometry, topology, and the multilayered space model. The indoor navigation extension model, which is the semantic extension model for indoor navigation, is so far the only extension module defined on the top of the core model. Many other extension modules may be defined for each application area such as indoor cadastral extension (Alattas et al., 2017), indoor geo-referenced multimedia extension, indoor facility management extension.

4.1 IndoorGML Core Module

The Core Module of IndoorGML defines the common framework of indoor spatial data model, which includes the cell and cell boundary geometry model, topological model, and multilayer space model. The primitive spatial types of this module come from those defined by ISO 19107. The UML class diagram of the Core Module is given as Fig. 5. Note that the Core Module is also given as a XML schema to express indoor maps in XML documents.

The Core Module includes four basic types: State, Transition, Cell Space, and Cell Space Boundary. While Cell Space and Cell Space Boundary belong to the indoor topographic space representing cell and cell boundary, respectively, State and Transition belong to the topological graph derived from Cell Space and Cell Space Boundary by Poincaré duality.

Cell Space defines a basic unit type of the cellular indoor space model, such as room, corridor, and hall. It basically contains a GML identifier (OGC, 2007) with proper attributes.

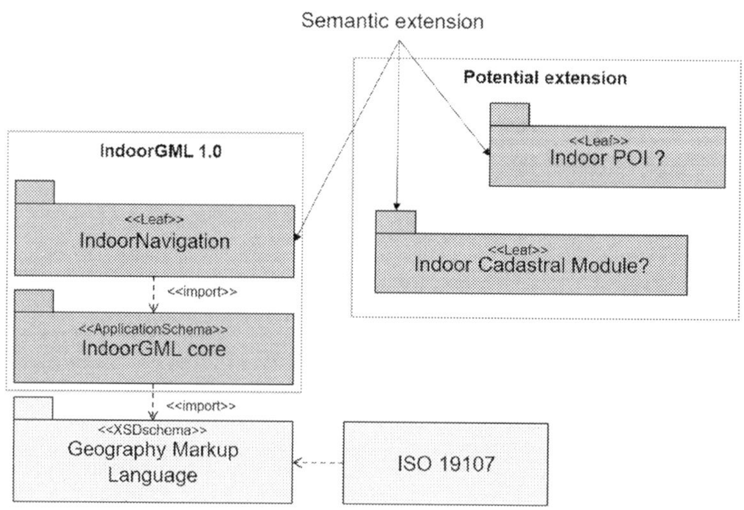

FIG. 4 Modular structure of IndoorGML (OGC, 2012b).

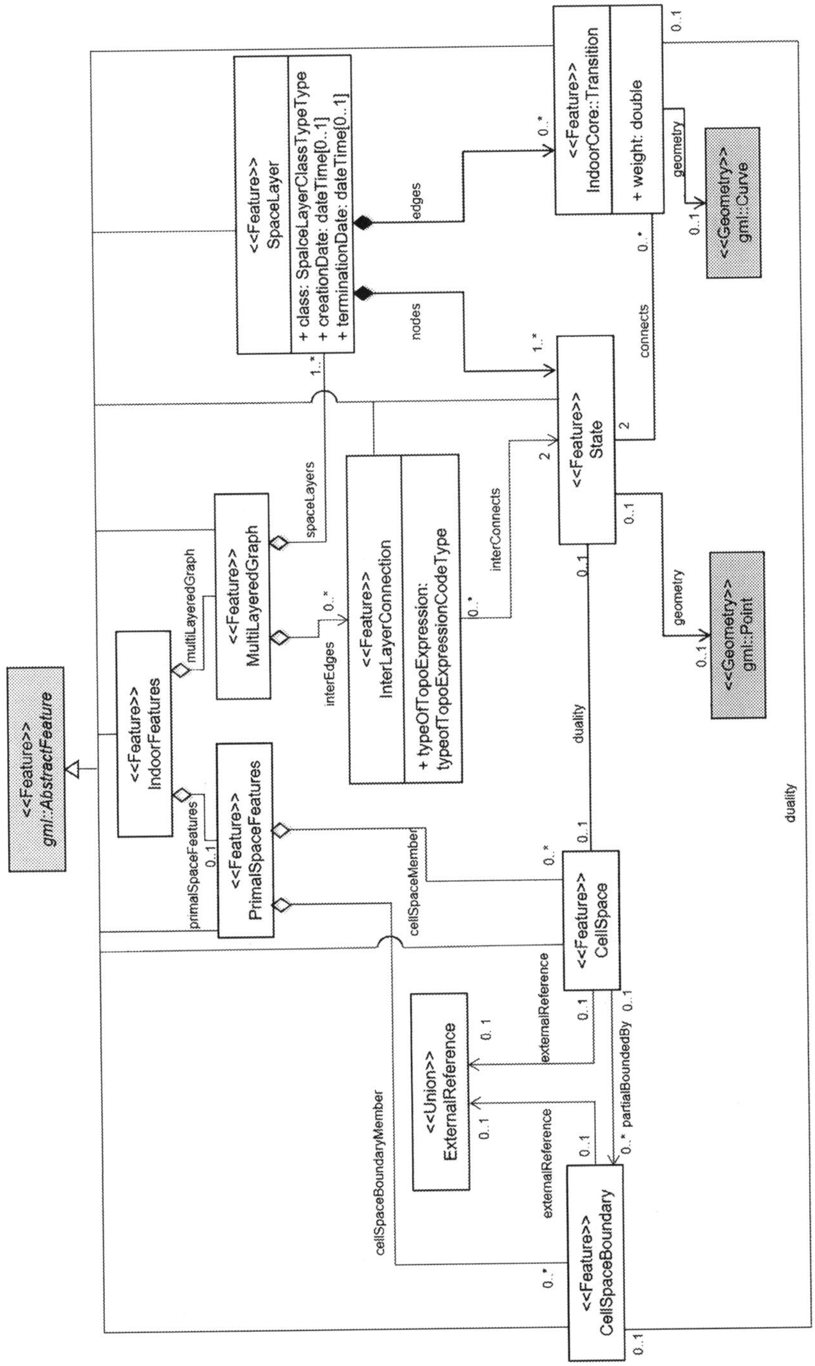

FIG. 5 OGC IndoorGML Core Module (OGC, 2012b).

It may also contain a reference to an external object, which may provide additional information. The geometric type of Cell Space may be either a solid or surface depending on the dimensionality of the space. It may not contain any geometry as discussed in the previous section. While Cell Space represents a cell in indoor space, Cell Space Boundary defines the boundary geometry of a Cell Space object, and its geometry may be a surface or curve depending on the dimensionality. State and Transition define the feature types of the topological graph corresponding with Cell Space and Cell Space Boundary in terms of connectivity topology.

4.2 IndoorGML Navigation Module

While the core module defines the basic feature types in indoor space, we may extend it for specific application areas. As indoor navigation is a typical and one of the most demanded application, the current version of IndoorGML includes the first extension for indoor navigation. Fig. 6 shows the UML class diagram of the indoor navigation module.

The feature types of the indoor navigation module are divided into two categories; cells space and cell boundary. The feature types belonging to cell spaces are illustrated in Fig. 7. Among the feature types in the navigation module, it is worthwhile to pay attention to Anchor Space and Anchor Boundary. It allows the connection between indoor and outdoor spaces as depicted in Fig. 8 and the seamless navigation can be implemented using the anchor. In addition to the connection between indoor and outdoor spaces, we may define extra attributes such as building address, the URL of the building, or the transformation parameters for two spatial reference systems of indoor and outdoor spaces.

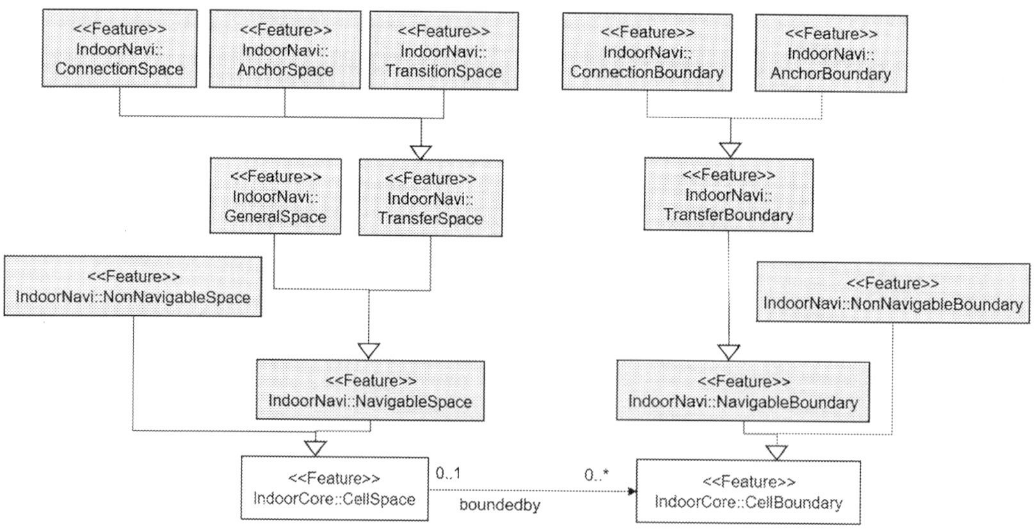

FIG. 6 IndoorGML Navigation Module (OGC, 2012b).

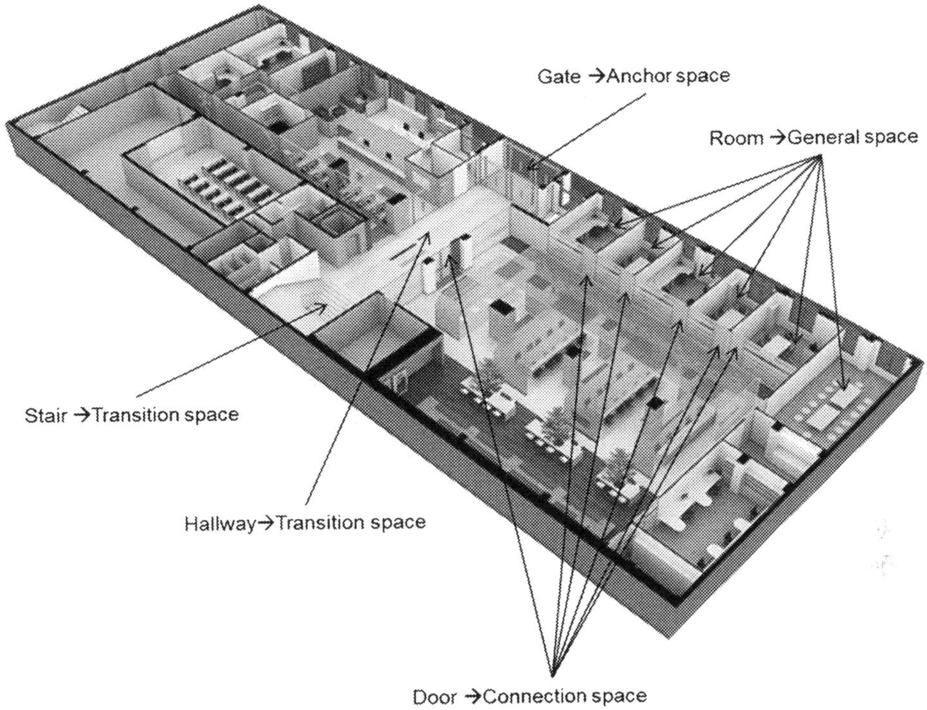

FIG. 7 Features in Indoor Navigation Module (OGC, 2012b).

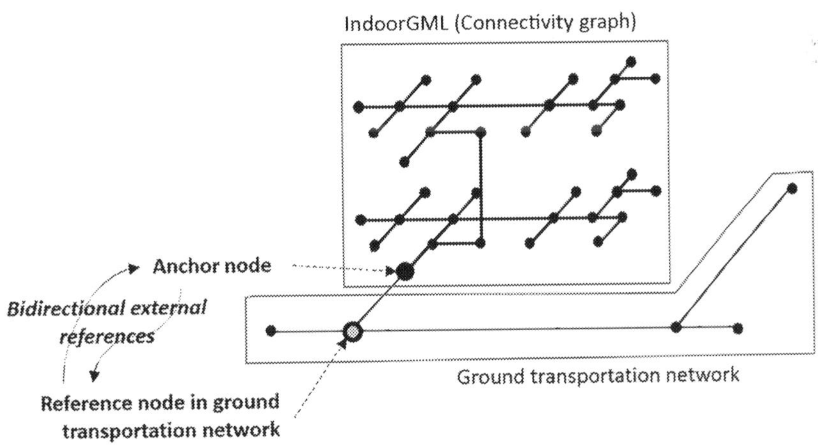

FIG. 8 Anchor (OGC, 2012b).

5 Implementation Issues

In this section, we discuss major issues on implementing IndoorGML. A more detail discussion is also found in Kang and Li (2017).

CELL DETERMINATION AND DECOMPOSITION

Since cells are the basic units of indoor space, the construction process of indoor maps in IndoorGML starts from the determination of cells. In many applications, we often find cells as shown in Figs. 9 and 10 . In order to assign semantics to each cell, it is required to decompose them in proper ways. The decomposition of a big cell depends on the type of indoor spaces and applications. For example, in airports, the space is divided from functional perspectives such as arrival, departure halls, office area, and so on, where the arrival hall is also divided into transit areas, immigration and passport control areas, and baggage claims. The decomposition rules in airports should be much different from

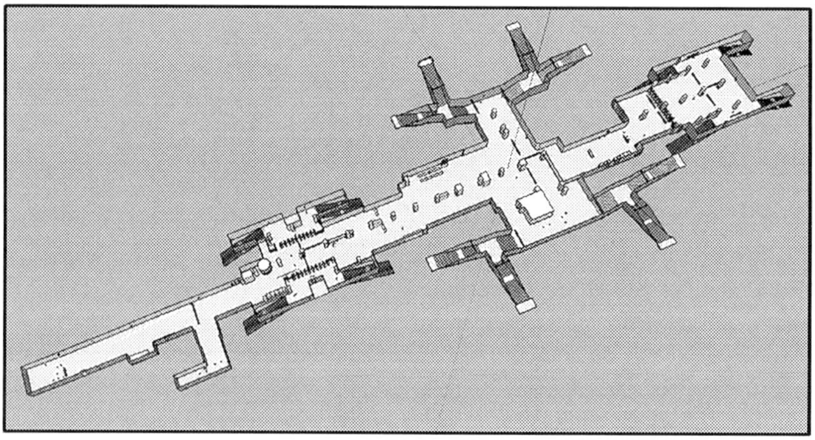

FIG. 9 A big cell at a subway station in Seoul (Ryoo et al., 2015).

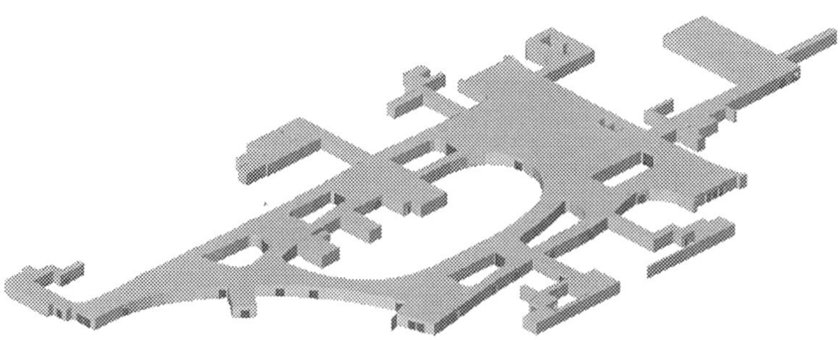

FIG. 10 A big cell at a shopping mall in Seoul (Ryoo et al., 2015).

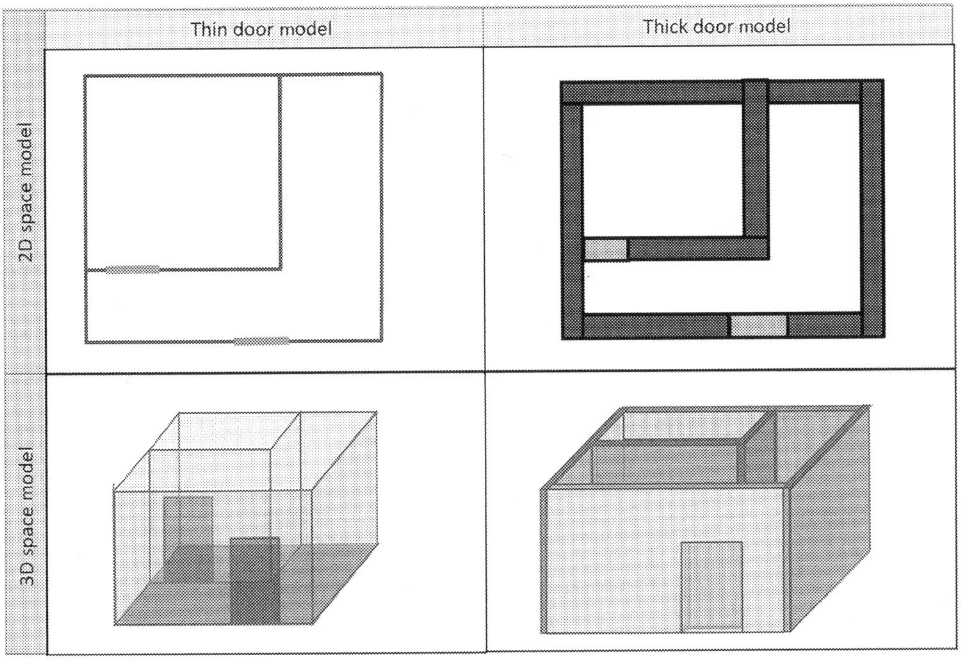

FIG. 11 Thin door model versus thick door model (OGC, 2012b).

subway station like Fig. 9 and shopping malls as Fig. 10. It is therefore necessary to prepare a set of decomposition rules for each indoor type (Diakité and Zlatanova, 2008).

THICK DOOR MODEL VS. THIN DOOR MODEL

There are two approaches to model an indoor space using IndoorGML—thin wall model and thick wall model. While the wall or door are represented as a surface in thin door model, they have a certain thickness in thick wall model as shown in Fig. 11. Note that the wall itself is also regarded as a nonnavigable cell in thick door model.

PATH GEOMETRY

The edge in connectivity graph does not only indicate the connectivity between two nodes but also its geometry, where the edge may be either a straight line or curve. Then the geometry of edge is useful to compute indoor distance and to represent paths of indoor transportations such as robots.

SPACE CLOSURE

In CityGML, a cell can be represented only as an instance of Room, where its geometry of Room is either a Solid as closed space or as a Multi-Surface, which is not necessarily closed. For example, stairs are considered as Interior Building Installation, whose geometry is a MultiSurface as shown in Fig. 12. If we want to close the space, then we need to define an additional feature of Closure Surface or Floor Surface of CityGML, which may

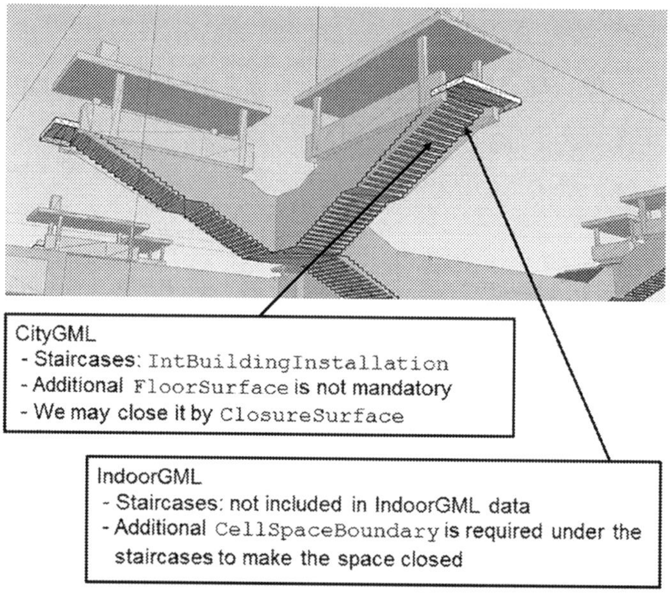

FIG. 12 Closed space—Staircase (Ryoo et al., 2015).

be very easily missing in practice. However, stairs are not included in IndoorGML since the representation of indoor installations is not within the scope. But stair case is represented as a cell of closed space, which is a Solid in 3D space or a Surface in 2D space. In both cases, the geometry of cell is always closed and it facilitates spatial operations such as point-in-polygon or point-in-polyhedron.

HIERARCHICAL STRUCTURE

Most indoor spaces have hierarchical structures. An efficient way to represent hierarchical structures of indoor space was introduced in Stoffel et al. (2008). It is possible to redefine the hierarchical graph for indoor space by the multilayered space model of IndoorGML (Kang and Li, 2017; Kim and Li, 2016). Each level of hierarchy is defined as a single space layer of the multilayered space model, and the relationships between two levels are represented by the interlayer connection of IndoorGML. An example is shown in Fig. 13. Therefore, G_0 is the single-layered graph of the bottom level. The highest level G_h, namely the root graph of the hierarchical graph, contains only a node without edge, where h is the height of the hierarchical graphs. Fig. 13 shows an example of the hierarchical structure for an indoor space by the multilayered space model of IndoorGML. S_1 and S_2 in Level 1 are the aggregations of $\{R_1, R_2, C_5, R_6\}$ and $\{C_3, C_4, R_7, R_8\}$ in Level 0, respectively. T_1 is the entire indoor space as the aggregation of $\{S_1, S_2\}$. Then, the G_0 layer indicates the base graph of the indoor space; G_1 is the next level layer of the hierarchy; and G_2 is the layer for the root level. The relationships between layers are given via interlayer connections. Note that the topological properties of interlayer connections in Fig. 13 are INSIDE.

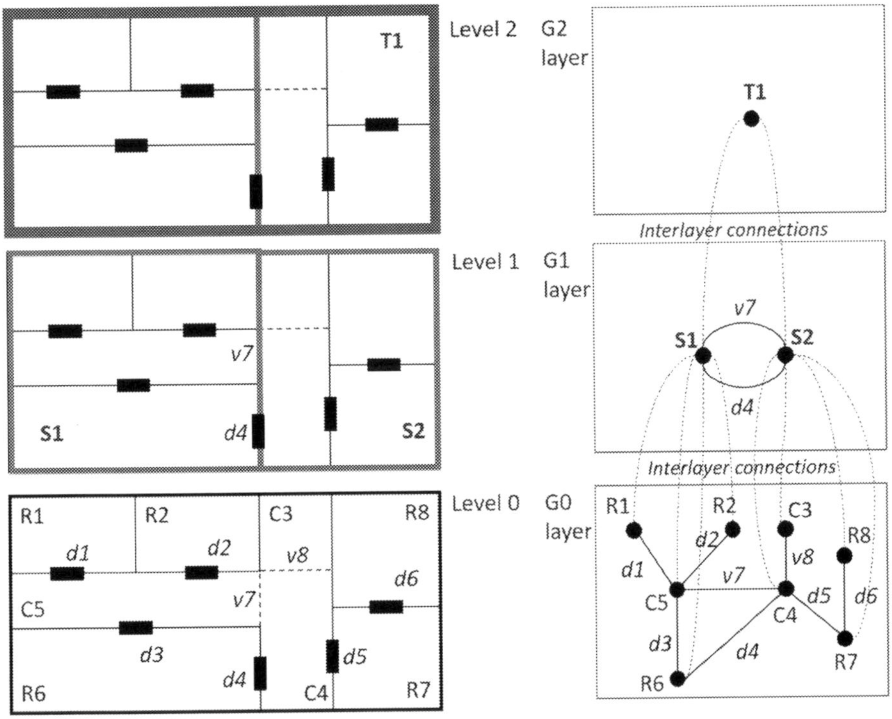

FIG. 13 Hierarchical structure and multilayered space model of IndoorGML (d_i and v_j indicate the connections via the *i*th door and the *j*th virtual boundary, respectively) (Kang and Li, 2017).

WALL TEXTURE

Unlike CityGML where every wall surface is considered as a feature with attributes and textures (OGC, 2012a), it is optional in IndoorGML to define wall surface as an independent feature. Furthermore, we cannot assign texture to wall surface in IndoorGML, since the visualization is not among the purposes of IndoorGML and more precisely no orientation is defined for wall surface.

VERTICAL CONNECTION

The indoor space has a set of connections between floors such as lifts, escalators, and stairs as well as horizontal connections. However, they may be differently implemented by IndoorGML as shown in Fig. 14. In the case of elevators, the elevator shaft is regarded as a cell and the vertical connections are established via the elevator shaft cell. In the contrary, the staircase can be divided into small cells, each of which has a connection to the corresponding floor. The vertical connections via elevators are differently represented in IndoorGML. Since each escalator does not own its space extent in indoor space unlike elevators, it is simply considered as a transportation mode between floors. The connection via escalator is therefore treated as vertical doors shown in right-hand side of Fig. 14.

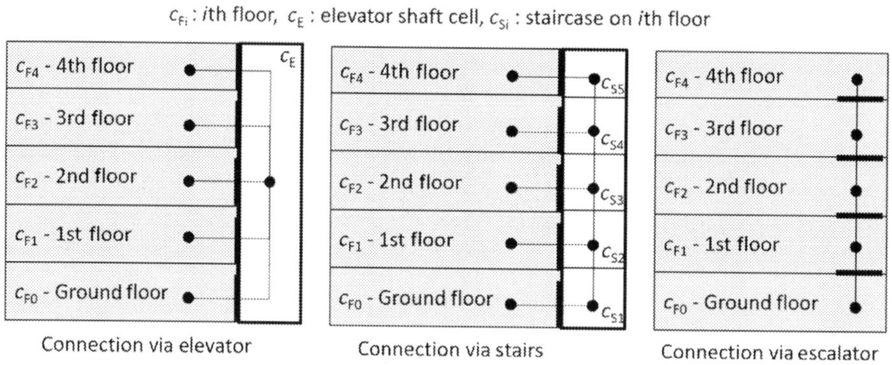

FIG. 14 Implementations of vertical connections by IndoorGML.

6 Use Cases

The robustness and soundness of the previously illustrated conceptual model underpinning IndoorGML has been already validated in the context of several use cases. To this extent, an interesting set of use cases have been validated in real-life conditions by the project i-locate "i-locate—Indoor/outdoor LOCation and Asset management Through open gEodata" (Conti and Eccher, 2014). The project, which has been funded by the European Commission (EC), delivered one of the first implementations of IndoorGML (Conti, 2015b) together with a suite of associated tools, including a middleware, an online portal (i-locate project, 2015), and a IndoorGML plug-in for JOSM the popular authoring tool for OpenStreetmap (Conti, 2015a). Perhaps equally interesting, the project has also assessed the use of IndoorGML in the context of two key use cases, namely user navigation and asset management, within different verticals such as healthcare, museums, etc. The assessment has been carried on in the context of 13 pilot sites across Europe, for a duration of a year in real operational scenarios. All the pilot sites shared two macro-use cases, albeit applied in different vertical domains, that is, indoor and outdoor guidance of people and real-time asset management.

The 13 sites included four major hospitals (in Malta, Romania, Italy, and Greece), two university campus (in the Netherlands and Germany), a major international museum in Romania, several public buildings from four municipalities (in Croatia, Romania, and Italy), a nursing home in Romania, and business park in Luxembourg. While a comprehensive description of each pilot, Morganto et al. (2015); Napoleoni et al. (2016) is clear beyond the scope of this chapter, nevertheless it is worth highlighting that the size, heterogeneity and duration of the pilot (12 months) allowed for a very significant assessment of the potential of IndoorGML in real operational scenarios, ranging from creation of multiple graphs using the online portal or JOSM plugin (see Fig. 15) to their use across the i-locate service stack in order to provide tailored services to final users using a variety of location technologies using, for instance, ZigBee, Ultra-Wide Band, Bluetooth, Wi-Fi, cell-ID, GNSS (EGNOS), as well as camera-based systems.

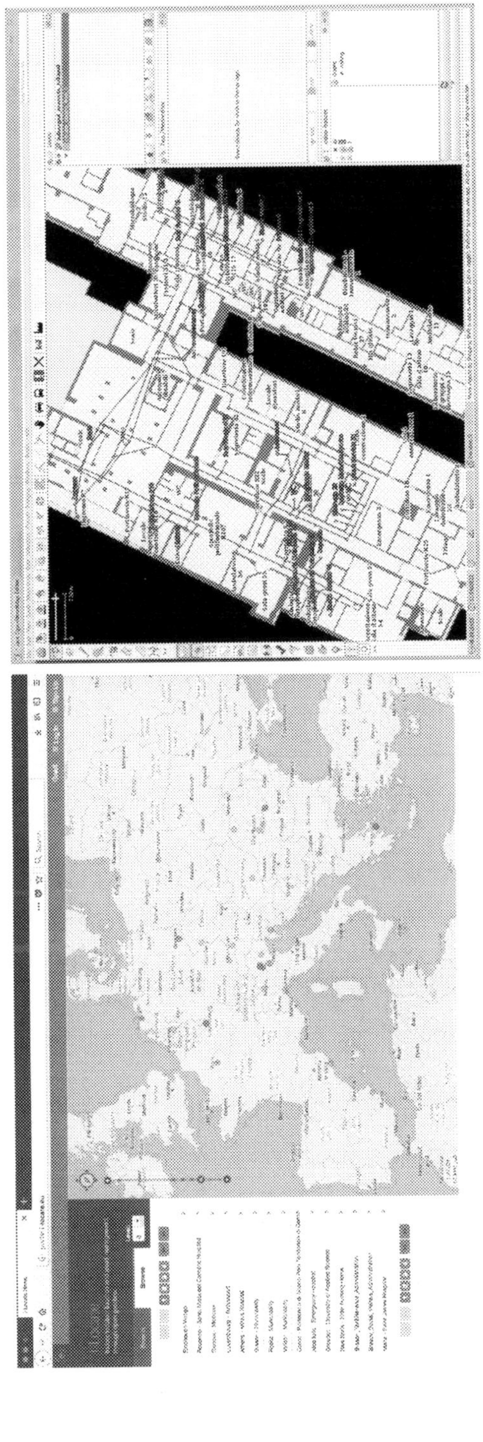

FIG. 15 Images of the i-locate portal (*left*) (i-locate project, 2015) and JOSM featuring the i-locate plugin (*right*) (Conti, 2015a) showing the connectivity graph, encoded as IndoorGML (i-locate project, 2015), of the hospital where i-locate technology has been used.

All the pilots have shared two macro use cases, that is indoor way finding and management of both of people or asset. The typical wayfinding use case has seen the final users being routed indoor and outdoor, through the i-locate App, to reach a given destination. To do so each IndoorGML graph has been further connected, via an anchor node, through the OpenStreetMap outdoor road network, in order to allow for hybrid indoor and outdoor navigation. The resulting hypergraph, that is a graph of graphs, has then been compiled and used by a specifically developed web services providing navigation and routing capabilities. Most interestingly, the serviced has designed to allow for indoor to indoor, outdoor to outdoor, and, most interestingly, outdoor/indoor navigation, calculating optimized paths according to a set of customizable criteria which, in turn, defined the weight functions used by the routing system (Feng et al., 2015; Konstantinidis et al., 2017).

While the following examples have shown use of IndoorGML in traditional use cases, albeit applied to a variety of different verticals, an additional example may help show the potential that be unleashed by leveraging on the IndoorGML concept of cell. In most use cases, cells are used to refer to rooms, or indoor spaces of a building with a specific functioning connotation. An example could be the different check-in desks of airport departure hall, that, although not physically separated each other, could be modeled as cell and used to provide indoor guidance to each of the different desk. The work carried on by two ongoing EC projects, UNCAP and then CAPTAIN, is instead emergence of predictors of cognitive decline among elderly adults.

Body movements of elder adults are continuously detected by using 3D sensors (Nively, 2015). By deriving their body spatial configuration (position in space of different body parts) in real time, it is possible to define, among other activity indicators, trajectories, and speed over time of the senior. In turn, this information can be used to extract important information on behavioral patterns related to the their Activities of Daily Living (ADL). By applying Density-Based Spatial Clustering of Applications with Noise (DBScan), it has been possible to cluster position data over time according to the minimum distance between two neighboring points (eps parameter) and the minimum neighborhood points that are sufficient to constitute a cluster (minPts parameter). The resulting clusters, technically referred to as High-Density Regions (HDRs), are further processed to calculated the smallest convex polygon that surrounds the points of each HDR, which is finally used to define the IndoorGML cell. In this case the cell represents a portion of space associated with a specific behavioral pattern.

In the work illustrated in Konstantinidis et al. (2016, 2017), the methodology was applied to user's activity data collected over a 12-month period while cells were generated to highlight possible behavioral variations over time at a monthly frequency. While each cell is related to a specific spatial behavior within a given timeframe, the connectivity between different cells, derived through statistical analysis of the activity data, can be used to model transitions with highest occurrence between cells, whose variation over time can in turn be used to infer changing of usual behaviors. The initial encouraging results (Konstantinidis et al., 2016, 2017), are being extended to include outside data analysis collected by leveraging on high-precision GNSS (Galileo) and by leveraging, as in the previous example, on the concept of anchor node, to expand the scope of the aforementioned method to indoor and outdoor spaces.

7 Conclusion

As indoor maps are a fundamental component of indoor spatial information services including indoor LBS, a number of different map formats has been developed by standard development institutes and vendors. Each approach reflects its requirements and viewpoints. However, indoor spatial information services consist of several building blocks from spatial technologies as outdoor information services. In order to provide the interoperability between different building blocks and setup an ecosystem, we need a common data model and format of indoor maps. The OGC has published IndoorGML as a standard for indoor spatial information.

The main features of IndoorGML include the cellular space model, the geometric, topological, and semantic models of cells, and the multilayered space model. IndoorGML provides only the features that are not found in other standard data models and formats, and particularly emphasizes the indoor cellular space model and network topology in indoor space. However, it does not form a complete set of all features for indoor maps and therefore has to be integrated with other standards such as OGC CityGML, IFC, KML, etc., to develop an indoor spatial information application.

In this chapter, we also discussed the implementation issues of IndoorGML from different viewpoints as well as use cases, such as cell determination and decomposition, thick-door model, path geometry, cell space closure, hierarchical structures, and vertical connections, which we currently believe the most important ones. However, it is not possible to limit the application scopes of indoor maps and IndoorGML, and a number of implementation issues may be expected. Particularly we may need more specific implementation specifications for each type of applications, such as robotics, indoor asset managements, indoor security controls, and so on. These specifications can be added as extensions modules of IndoorGML, which are also future works of IndoorGML community.

Acknowledgments

Part of the work described in this chapter has been co-funded by BK21PLUS, Creative Human Resource Development Program for IT Convergence, the European Commission through the projects i-locate (Grant Agreement no. 621040), UNCAP (Grant Agreement no. 643555), and CAPTAIN (Grant Agreement no. 769830), and by the Ministry of Land, Infrastructure and Transport of Korean Government (Grant agreement no. 17NSIP-B135746-01). This document reflects only the views of the authors and the European Community is not liable for any use that might be made of the information contained within the document.

References

Afyouni, I., Cyril, R., Christophe, C., 2012. Spatial models for context-aware indoor navigation systems: a survey. J. Spat. Inf. Sci. 1 (4), 85–123.

Alattas, A., Zlatanova, S., Van Oosterom, P., Chatzinikolaou, E., Lemmen, C., Li, K.J., 2017. Supporting indoor navigation using access rights to spaces based on combined use of IndoorGML and LADM models. ISPRS Int. J. Geoinf. 6 (12), 384.

Becker, T., Nagel, C., Kolbe, T.H., 2009. A multilayered space-event model for navigation in indoor spaces. In: 3D Geo-Information Sciences. Springer, pp. 61–77.

BuildingSmart, 2009. IFC Standard. Available from: http://www.buildingsmart-tech.org/specifications/ifc-overview (Accessed July 2018).

Conti, G., 2015a. i-locate extends OpenStreetMap editor for indoor navigation graph editing. Available from: http://www.opengeospatial.org/blog/2270 (Accessed July 2018).

Conti, G., 2015b. Indoor/outdoor location and asset management through open geodata—i-locate. Available from: http://www.opengeospatial.org/blog/2263 (Accessed July 2018).

Conti, G., Eccher, C., 2014. Indoor/outdoor LOCation and Asset management Through open gEodata. Available from: http://mediageo.it/ocs/index.php/esri/15cue/paper/viewFile/41/83 (Accessed July 2018).

Diakité A. A., Zlatanova, S., 2018. Spatial subdivision of complex indoor environments for 3D indoor navigation, International Journal of Geographical Information Science 32(2), 213–235.

Feng, T., Jessurun, J., Arentze, T., 2015. Cohesive routing service for indoor and outdoor navigation. In: Proceedings of FOSS4G Europe 2015.

i-locate project, 2015. i-locate online portal. Available from: http://portal.i-locate.eu (Accessed July 2018).

ISO TC211, 2003. Geographic Information-Spatial Schema, ISO 19107:2003. ISO, Geneva, Switzerland.

ISO/TC59/SC13, 2015. Building Construction-Organization of Information About Construction Works—Part 2: Framework for Classification, ISO 12006-2:2015. ISO, Geneva, Switzerland.

Kang, H.K., Li, K.J., 2017. A standard indoor spatial data model—OGC IndoorGML and implementation approaches. ISPRS Int. J. Geoinf. 6 (4), 116.

Kim, J.S., Li, K.J., 2016. Location K-anonymity in indoor spaces. GeoInformatica 20 (3), 415–451.

Konstantinidis, E.I., Billis, A.S., Plotegher, L., Conti, G., 2016. Indoor location IoT analytics "in the wild": active and healthy ageing cases. In: Proceedings of MEDICON 2016.

Konstantinidis, E.I., Billis, A.S., Dupre, R., Montenegro, J.M.F., Conti, G., Argyriou, V., Bamidis, P.D., 2017. IoT of active and healthy ageing: cases from indoor location analytics in the wild. Health Technol. 7 (1), 41–49.

Lee, J., 2004. A spatial access-oriented implementation of a 3-D GIS topological data model for urban entities. GeoInformatica 8 (3), 237–264.

Li, D., Lee, D.L., 2008. A topology-based semantic location model for indoor applications. In: Proceedings of the 16th ACM SIGSPATIAL International Conference on Advances in Geographic Information Systems, p. 6.

Lu, H., Cao, X., Jensen, C.S., 2012. A foundation for efficient indoor distance-aware query processing. In: 2012 IEEE 28th International Conference on Data Engineering (ICDE), pp. 438–449.

Morganto, E., et al., 2015. i-locate project deliverable: pilot scope description and system requirements. Available from: http://www.i-locate.eu/?smd_process_download=1&download_id=8103 (Accessed July 2018).

Napoleoni, F., et al., 2016. i-locate project deliverable: pilot scope description and system requirements. Available from: http://www.i-locate.eu/data/uploads/2016/09/D.5.5-Final-Pilots-Evaluation.pdf (Accessed July 2018).

Nively, 2015. MentorAge. Available from: http://www.mentor-age.com (Accessed 15 December 2017).

OGC, 2007. OGC GML, Document No. 07-036. Available from: http://www.opengeospatial.org/standards/gml (Accessed July 2018).

OGC, 2012a. OGC CityGML Encoding Standard, Document No. 12-019. Available from: http://www.opengeospatial.org/standards/citygml (Accessed July 2018).

OGC, 2012b. OGC IndoorGML, Document No. 14-005r4. Available from: http://www.opengeospatial.org/standards/indoorgml (Accessed July 2018).

Ryoo, H.G., Kim, T., Li, K.J., 2015. Comparison between two OGC standards for indoor space: CityGML and IndoorGML. In: Proceedings of the Seventh ACM SIGSPATIAL International Workshop on Indoor Spatial Awareness, p. 1.

Schilit, B., Adams, N., Want, R., 1994. Context-aware computing applications. In: First Workshop on Mobile Computing Systems and Applications, 1994. WMCSA 1994, pp. 85–90.

Stoffel, E.P., Schoder, K., Ohlbach, H.J., 2008. Applying hierarchical graphs to pedestrian indoor navigation. In: Proceedings of the 16th ACM SIGSPATIAL International Conference on Advances in Geographic Information Systems, p. 54.

Xie, X., Lu, H., Pedersen, T.B., 2015. Distance-aware join for indoor moving objects. IEEE Trans. Knowl. Data Eng. 27 (2), 428–442.

Yuan, W., Schneider, M., 2010. Supporting continuous range queries in indoor space. In: 2010 Eleventh International Conference on Mobile Data Management (MDM), pp. 209–214.

11

The EvAAL Evaluation Framework and the IPIN Competitions

Francesco Potortì, Antonino Crivello, Filippo Palumbo

INSTITUTE OF INFORMATION SCIENCE AND TECHNOLOGIES "A. FAEDO" NATIONAL RESEARCH COUNCIL, ISTI-CNR, PISA, ITALY

1 Motivation and Challenges

Indoor localization systems have yet a long way to go before becoming an off-the-shelf service like outdoor localization is. Several roadblocks exist that hinder the possibility of ubiquitous and seamless positioning and navigation applications on our mobile devices. Next to technological, privacy, and standardization issues, evaluation of localization systems is one of the challenges that is currently being tackled by researchers in this field.

As in any mature technology field, common evaluation criteria are fundamental in order to add transparency to the market by defining a common performance language and eventually to build and nurture stakeholders' trust.

The problem with indoor localization systems is that they are generally complex. While in the laboratory the base techniques are individually analyzed and optimized, real working systems use many techniques that work synergically, thanks to the use of data fusion methods. At the base of these techniques, a wide spectrum of sensors work to provide raw data. On top of these techniques, applications are dedicated to a wide variety of use cases.

It is therefore not straightforward to devise ways to evaluate indoor localization systems through a series of parameters. It is not even easy to just compare two of them, because comparison is possible and meaningful on many dimensions, depending on the particular use case.

2 Background

In 2010, indoor localization had become a significant research field on its own, but it lacked of a dedicated forum. The Indoor Positioning and Indoor Navigation (IPIN) conference was born in Zurich (CH) to fill this gap. The first edition gathered about 200 attendees.

Geographical and Fingerprinting Data for Positioning and Navigation Systems. https://doi.org/10.1016/B978-0-12-813189-3.00011-3

In that same year, the EU FP7 universAAL project started its work toward creating a universal framework for developing applications for Ambient Assisted Living (AAL) and, more generally, for smart homes and smart environments, building on advances in ubiquitous computing, distributed middleware and pervasive computing and communication (Ram et al., 2013). The universAAL framework is intended to support an ecosystem of independent applications, so the problem of comparing and evaluating their performance naturally came forward.

As an answer to this demand, the universAAL project started EvAAL, with the purpose of evaluating AAL systems through competitive benchmarking (Barsocchi et al., 2013). The idea was to gather together working systems, both prototypal and mature, and independently compare their performance in one or several specific areas, with the long-term objective of creating a set of evaluation benchmarks for indoor pervasive systems. In fact, two areas were considered during EvAAL competitions, starting in 2011 in Valencia (ES): indoor localization and indoor activity recognition.

EvAAL competitions were organized yearly during the lifespan of the universAAL project, until 2013. In 2014, the IPIN conference decided to start an indoor competition on its own, building on EvAAL's experience, and the first IPIN competition was born.

In the same year, the Microsoft Indoor Localization Competition was launched, in association with the International Conference on Information Processing in Sensor Networks (IPSN) (Lymberopoulos et al., 2015). Rather than focusing on rigorous evaluation of working systems as the EvAAL and IPIN competition did, it has focused on simplicity and comparison of basic functionality, even for very prototypal systems, with the result of attracting a higher number of contestants with respect to EvAAL and IPIN.

The three initiatives previously mentioned are described in some more detail in the following.

2.1 The IPIN Conference

IPIN is the only long-lasting conference specifically dedicated to indoor localization. It is essentially dedicated to "hard-core" topics, that is, to specific low-level technology-oriented hardware and software localization techniques. It is interesting to look at the session titles along the history of IPIN conferences to look for the evolution of topics not strictly connect to low-level techniques.

The eight IPIN conferences in the years 2010–2017 had an average of 24 sessions. Of these, an average of three sessions were devoted to a topic not specifically centered on low-level software or hardware localization techniques. Table 1 lists the topics of these mid- to high-level topics.

It is interesting to note that, while since 2014 IPIN has dedicated specific sessions to discuss the results of the colocated IPIN competition, only in 2017 a session was explicitly devoted to the evaluation of indoor localization systems. This is a strong indication that the topic of system evaluation on its own has drawn attention only recently among researchers

Table 1 Mid- to High-Level Session Topics in IPIN Conferences

Year	FW	req.	apps	cntxt	comp.	maps	motion	eval.
2010	x	x	x					
2011			x	x				
2012		x	x					
2013	x	x		x				
2014	x			x	x			
2015				x	x	x	x	
2016	x	x			x	x		
2017			x	x	x	x		x

Notes: apps, applications; *cntxt*, context-aware systems or applications; *comp.*, IPIN competition-dedicated sessions; *eval.*, evaluation of localization systems; *FW*, frameworks and libraries; *maps*, map generation and rendering; *motion*, human motion models and monitoring; *req.*, user requirements.

and industry. This emerging attention can be partly attributed to the pioneering activity of indoor localization competitions, starting with EvAAL, and partly to the fact that while localization systems are starting to approach the market, the need for standard methods of evaluation is becoming apparent.

2.2 The EvAAL Indoor Localization Competition

The EvAAL initiative was launched by the European FP7 universAAL project in 2010 as a way of "Evaluating Ambient Assisted Living systems through competitive benchmarking." In 2011, the first EvAAL competition was held at the CIAmI Living Lab in Valencia (ES), with a single track devoted to Indoor Localization and Tracking (Barsocchi et al., 2012).

When EvAAL was born, its long-term goal was to build one or more frameworks for evaluating entire AAL systems, a huge task which was tackled step by step by considering single system modules. The first such module was in fact indoor localization. In 2012, a second track was added, namely Activity Recognition for AAL. Both tracks were present in the 2013 edition too. Due to lack of funding from the universAAL project, which ended at the beginning of 2014, EvAAL suspended its activity as competition organizer, but was careful to preserve its heritage through its web site,[1] which hosts extensive documentation of the three EvAAL competitions and of the subsequent IPIN competitions based on the EvAAL framework (Potortì et al., 2017).

During the years 2011–2013, the Indoor Localization and Tracking competition has been based on the same idea: inside a living lab, that was a small house instrumented with various sensors, a path, unknown to competitors, was drawn in advance; competing systems were given a fixed time for installing their devices in the smart home and

[1] See http://evaal.aaloa.org.

estimating in real time the position of an *actor* walking the path. The basic criteria used for the setup were:

Accommodating any Technology Competitors were free to use any technology that could be installed in the living lab and on the *actor*'s body in 1 h time.

Natural movements and environment Measurements were done in real time on an *actor* moving in a natural way, in a natural environment: he walked around the house, sit on the bed or the coach, looked for a book in a bookshelf, turned on the TV set, or the shower tap.

Reproducible path, equal for all competitors The path walked by the *actor* was precisely known (in fact, drawn step by step on the floor) and walked at precisely known speed following a chime marking each step. This arrangement allowed for an estimated path reproducibility with 10 cm error in space and 100 ms error in time, well below the accuracy required for human indoor localization.

Secret path Competitors got to know the path shape only after their own installation was complete and measurement was going to begin, because the markers on the floor were hidden by carpets before the measurement phase and only one competitor at a time was admitted to the area.

Independent measurements Competing systems had to send location estimates in real time to a central database, twice per second.

Accurately controlled timing Each competitor had 1 h for installing their hardware in the living lab and checking the communication with the measurement system provided by organizers.

Different scenarios Three scenarios were used: first, a person was located as being inside one of several Areas of Interest (AoI) or outside any AoI; second, a person was located with absolute coordinates inside the living lab; third was like the previous case but a second *disturbing actor* moved on a predefined path different from the main path.

Evaluation was based on a set of predefined metrics, both objective and subjective, the latter based on scores given by a small committee after an interview to the competitors. The final score was a weighted average of the metric scores:

Accuracy (objective, weight 0.35) The third quartile of point localization error, where error is defined as the distance from the ground truth position (the mark on the floor) and the position estimated by the competing system, computed through linear time interpolation.

Availability(objective, weight 0.20) The quote of real-time samples, produced by the competing system, that were at a distance of 500 ms from each other.

Installation complexity (objective, weight 0.10) The time taken by competitors to install their system, with a min time of 10 min and maximum of 1 h.

User acceptance (subjective, weight 0.2) Interview scoring based on characteristics like battery duration, possibility of hiding the installation in a house, need of cabling, need of periodic recalibration, and so on.

Interoperability (subjective, weight 0.15) Interview scoring based on characteristics like presence of documented API, use of a free software license, use of standard protocols and libraries, operating systems supported, and so on.

The score with the highest weight was the accuracy performance, as should be expected from evaluation metrics of a positioning and tracking system. The choice of third quartile favors result stability and credibility (Barsocchi et al., 2013), and was a prominent distinguishing characteristic of the EvAAL competitions.

The setup and evaluation criteria made EvAAL a rigorous and difficult competition, and in fact the number of attendants for the localization track was seven or eight in all three editions. Competing systems not only had to show good performance, but they had to be installed from scratch in an unknown environment in 1 h time, had to interact with an external logging system, had to work without interruption for the 10 min or so of the longest path walked by the actor. All these requirements were hard to meet for prototypal or unstable systems.

The upside was that EvAAL competitions were *realistic*. The actor moved in a realistic way in a real domestic environment and the results were gathered and displayed in real time. As a consequence, the accuracy performance was significantly lower than what you can read in academic papers, as they reflected real-life situations. From this point of view, the EvAAL competitions were a breakthrough, as for the first time they provided realistic performance measurements of indoor localization systems.

2.3 The Microsoft Indoor Localization Competition

In 2014, the International conference on information Processing in Sensor Networks (IPSN) hosted the first edition of the Microsoft indoor localization competition. The competition favors inclusion by setting a measurement environment typical of laboratory conditions, thus allowing for participation of prototypes even at a very preliminary stage. Specifically, competitors are asked to place their positioning system on a series of key points in sequence, and statically estimate the keypoints' coordinates.

The environment is not meant to represent any specific use case, and in 2014–2017 years varied from few rooms on a single floor to a 600 m^2, two-floor area. Scoring is based on accuracy only, consistently with the technology-oriented nature of the competition. The final score is based on the mean of point errors at keypoints. In the latest years, 2D and 3D tracks were considered.

On the upside, the number of participants to the Microsoft competition has been significantly higher than EvAAL's, almost always exceeding 20 participants.

3 The EvAAL Framework

As a result of the experience gained from the EvAAL competitions and the feedback obtained from the organizers and competitors, the EvAAL committee has formalized an evaluation framework (Potortì et al., 2017) to be applied to indoor localization competitions in order to measure and compare the performance obtained by the competing systems. The EvAAL framework is characterized by several *core* (the distinguishing features of the EvAAL framework) and *extended* (all adopted by the EvAAL competitions) criteria. The *core* criteria are the following:

1. *Natural movement of an actor*: The agent testing a localization system walks with a regular pace along a predefined path. The actor can rest in a few points and walk again until the end of the path.
2. *Realistic environment*: The path the actor walks is defined in a realistic setting.
3. *Realistic measurement resolution*: The minimum time and space error considered are relative to people's movement. The space resolution for a person is defined by the diameter of the body projection on the ground, which is set to 50 cm. The time resolution is defined by the time a person takes to walk a distance equal to the space resolution. In an indoor environment, considering a maximum speed of 1 m/s, the time resolution is 0.5 s.
4. *Third quartile of point Euclidean error*: The accuracy score is based on the third quartile of the error, which is defined as the 2D Euclidean distance between the measurement points and the estimated points. More discussion on this can be found in Potortì et al. (2017) and Barsocchi et al. (2013) and at the end of Section 4.2.

The *extended* criteria additionally adopted by the first EvAAL competitions are the following:

5. *Secret path*: The final path is disclosed immediately before the test starts, and only to the competitor whose system is under test. This prevents competitors from designing systems exploiting specific features of the path.
6. *Independent actor*: The actor is an agent not trained to use the localization system.
7. *Independent logging system*: The competitor system estimates the position at a rate of twice per second, and sends the estimates to a logging application provided by the EvAAL committee. This prevents any malicious actions from the competitors. The source code of the logging system is publicly available.[2]

[2]See http://evaal.aaloa.org/2017/software-for-on-site-tracks.

8. *Identical path and timing*: The actor walks along the same identical path with the same identical timing for all competitors, within time and space errors within the above-defined resolutions.

4 The IPIN Competitions

The experience of the EvAAL competitions was transferred to IPIN. The first edition of the IPIN competition was held at the IPIN 2014 Conference, located in Busan, South Korea. There were some significant differences with respect to EvAAL competitions:

No instrumentation Competitors were not allowed to instrument the competition area with their own devices.

Single technology Competition was restricted to a single technology per track.

Large area The size of the competition area was significantly larger than a small apartment.

Simple scoring Only point error accuracy was considered for scoring.

Use of keypoints Point errors where computed at a number of keypoints along the path.

Table 2 shows an overview of the number of tracks and competitors in past EvAAL and IPIN indoor localization competitions. While the numbers may look small, it is interesting to observe how such a challenging competition, requiring significant preparation effort—and significant on-site effort for on-site competing teams—keeps attracting an essentially constant number of competitors, meaning that the IPIN competition maintains its attractiveness while technology advances.

The IPIN 2014 competition, which was held in Busan (KR), was composed of two tracks: positioning through smartphone-based solution and foot-mounted pedestrian dead reckoning. Many characteristics and criteria were in common with the previous EvAAL competitions, as the IPIN competitions are based on the EvAAL framework described in Section 3.

Table 2 Tracks and Competitors in Past Indoor Localization Competition Tracks

Edition	Tracks	Competitors Real-Time	Competitors Offline
EvAAL 2011	1	7	–
EvAAL 2012	1	8	–
EvAAL 2013	1	7	–
IPIN 2014	2	7	–
IPIN 2015	3	6	4
IPIN 2016	4	14	5
IPIN 2017	4	7	9

Competitors were able to perform their own survey of the public area where the competition was held, for a whole day. This was especially useful for systems using fingerprint techniques. Since no instrumentation on the competition area was allowed, only the already deployed Wi-Fi access points could be used by competitors. Competing systems had to be carried by an actor without impairing her or his movements.

The area was a three-floor building used for conferences and big events, and the path spanned few floors, going through stairs. In such an environment, it would have been impossible to measure point localization error at each step, as it was done in EvAAL competitions. Rather, point error was measured at a series of *keypoints*, marked on the floor with adhesive plastic. The actor, rather than following a precisely defined path with steps following a chime, was free to walk in the environment, with the only constraint of passing over all keypoints in the right order. This behavior made it possible to host the competition in a public area, where other people's path could collide with the actor's one. A timestamp was collected at each keypoint to allow for independent error measurement, as detailed in Potortì et al. (2015).

The IPIN 2015 competition was held in Banff (CA). The competition consisted of two on-site and one off-site tracks: smartphone-based positioning, foot-mounted pedestrian dead reckoning positioning, and Wi-Fi fingerprinting in large environments (off-site). Tracks 1 and 2 (smartphone-based and pedestrian dead reckoning) were similar to previous year's ones. In the off-site Track 3 "Wi-Fi fingerprinting in large environments," competitors had access to a large Wi-Fi fingerprint database, to which they can apply their algorithms off-line. During this edition, 10 different teams participated in the three different tracks.

The IPIN 2016 competition was held in Alcalá de Henares (ES). The competition consisted of three on-site and one off-site tracks: smartphone-based positioning, foot-mounted pedestrian dead reckoning positioning, smartphone-based (off-site), and indoor mobile robot positioning. Tracks 1 and 2 were similar to previous years' ones. Track 3 had the goal to evaluate the performance of different indoor localization solutions based on the signals available to a smartphone (such as Wi-Fi readings, inertial measurements, etc.) that were received while a person was walking along few multifloor buildings. Track 4 was dedicated to robot positioning. The goal was monitoring the trajectory followed by a mobile robot, along a predetermined track inside an indoor area, by using a localization system installed by competitors in the navigation area and on board the robot (without interaction with the mobile robot systems). Competitors would be provided with a map of the area, while the predefined path followed by the robot would not be disclosed until the day of the competition.

The IPIN 2017 competition was held in Sapporo (JP). The competition consisted of two on-site and two off-site tracks: smartphone-based positioning, foot-mounted pedestrian dead reckoning positioning, smartphone-based (off-site), and PDR for warehouse picking. Tracks 1, 2, and 3 were similar to previous year ones. Track 4 was devoted to warehouse picking solutions based on PDR technology. It was an off-line competition based on picking data measured in a real warehouse.

4.1 Applying the EvAAL Framework to IPIN Competitions

IPIN competitions adopted the EvAAL framework by applying its *core* criteria to all tracks and part of its *extended* criteria in some tracks. We start by detailing how the *core* criteria were applied.

- *Natural movement of an actor* and *realistic environment*: In Tracks 1 (real-time smartphone-based) and 2 (real-time dead reckoning), present in all editions, the actor moves naturally in a realistic and complex environment spanning several floors of one big building. In Track 3 ("Wi-Fi fingerprinting in large environments" in the 2015 edition and "off-site smartphone-based" in 2016 and 2017 editions), the actor walks along floors of few big buildings. In Track 4, the robot moves at the best of its capabilities in a complex single-floor track in the 2016 edition, while, in the 2017 edition, the actor moves naturally in a realistic warehouse.
- *Realistic measurement resolution*: The space-time error resolution for each year's Tracks 1–3, where the agent is a person, are 0.5 m and 0.5 s, while space-time resolution for 2016 Track 4, where the agent is a robot, are 1 mm and 0.1 s (only adherence to the trajectory is considered given the overwhelming importance of space accuracy with respect to time accuracy as far as robots are concerned) and for 2017 Track 4, different resolutions are considered for each task (i.e., PDR, picking work, human moving, obstacle interference).
- *Third quartile of the point Euclidean error*: The accuracy score obtained by competitors of each track was evaluated according to the *core* criteria of the EvAAL framework related to the third quartile of point Euclidean error. It is measured using the *xy* coordinates (longitude and latitude) provided by competitors as output. Also, a penalty of 15 m is added for each floor error.

The *extended* criteria of the EvAAL framework only make sense for the real-time tracks. Here is how they were used through the IPIN competitions:

- *Secret path*: In Tracks 1 and 2, the path is kept secret only until 1 h before the competition begins, because it would be impractical to keep it hidden from the competitors after the first one in a public environment. Competitors were trusted not to add this knowledge to their systems. In 2016 Track 4 (real-time robotic), a cover was used to avoid any visual reference of the path and other visual markers, so the path was kept secret even during the competition.
- *Independent actor*: This was always used in Track 1 (smartphone-based). For Track 2, competitors themselves took the role of actors, but in 2014 and 2018 results were obtained both with competitor actor and independent actor.
- *Independent logging system*: The logging system is independent only in Track 1, while competitors in Track 2 are asked to provide a log file themselves.
- *Identical path and timing*: In Tracks 1 and 2 the paths and timing are similar but not equal, because the actor is only required to step over the key points in the right order,

without any specific constraint on the path to follow between points and the stride rhythm. In 2016 Track 4 (real-time robotic), timing and path were strictly identical as the agent was a robot, thus not affecting the final accuracy of the competitor in that context.

4.2 Discussion on the Error Statistics

From a scoring point of view, the most characteristic of the *core* EvAAL criteria is the use of *third quartile of point Euclidean error* as the metric for ranking the competing systems. This was the method used during the EvAAL competitions, which were performed on a single floor.

During IPIN competitions, which were performed on multifloor buildings, the Euclidean error was evaluated in 2D, and a penalty of 15 m was added for each wrong floor detection.

The reason behind using a point error instead of comparing trajectories (i.e., the Fréchet distance (Mathisen et al., 2016; Schauer et al., 2016)) is that the latter is less adequate to navigation purposes, for which the real-time identification of the position is more important than the path followed. The only exception was Track 4 in 2017, where the final score was the sum of several metrics, namely integrated positioning error evaluation; PDR error evaluation; picking work evaluation; human moving velocity evaluation; obstacle interference evaluation; and update frequency evaluation.

The reasons behind using the third quartile as the error metric are discussed in Potortì et al. (2017) and Barsocchi et al. (2013): first, the choice of a quantile statistics grants measurement robustness and answers the practical question of what is the maximum error for a given quote of samples; second, the choice of 0.75 as the quantile is consistent with the experimental nature of the competition, where most competing systems are not engineered well enough to be ready for the market. This is in contrast with the choice made in the ISO/IEC 18305:2016 Standard (ISO/IEC 18305:2016(en), 2016), where the considered quantile is 0.95, which is appropriate for a well-engineered, ready-to-market system, while it is excessively severe with respect to the current state of the art in indoor localization systems.

The reason for using an additive error penalty proportional to the floor identification error is a compromise between simplicity and realism. An even simpler solution would have been to adopt a spherical error, one of the metrics considered in the ISO/IEC 18305:2016 Standard. However, this is not appropriate for common multifloor buildings, where the weight of a Euclidean error is much more significant vertically than horizontally. An accurate, but much more complex solution, involves disposing altogether of the Euclidean distance and computing on a map the length of the path from the real point to the wrongly estimated point, as discussed in Mendoza-Silva et al. (2017). This solution will be considered for use in the future IPIN competitions.

5 IPIN Competing Systems

The IPIN competition is aimed at bringing together academic and industrial research communities for evaluating different approaches and envisioning new research opportunities in the indoor localization arena, where no accepted standards do yet exist. In this section, we introduce an overview of several real-time competing system, focusing our discussion on the different choices and technologies implemented by the competitors.

Along the various editions, many systems and techniques have been proposed. However, some common characteristics can be observed. We divide these similarities in two main categories: (i) raw data processing and (ii) filtering/data fusion strategy. Creating a system able to work in a real-world scenario is a big challenge and it involves several different parts, including inertial sensors for step detection purpose, map matching information, Wi-Fi and magnetic field data collection, and compass data processing. Produced data are then fused to output a series of estimated position coordinates.

Table 3 reports a selection of different real-time systems that participated in IPIN competitions. Both smartphone-based systems (Track 1) and pedestrian dead reckoning systems (Track 2) are listed, in order to highlight common modules and fusion strategies used. All systems in Table 3 were able to complete the whole path during the real-time competition. The two main fusion strategies chosen are Particle Filter and Kalman Filter,

Table 3 Some Representative Competing Systems Along Different IPIN Competitions

Edition	Competitor	Tracks	Raw-Data Modules	Fusion Strategy
2014	Kailos (Han et al., 2014)	1	Map, Wi-Fi, PDR	Hidden Markov model
2014	Hubilon (Park, 2014)	1	Map, Wi-Fi, PDR	Particle filter
2014	Spirit (Berkovich, 2014)	1	Map, Wi-Fi, PDR, Magnetometer	Particle filter
2015	MMSS (Li et al., 2015)	1	Map, Wi-Fi, PDR, Magnetometer	Kalman filter
2015	NESL (Ju et al., 2014)	2	PDR	Kalman filter
2016	Navindoor Fetzer et al. (2016)	1	Map, Wi-Fi, PDR, Magnetometer	Particle filter
2016	WiMag (Guo et al., 2016)	1	Map, Wi-Fi, PDR, Magnetometer	Particle filter
2016	NESL (Ju et al., 2014)	2	Magnetometer, Barometer, Gyroscope	Zero velocity update
2016	Sysnav (Chesneau et al., 2016)	2	PDR	Extended Kalman filter
2017	NESL (Ju et al., 2014)	1	Map, PDR, Magnetometer	Extended Kalman filter
2017	MCL[a]	1	Map, Wi-Fi, Magnetometer	–
2017	Magneto (Chesneau et al., 2017)	2	PDR, Magnetometer	Extended Kalman filter

[a]See http://evaal.aaloa.org/2017/competitors.

Table 4 Competing Smartphone-Based Systems Which Are Not Reported a Final Result

Edition	Modules	Strategies
2014	Map, Wi-Fi, PDR	–
2015	Map, Imaging	–
2016	Map, Wi-Fi, PDR	Particle filter
2017	Map, Wi-Fi, PDR	No standard fusion strategy

while most raw data modules can be categorized as: Map Information, PDR (step detection and orientation), Magnetometer, and Wi-Fi.

Table 4 reports a selection of different smartphone-based systems, which are applied to the real-time competitions but that were not able to reach a final result. We observe that the best Track 1 (smartphone-based) systems use many raw-data modules and adopt a reliable and well-known fusion strategy.

5.1 An Overview on the Internals of Real-Time Systems

Fig. 1 shows a graphical simplified overview of how the raw-data modules interact with the fusion strategy to produce positioning estimates.

5.1.1 Raw-Data Modules

Some typical raw-data subsystems are here given an overview: pedestrian dead reckoning, orientation, Wi-Fi, magnetic field, and map matching.

Pedestrian dead reckoning is a relative positioning module useful to estimate the traveling distance and the users' direction. In general, this module is based on the use of a combination of three sensors: magnetometer, accelerometer, and gyroscope. Accelerometer is used to detect the step event, from which speed can be evaluated. Step

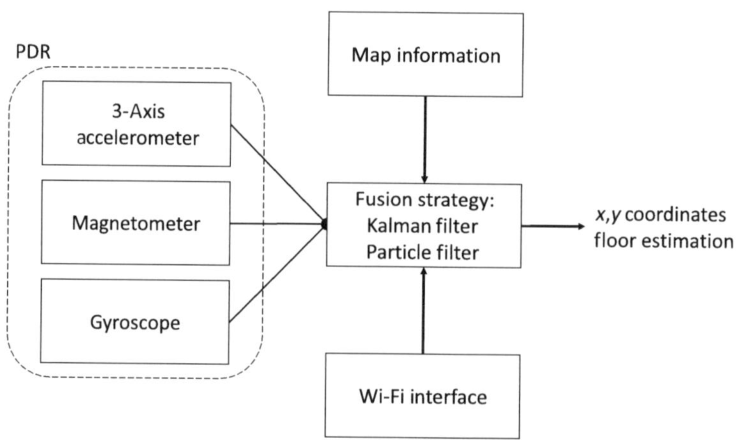

FIG. 1 A simplified overview of the interaction between raw-data modules and fusion strategy.

detection is implemented differently in the two main scenarios, that is, foot-mounted sensor and hand-held sensor. For the foot-mounted case, there is a phase when the foot is in contact with the floor for a fraction of a second, which is relatively simple to identify using a technique known as zero velocity update (ZUPT) (Foxlin, 2005). When the sensor is held in hand, as in the case of a smartphone, a spectral analysis of acceleration is used to detect low frequencies of acceleration to identify steps.

The moving direction of a pedestrian can be evaluated considering the difference between the magnetic north and the direction of the smartphone. The user orientation can be estimated using magnetic and gyroscope sensor. These values have to be corrected when the deviation errors are accumulated, due to the quality of the sensors and the behavior of the user. The main problem of these modules is the a priori knowledge of the absolute initial position. Otherwise, the usage of magnetometer and gyroscope can only produce relative position coordinates.

Wi-Fi scanning can be used for fingerprinting or for range-based methods. Both are based on RSSI measurements. Fingerprinting is based on a priori knowledge of a fingerprint database built during a site survey phase. During the positioning phase the Wi-Fi fingerprint module finds the vector of RSSI measurements for an unknown position, which is nearest to measurements stored in the database. Range-based methods use a combination of geometric techniques, essentially based on triangulation or multilateration and error minimization methods.

Similarly to Wi-Fi fingerprinting, a system can benefit from the magnetometer sensor implementing magnetic fingerprinting based on magnetic field vectors. The magnetometer of smartphones measures the magnetic field in the device coordinate system. As smartphones may be oriented arbitrarily in the user's hand, the measurements are transformed to horizontal coordinate system of the floor plan. The device orientation angles required for the transformation are estimated using the gravity vector coming from the accelerometer and the orientation coming from PDR. Magnetic fingerprinting is based on comparing the magnetic field vector measured in real-time in an unknown position with data in a fingerprint map that contains magnetic field data in known locations.

The last raw data source considered in this brief component description is the map information, being a fundamental information for navigation purpose. An efficient map matching algorithm allows to define a route of the user by matching the actual position into a building floor plan. Many improvements can be done in drawing the trajectory, for example, computing the possibility of a transition from a zone to another, to avoid crossing walls and closed door (Potortì and Palumbo, 2015; Palumbo and Barsocchi, 2014).

5.1.2 Fusion Strategies

Two different approaches are mainly used for fusing raw data: Kalman filter and particle filter.

The Kalman filter is a recursive Bayesian filter, which is optimal for Gaussian linear systems. Thanks to its easy implementation, it has been applied to many different fields for data fusion. It is an optimal estimator, assuming the initial uncertainty is Gaussian and the observation model and system dynamics are linear functions of the state. Because most systems are not strictly linear, researchers typically use the extended Kalman filter (EKF), which linearizes the system using a first-order Taylor series expansions. Kalman filter is the best option if the uncertainty in the state is not too high, which limits them to location tracking using either accurate sensors or sensors with high update rates.

A particle filter goes trough four steps, which are continuously repeated during its execution: cloud particles initialization, propagation, update or correction, and re-sampling. Initially a cloud of particles is generated in random places using a priori distribution probability assumptions. Subsequently, in the propagation stage, the coordinates and the heading of each particle are perturbed using a pedestrian motion model, using data from PDR. Other sources of information, such as magnetic data, map matching, and Wi-Fi, are then used to remove particles whose position is unlikely. Then new particles are generated based on the current distribution probability estimate to repopulate the particle cloud. Unlike Kalman filters, particle filters can converge to the true posterior even in non-Gaussian, nonlinear dynamic systems, at the price of much higher computation load (He and Chan, 2016).

6 Conclusion and Future Directions

As soon as research on indoor localization and tracking reached a sufficient number of interested research groups and industries, the need for common benchmarks has started to emerge. This need has been met by EvAAL first, then by the Microsoft and the IPIN competitions.

Comments of competitors were homogeneous during the EvAAL competitions first and during the IPIN ones next: they were impressed by the rigorous methodology used for the measurement and most said to have gained significant insight in the inner working and the potential of their own systems.

The IPIN competition is particularly interesting in that it concentrates on working systems in realistic situations, and provides realistic measures of what can be expected from a real-life system, which was shown to be significantly different from the generally optimistic figures that one can read in laboratory papers.

IPIN sessions dedicated to EvAAL have raised significant interest among IPIN attendees, especially in 2015, when a plenary session was dedicated to the competition and the general principles were illustrated.

Now that this research area approaches the market, IPIN competition will need to accompany the process and to grow by supporting modern localization systems, which exploit a variety of sensor data. For example, IPIN competitors until now have vastly ignored BLE beacons, while it is to be expected that future commercial systems will exploit their potential (Faragher and Harle, 2015; Palumbo et al., 2015; Barsocchi et al., 2017): future IPIN competitions will likely address this issue by encouraging use of BLE beacons as an important source of information in areas where little Wi-Fi coverage is available.

Another area where IPIN can experiment with new solutions is the use of a more useful metric for computing the positioning error, such as the one mentioned in Section 4.2 and presented in Mendoza-Silva et al. (2017). In general, the objective of IPIN is to define standard procedures for the evaluation of indoor localization systems, in an effort to improve over the recent ISO/IEC 18305 Standard (ISO/IEC 18305:2016(en), 2016). This effort is being coordinated by the newborn IPIN Indoor Positioning Indoor Navigation (IPIN) International Standards Committee (ISC), and should produce its first results by 2018.

The challenging nature of the IPIN competition is its most precious asset, and probably the main reason for the relatively small and constant number of competitors over the years (see Table 2). The competition tracks have adapted to technological advances since 2011, thus maintaining attractiveness, and we will keep tracking new developments. Unless confronted with significant technological or market changes, such that research interest shifts away from indoor localization, we would consider it a success to witness a similar participation in the future.

References

Barsocchi, P., Potortì, F., Furfari, F., Gil, A.M.M., 2012. Comparing AAL indoor localization systems—indoor localization and tracking. In: Chessa, S., Knauth, S. (Eds.), Evaluating AAL Systems Through Competitive Benchmarking, vol. 309. Communications in Computer and Information Science. Springer, Lecce (IT), pp. 1–13.

Barsocchi, P., Chessa, S., Furfari, F., Potortì, F., 2013. Evaluating AAL solutions through competitive benchmarking: the localization competition. IEEE Pervasive Comput. Mag. 12 (4), 72–79. ISSN 1536-1268. https://doi.org/10.1109/MPRV.2013.23.

Barsocchi, P., Crivello, A., Girolami, M., Mavilia, F., Palumbo, F., 2017. Occupancy detection by multi-power Bluetooth low energy beaconing. In: 2017 International Conference on Indoor Positioning and Indoor Navigation (IPIN), pp. 1–6.

Berkovich, G., 2014. Accurate and reliable real-time indoor positioning on commercial smartphones. In: 2014 International Conference on Indoor Positioning and Indoor Navigation (IPIN), pp. 670–677.

Chesneau, C.I., Hillion, M., Prieur, C., 2016. Motion estimation of a rigid body with an EKF using magneto-inertial measurements. In: 2016 International Conference on Indoor Positioning and Indoor Navigation (IPIN), pp. 1–6.

Chesneau, C.I., Hillion, M., Hullo, J.F., Thibault, G., Prieur, C., 2017. Improving magneto-inertial attitude and position estimation by means of a magnetic heading observer. In: 2017 International Conference on Indoor Positioning and Indoor Navigation (IPIN), pp. 1–8.

Faragher, R., Harle, R., 2015. Location fingerprinting with Bluetooth low energy beacons. IEEE J. Sel. Areas Commun. 33 (11), 2418–2428.

Fetzer, T., Ebner, F., Deinzer, F., Köping, L., Grzegorzek, M., 2016. On Monte Carlo smoothing in multi sensor indoor localisation. In: 2016 International Conference on Indoor Positioning and Indoor Navigation (IPIN), pp. 1–8.

Foxlin, E., 2005. Pedestrian tracking with shoe-mounted inertial sensors. IEEE Comput. Graph. Appl. 25 (6), 38–46.

Guo, X., Shao, W., Zhao, F., Wang, Q., Li, D., Luo, H., 2016. WiMag: multimode fusion localization system based on magnetic/WiFi/PDR. In: 2016 International Conference on Indoor Positioning and Indoor Navigation (IPIN), pp. 1–8.

Han, D., Lee, S., Kim, S., 2014. KAILOS: KAIST indoor locating system. In: 2014 International Conference on Indoor Positioning and Indoor Navigation (IPIN), pp. 615–619.

He, S., Chan, S.-H.G., 2016. Wi-Fi fingerprint-based indoor positioning: recent advances and comparisons. IEEE Commun. Surv. Tutorials 18 (1), 466–490.

ISO/IEC 18305:2016(en), 2016. Information Technology—Real Time Locating Systems—Test and Evaluation of Localization and Tracking Systems. Technical Committee ISO/IEC JTC 1/SC 31—Automatic Identification and Data Capture Techniques, Geneva (CH).

Ju, H.J., Lee, M.S., Park, C.G., Lee, S., Park, S., 2014. Advanced heuristic drift elimination for indoor pedestrian navigation. In: 2014 International Conference on Indoor Positioning and Indoor Navigation (IPIN), pp. 729–732.

Li, Y., Zhang, P., Niu, X., Zhuang, Y., Lan, H., El-Sheimy, N., 2015. Real-time indoor navigation using smartphone sensors. In: 2015 International Conference on Indoor Positioning and Indoor Navigation (IPIN), pp. 1–10.

Lymberopoulos, D., Liu, J., Yang, X., Choudhury, R.R., Sen, S., Handziski, V., 2015. Microsoft indoor localization competition: experiences and lessons learned. GetMobile Mob. Comput. Commun. 18 (4), 24–31.

Mathisen, A., Sørensen, S.K., Stisen, A., Blunck, H., Grønbæk, K., 2016. A comparative analysis of indoor WiFi positioning at a large building complex. In: 2016 International Conference on Indoor Positioning and Indoor Navigation (IPIN), pp. 1–8.

Mendoza-Silva, G.M., Torres-Sospedra, J., Huerta, J., 2017. A more realistic error distance calculation for indoor positioning systems accuracy evaluation. In: 2017 International Conference on Indoor Positioning and Indoor Navigation (IPIN), pp. 1–8.

Palumbo, F., Barsocchi, P., 2014. Salt: source-agnostic localization technique based on context data from binary sensor networks. In: European Conference on Ambient Intelligence, pp. 17–32.

Palumbo, F., Barsocchi, P., Chessa, S., Augusto, J.C., 2015. A stigmergic approach to indoor localization using Bluetooth low energy beacons. In: 2015 12th IEEE International Conference on Advanced Video and Signal Based Surveillance (AVSS), pp. 1–6.

Park, Y., 2014. Smartphone based hybrid localization method to improve an accuracy on indoor navigation. In: 2014 International Conference on Indoor Positioning and Indoor Navigation (IPIN), pp. 705–708.

Potortì, F., Palumbo, F., 2015. CEO: a context event only indoor localization technique for AAL. J. Ambient Intell. Smart Environ. 7 (6), 745–760.

Potortì, F., Barsocchi, P., Girolami, M., Torres-Sospedra, J., Montoliu, R., 2015. Evaluating indoor localization solutions in large environments through competitive benchmarking: the EvAAL-ETRI competition. In: Proceedings of the International Conference on Indoor Positioning and Indoor Navigation (IPIN). Available from: http://ieeexplore.ieee.org/xpls/abs_all.jsp?arnumber=7346970.

Potortì, F., Park, S., Jiménez Ruiz, A.R., Barsocchi, P., Girolami, M., Crivello, A., Lee, S.Y., Lim, J.H., Torres-Sospedra, J., Seco, F., Montoliu, R., Mendoza-Silva, G.M., Pérez Rubio, M.D.C., Losada-Gutiérrez, C., Espinosa, F., Macias-Guarasa, J., 2017. Comparing the performance of indoor localization systems through the EVAAL framework. Sensors 17 (10). ISSN 1424-8220. https://doi.org/10.3390/s17102327.

Ram, R., Furfari, F., Girolami, M., Iba nez-Sánchez, G., Lázaro-Ramos, J.P., Mayer, C., Prazak-Aram, B., Zentek, T., 2013. UniversAAL: provisioning platform for AAL services. In: Ambient Intelligence-Software and Applications. Springer, pp. 105–112.

Schauer, L., Marcus, P., Linnhoff-Popien, C., 2016. Towards feasible Wi-Fi based indoor tracking systems using probabilistic methods. In: 2016 International Conference on Indoor Positioning and Indoor Navigation (IPIN), pp. 1–8.

12

IndoorLoc Platform: A Web Tool to Support the Comparison of Indoor Positioning Systems

Raul Montoliu, Emilio Sansano-Sansano, Joaquín Torres-Sospedra, Óscar Belmonte-Fernández

INSTITUTE OF NEW IMAGING TECHNOLOGIES, JAUME I UNIVERSITY, CASTELLÓN, SPAIN

1 Introduction

Geolocation systems have been present during decades allowing navigation services that guide users by car, foot or bicycle, and many others such as evacuation services and social network services. The Global Navigation Satellite Systems (GNSS) are able to provide these services in outdoor environments, but in most of the situations people spend a significant portion of their time in indoor environments such as offices, undergrounds, shopping malls, airports, etc., where these satellite-based positioning systems do not work. This is one of the reasons why the development of new indoor positioning and navigation systems has attracted the attention of many researchers in the last years.

This research effort has achieved the development of many different indoor positioning technologies, being the ones based on Received Signal Strength Indicator (RSSI) fingerprinting (He and Chan, 2016; Wu et al., 2013; Han et al., 2014) among the most popular. This technique is based on the measurement of the intensity of the received radio signals of the emitting devices (beacons) that are available at a particular place, and on the comparison of this measurement with a previously built RSSI dataset (also known as radio map). In this scenario, a fingerprint is an RSSI feature vector composed of received signal values from different emitting devices or beacons, associated with a precise position. The similarity of the received signals (fingerprint) with some of the stored fingerprints can be used to guess the approximate position on the subject. This technique is becoming increasingly important for indoor localization, since Wi-Fi is generally available in indoor environments where GPS signals cannot penetrate, and the wireless access points (WAPs) can be used as opportunistic beacons. Other types of indoor localization beacons (Bluetooth, RFID, etc.) can also be used in conjunction with Wi-Fi access points or as a standalone positioning system.

Geographical and Fingerprinting Data for Positioning and Navigation Systems. https://doi.org/10.1016/B978-0-12-813189-3.00012-5

Many different approaches have been object of research and many papers have been published trying to solve this indoor localization problem. However, it is very difficult to compare results from different approaches, since every research presents its estimated results using its own experimental setup and measures, and it is very difficult to reproduce the particularities of every single experiment. In the *Pattern Recognition and Machine Learning* research fields, the common practice is to test the results of each proposal using several well-known datasets (García et al., 2009). This allows researchers to fairly compare different methodologies in the literature. For instance, the UCI Machine Learning Repository (Lichman, 2013) and the web *Kaggle* (Goldbloom et al., 2017) are two well-known examples in this sense. However, in the fingerprint-based indoor localization research field, there is a limited number of such kind of databases (Nahrstedt and Vu, 2012; Torres-Sospedra et al., 2014, 2015b; Talvitie et al., 2014; Barsocchi et al., 2016; Moayeri et al., 2016).

Many different approaches have been object of research and many papers have been published trying to solve this indoor localization problem. However, it is very difficult to compare results from different approaches, since every research presents its estimated results using its own experimental setup and measures, and it is very difficult to reproduce the particularities of every single experiment.

This chapter consists of an introduction to the *IndoorLoc Platform*,[1] a web tool to support the comparison and the evaluation of indoor positioning algorithms. The platform is a centralized website where researchers can do the following actions:

1. Access to a public repository of datasets for RSSI fingerprinting.
2. Upload indoor positioning estimations on experimental setups included in the platform.
3. Include the estimation results in a ranking.
4. Analyze positioning methods.
5. Interact with the platform in a user-friendly environment to test the algorithms and datasets included.

In order to show a real example of the platform usage, this chapter also presents a comparative study of the performance of two fingerprinting-based indoor localization methods included in the platform when using four of the datasets also included in the platform. All the experiments presented are easily reproducible using the tools included in the platform.

The two methods shown differ in the methodology used to solve the indoor localization problem. They are a deterministic-based and a probabilistic-based method. The four datasets differ in the type of scenario where data has been captured, as for instance: the number of samples, the size of the scenario, the density of the samples, etc.

[1] http://indoorlocplatform.uji.es.

A preliminary version of this chapter was published (as a conference paper) in Montoliu et al. (2017). This chapter provides additional details of the *IndoorLoc Platform*, to help the reader to be aware of the different possibilities of the proposed web platform.

The rest of the chapter is organized as follows. Section 2 reviews related work. Section 3 describes the main sections in which the platform is divided. Sections 4 and 5 explain the datasets and methods, respectively, included in the platform. Section 6 presents a set of experiments performed using the algorithms and datasets included in the proposed platform. Section 7 describes a real case of use of the usage of the platform during a fingerprinting-based indoor positioning course. Finally, the most important conclusions arisen from this work are presented in Section 8.

2 Related Work

As it has been said in Section 1, most of the indoor positioning methods found in the literature present the experiments using their own experimental setup. A second related problem is that those datasets are not made available to the research community, making it impossible to reproduce the presented results. Both issues make a fair comparison of localization methods developed by different groups not feasible in a rigorous manner, since scenarios may change in an uncontrolled way.

A better way to compare positioning algorithms is to use the same experimental setup, and for that purpose, the use of a repository of prerecorded data in a large variety of buildings and contexts can be very useful. Some good examples of data repositories in the machine learning community are the UCI Machine Learning Repository (Lichman, 2013) and *Kaggle* (Goldbloom et al., 2017), both created for evaluating machine learning algorithms with common databases.

Another alternative are competitions where several research groups should prepare their methods to obtain the best results using a common experimental setup, or even the same prerecorded data. Some examples of competition are: Microsoft-IPSN (Lymberopoulos et al., 2014, 2015, 2016, 2017), EvAAL (Potortì et al., 2015), and EVARILOS (Lemic et al., 2015). The first off-site indoor location competition was the third track of the EvAAL-ETRI Indoor Location competition (Torres-Sospedra et al., 2017b), called *Wi-Fi fingerprinting in large environments*, which was held during the Sixth International Conference on Indoor Positioning and Indoor Navigation (IPIN'15). In this event, the competitors had access to the *UJIIndoorLoc* (Torres-Sospedra et al., 2014) dataset, that has been included in the proposed platform. A similar competition was held in the Seventh International Conference on Indoor Positioning and Indoor Navigation (IPIN'16) (Torres-Sospedra et al., 2017a), where the dataset used was more challenging, since data provided by all sensors embedded in typical smartphones was included, acquired by different people moving in different types of buildings.

One of the main problems of such competitions is that when they finish, researchers cannot continue improving their methods. In addition, the different datasets are located in different web pages. The proposed web platform is focused on providing a common place for researchers to access to fingerprint-related datasets. Another of the main objectives is to provide a continuous competition without deadlines. Therefore, researchers will not have time restrictions to test their methods and submit their results to the platform.

The most similar work to the *IndoorLoc Platform* is Lemic et al. (2015), where the authors presented a web platform for evaluation of RF-based indoor localization algorithms with two core services: one focused on the storage of raw data and the other focused on automated calculation of metrics for performance assessment. They also include an SDK for convenient access to the platform from MATLAB and python. The two first characteristics are included in the proposed web platform. The SDK is not needed in our case, since users can directly interact with the web platform to upload their results. The main differences of the proposed web platform with respect to Lemic et al. (2015) are as follows:

1. It is more focused on fingerprinting methods.
2. It also includes a dashboard section where researchers can make experiments using the methods and datasets included in the platform in a user-friendly environment.
3. There is a ranking section where researchers can check the accuracy of their method against the methods of other researchers. In addition, the proposed web platform has been designed in order to easily upload new methods and datasets.

The *IndoorLoc Platform* has been designed with a state-of-art visual style and with a user-centered interface making the access to all the elements of the platform very intuitive. For instance, the home web page (see Fig. 1) directly presents the main sections of the platform. Another example is that users can download a dataset or upload a result with just a few mouse clicks. The platform is also responsive, and will automatically adapt to the device screen used to access it.

In addition, the platform has a high formative component, because even a user without programming knowledge can interact with the algorithms and datasets included. Although, it will be the users with a high programming skill who will be able to get a better advantage of the platform because, probably, they will be able to improve the results that can be obtained with the algorithms included in the platform.

3 Overview of the Platform

Fig. 1 shows the homepage of the platform. The homepage displays a summary of the contents of the four main sections in which the platform is structured, while the menu in the upper section of the platform allows access to each one of these sections.

The four main sections of the platform are as follows:

- **Datasets**: This section is a repository of several datasets stored in the platform. These datasets are available to download so users can use them in their own experiments.

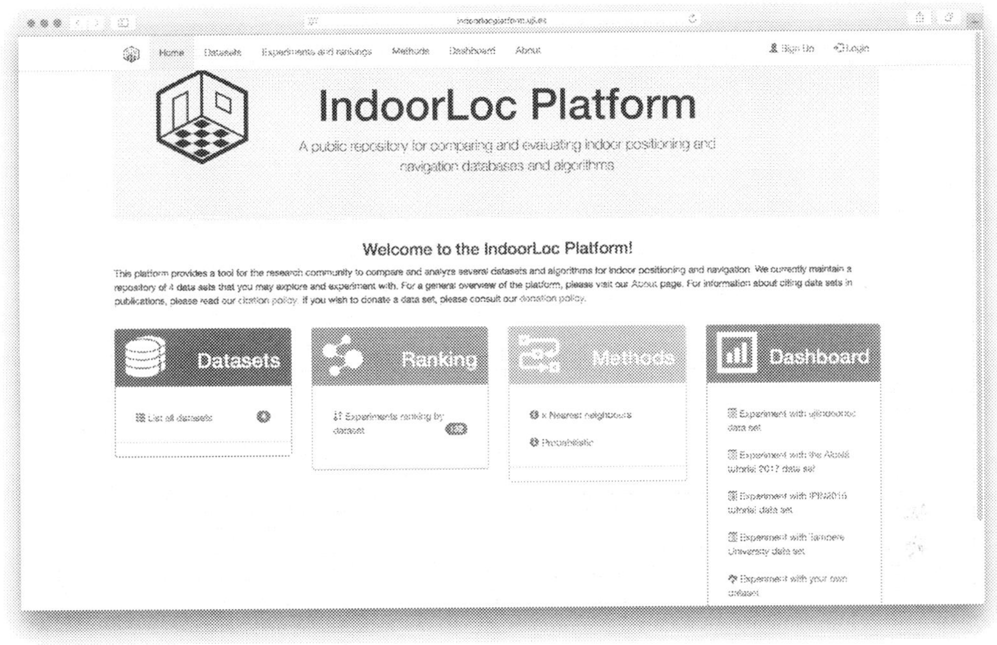

FIG. 1 Homepage of the IndoorLoc platform.

- **Ranking**: In this section, users can upload the results of their own algorithms to obtain an estimation of the accuracy of their methods when using the datasets included in the platform. In addition, the results can be included in the ranking, where the best results of each dataset are showed sorted by accuracy.
- **Methods**: This section presents a set of well-known algorithms so users can study their implementation.
- **Dashboard**: In this section, users can test the algorithms included in the platform, using some of the datasets included, in a user-friendly environment.

These sections are briefly described in the next subsections.

3.1 Datasets

Fig. 2 shows the datasets section of the platform. The *Datasets* section displays the basic information about all the datasets included in the platform. In addition, the links to download all the files related to each dataset are also included.

Each dataset can be composed of up to four files:

- **Dataset info**: A *pdf* file with information and features about the dataset. The description includes the name of the donors, the contact information, general

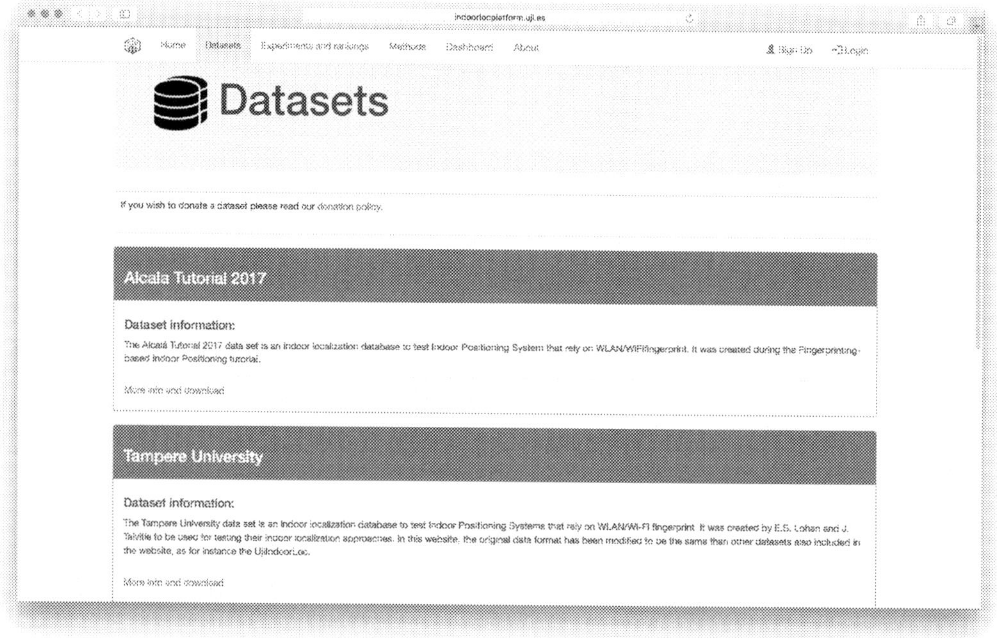

FIG. 2 Datasets section of the IndoorLoc platform.

information about the dataset, a description of the files included, the attributes description, the format of the result file, and the citation request.

- **Training set**: A file with the samples to be used to train the localization models. It includes the localization of the samples.
- **Validation set**: This file is similar to the training set file, and also includes the localization of the samples. Should be used to assess the performance of the localization model created using the training set data.
- **Test set**: This file is also used to assess the performance of the localization model, but does not include the actual localization of the samples, only its fingerprint. To obtain an estimation of the accuracy of the model, users can run their methods to obtain an estimation of the localization of the samples of the test set, and then upload their results to the platform to get an evaluation of the performance of the model. The true localization of the samples is stored in the platform.

The training, validation, and test files have a *comma-separated values* (CSV) file format. The three first files (info, training, and validation) are accessible to everyone. Only registered users are allowed to access the test file. Not all the datasets included in the platform have a validation set. In that case, users can use techniques as cross-validation (Bishop, 2006) to assess the performance of the localization model generated.

At the moment of writing this chapter, there are six different databases included in the platform. Four of them are related to the Wi-Fi fingerprinting indoor localization problem. They are briefly described in Section 4.

Registered users can upload their own datasets following the instructions provided by the platform. Before to be definitively added to the platform, each new dataset is rigorously examined by the administrators of the platform to ensure that it has the required quality.

3.2 Ranking

One of the main objectives of the *IndoorLoc* web platform is to provide a tool to the indoor localization community to compare their methods using well-known datasets. This section has been devoted to this purpose. For each dataset, a list of the best methods, according to a figure of merit, is shown.

Registered users are allowed to upload the results of their methods following the instructions included in the description of the dataset (see Section 3.1). Once the results file has been uploaded, the platform calculates the figure of merit for this dataset using the estimated locations provided by the user and the ground truth internally (and privately) stored in the platform. After the figure of merit is calculated and displayed, users have the choice of including or not the result in the ranking.

Each entry in the ranking has a description field, provided by the user, showing info about the experiment performed to obtain such result, e.g., the parameters used or the algorithm details.

Fig. 3 shows the ranking page for the *IPIN2016 Tutorial* dataset. At the moment of writing this text, the ranking is composed by two experiments performed by the same user. According to the notes written by the user, the result of the leader was obtained using the probabilistic method and the one in the second position using a knn algorithm.

Fig. 4 shows the ranking for the *UJIIndoorLoc* dataset. In this case, the four best results obtained in the third track of the EvAAL-ETRI Indoor Location competition (Torres-Sospedra et al., 2017b), where this dataset was used, have been manually introduced by the web creators to give a baseline reference.

3.3 Methods

This section shows some basic information about the methods included in the platform. This information consists on the explanation of the method though R[2] source code using comprehensible examples. In addition, links to the Dashboard section, where users can test these methods, are also included Fig. 5 shows the method webpage.

At the moment of writing this text, two methods have been included in the platform: deterministic-based (Bahl and Padmanabhan, 2000) and probabilistic-based (Youssef and Agrawala, 2005). They are described in Section 5.

[2]https://www.r-project.org/.

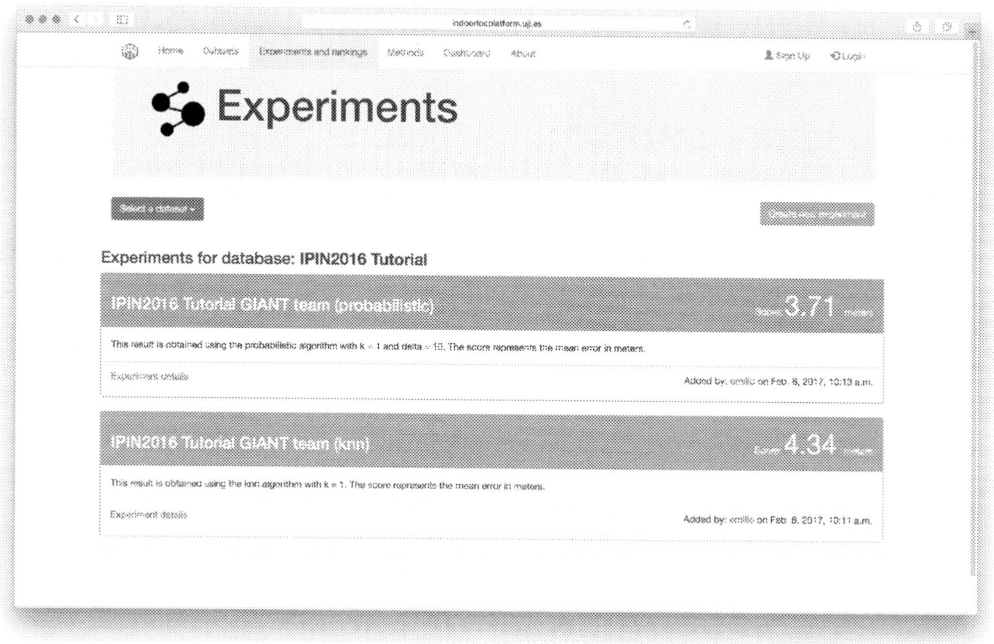

FIG. 3 Ranking webpage of the *IPIN2016 Tutorial* dataset. Two experiments have been included in the ranking, the first one (according to the notes written by the contributor) using a probabilistic-based algorithm and the second one using a knn-based method.

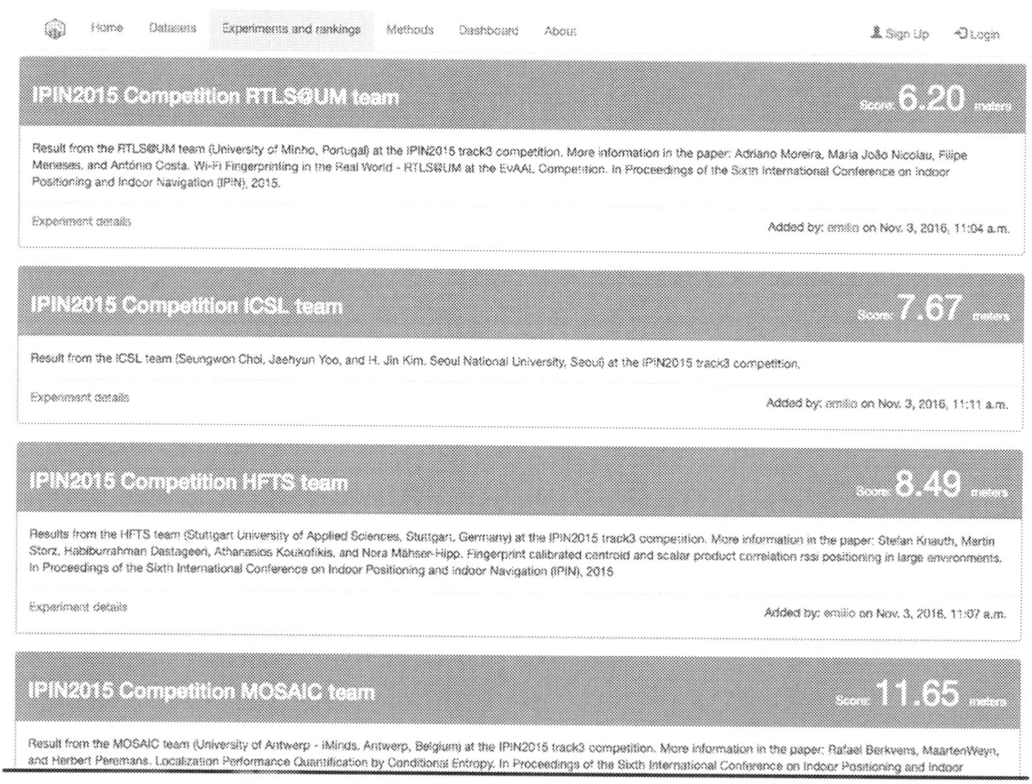

FIG. 4 Ranking webpage of the *UJIIndoorLoc* dataset.

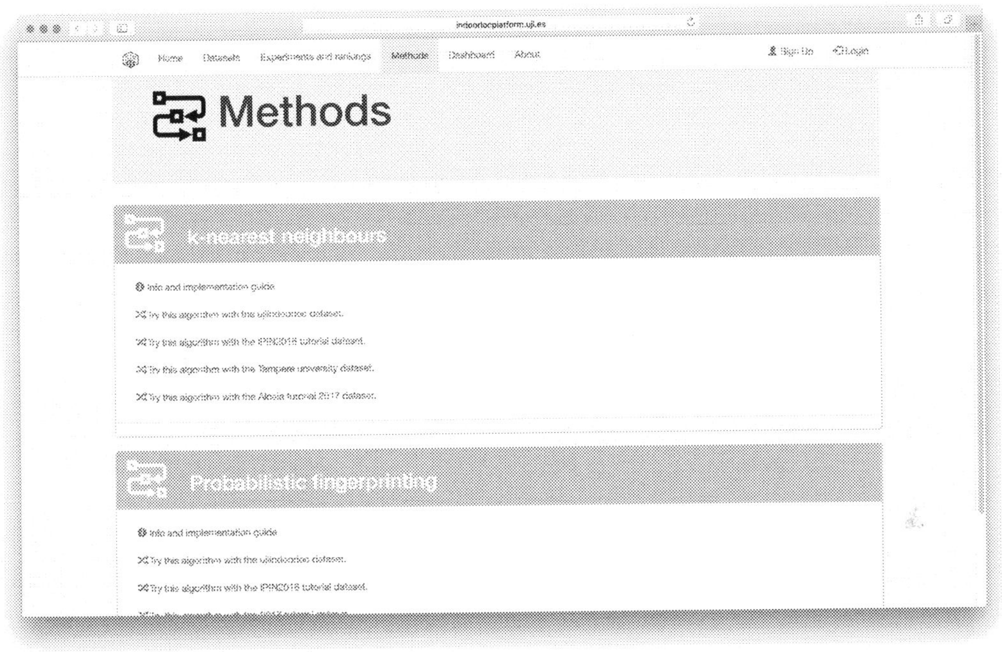

FIG. 5 Methods section of the IndoorLoc platform.

To add new methods to the platform, users must contact with platform administrators. Similarly to the dataset case, each new method is rigorously examined by the administrators of the platform to ensure that it has the required quality.

3.4 Dashboard

In the Dashboard section, users can test the methods included in the platform, using the datasets also included in the platform, in a user-friendly interface. Fig. 6 shows an example of a dashboard for the *UJIIndoorLoc* dataset. In particular, the user selected the building 0, the floor 1, the validation set (to estimate the locations), and the deterministic method. After clicking on the *Estimate error* button, the platform internally estimates the location of the validation samples and calculates some statistics, as the mean and the median of the estimation error. It also shows one figure with the error histogram and density, and another figure with the empirical cumulative density function.

Registered users are allowed to use their own dataset using the methods included in the platform. This dataset must be formatted using a set of rules specified in the web platform.

The platform can be easily improved adding more measurements to estimate the methods performance and with other kind of figures, thanks to the R Shiny environment.

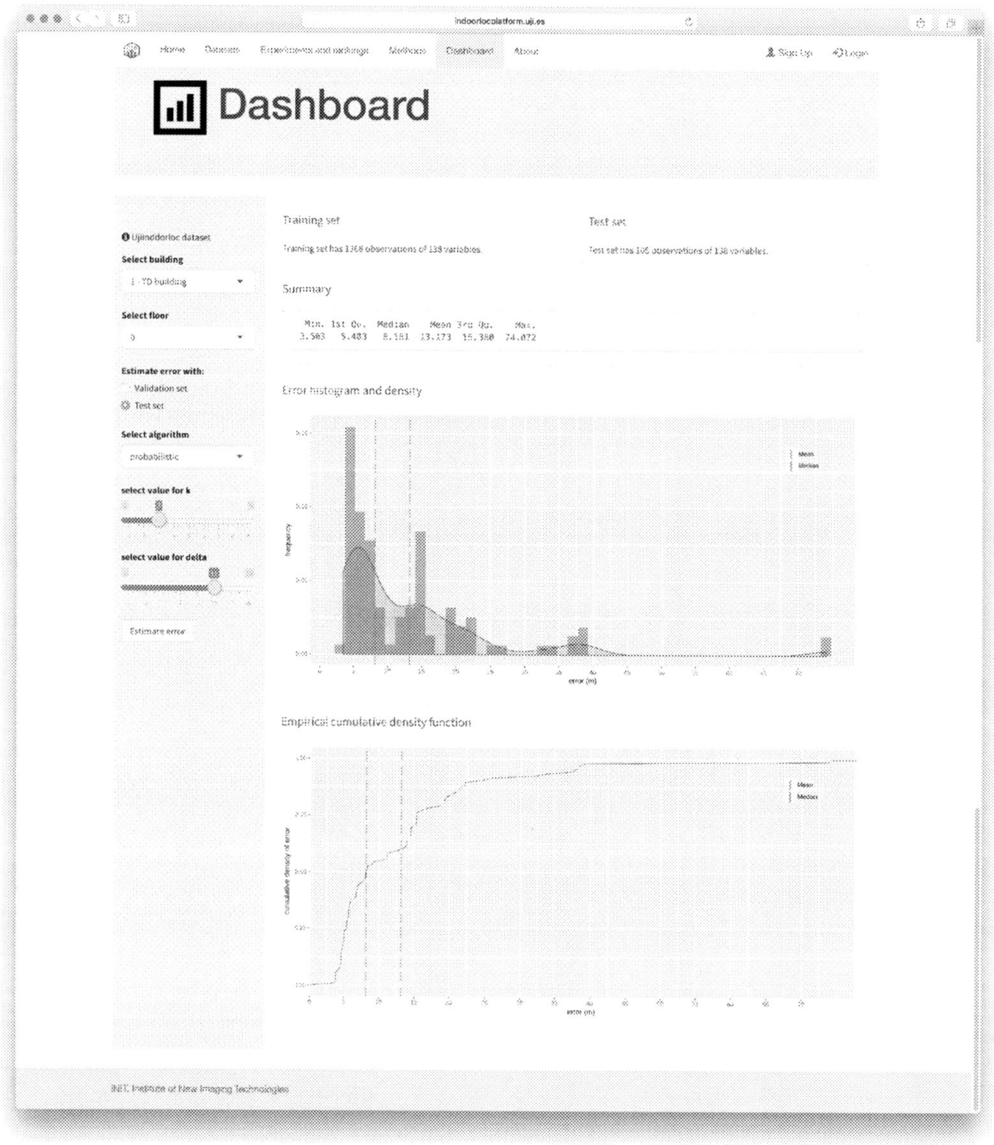

FIG. 6 Example of the *Dashboard* section of the platform.

3.5 Implementation Details

The platform has been implemented using these open source tools:

- **Django**[3]: A Python web framework to build the web application. Django follows the model-view-template (MVT) architectural pattern, and allows rapid development of database-driven websites.
- **Shiny**[4]: An R package that eases the building of interactive web *apps* using R code. The Shiny server hosts the *apps* on the platform that run embedded in the dashboard page.
- **RMarkdown**[5]: Documents created with the R Markdown technology are fully reproducible, and use a notebook interface to weave together text and code to produce elegantly formatted output. R Markdown uses multiple languages including R, Python, and SQL.
- **ipft R package**[6]: This R package includes algorithms and utility functions for indoor positioning using fingerprinting techniques. These functions are designed for manipulation of RSSI datasets, estimation of positions, comparison of the performance of different models, and graphical visualization of data.
- **Apache**[7]: Open source HTTP web server that works on Unix-like systems (BSD, GNU/Linux, etc.), Microsoft Windows, Macintosh, and other platforms, and provides HTTP services in sync with the current HTTP standards.

4 Datasets Included in the Platform

At the moment of writing this chapter, the platform includes six different datasets, four of them dedicated to the Wi-Fi fingerprinting problem. They are briefly introduced in the next sections.

4.1 Wi-Fi-Based Datasets

In the four Wi-Fi-based datasets, Wi-Fi fingerprints are characterized by the detected WAPs and the corresponding RSSI. The intensity values are represented as negative integer values near to −100 dBm (extremely poor signal) to 0 dBm. The positive value 100 is used to denote when a WAP was not detected. Tables 1–4 show a summary of the main characteristics of each dataset.

[3]https://www.djangoproject.com/.
[4]https://shiny.rstudio.com/.
[5]https://rmarkdown.rstudio.com/.
[6]https://cran.r-project.org/web/packages/ipft/.
[7]https://httpd.apache.org/.

Table 1 Main Characteristic of the *UJIIndoorLoc* Dataset

Number of buildings	3
Number of floors	4–5
Number of WAPs	520
Number of training samples	19,937
Number of validation samples	1111
Number of test samples	4900

Table 2 Main Characteristic of the *IPIN2016 Tutorial* Dataset

Number of buildings	1
Number of floors	1
Number of WAPs	168
Number of training samples	927
Number of validation samples	0
Number of test samples	702

Table 3 Main Characteristic of the *Tampere* Dataset

Number of buildings	2
Number of floors	3–4
Number of WAPs	390–354
Number of training samples	1478–583
Number of validation samples	0
Number of test samples	589–175

Table 4 Main Characteristic of the *Alcalá2017 Tutorial* Dataset

Number of buildings	1
Number of floors	1
Number of WAPs	152
Number of training samples	670
Number of validation samples	0
Number of test samples	405

4.1.1 UJIIndoorLoc

The UJIIndoorLoc (Torres-Sospedra et al., 2014) database covers three buildings of Universitat Jaume I[8] (Spain), with four or more floors and an area of almost 110,000 m^2. It can be used for classification, e.g., actual building and floor identification, or regression, e.g., actual longitude and latitude estimation. It was created in 2013 and 2014 by means of more than 20 different users and 25 Android devices. The database consists of 19,937 training/reference records and 1111 validation records. There is also a test file where the ground truth is not accessible.

The 529 attributes contain the Wi-Fi fingerprint, the coordinates (latitude, longitude, and floor) and Building ID, and other useful information such as the particular space (offices, labs, etc.) and the relative position (inside/outside the space) where the capture was taken, information about who (user), how (android device and version) and when (timestamp) Wi-Fi capture was taken, among other information. During the database creation, 520 different WAPs were detected. Thus, the Wi-Fi fingerprint is composed of 520 intensity values.

This dataset was used in the off-site track of the EvAAL-ETRI Indoor Localization Competition which was part of the Sixth International Conference on Indoor Positioning and Indoor Navigation (IPIN'15) (Torres-Sospedra et al., 2017b). The best results obtained in the competition have been included in the platform in the corresponding ranking.

Since the particular implementation of the localization methods included in the platform assumes that all the samples are in the same building and floor, the complete dataset has been divided into 11 different datasets.

4.1.2 IPIN2016 Tutorial

As an alternative of the *UJIIndoorLoc* dataset, the *IPIN2016 Tutorial* dataset is focused on the study of a small scenario. In particular, it covers a corridor of the School of Engineering of the University of Alcalá[9] (Spain). It is the place where a tutorial on Wi-Fi fingerprinting was held during the IPIN2016 conference. The database consists of 927 training/reference records and 702 test ones. The 177 attributes contain the Wi-Fi fingerprint (168 WAPs), the coordinates where it was taken, and other useful information.

4.1.3 Tampere University

This database (Cramariuc and Lohan, 2016) covers two building of the Tampere University of technology[10] (Finland), with four and three floors, respectively. In the first building, there are 1478 training/reference records and 489 test ones. The 312 attributes contain the Wi-Fi fingerprint (309 WAPs) and the coordinates (longitude, latitude, and height). In the second building, there are 583 training/reference records and 175 test ones. The 357 attributes contain the Wi-Fi fingerprint (354 WAPs) and the coordinates (longitude,

[8] http://www.uji.es.

[9] https://www.uah.es/es/.

[10] http://www.tut.fi/.

latitude, and height). An important difference of this dataset, with respect the *UJIIndoor-Loc*, is that in the former there is just one sample in each training location, while in the latter the number of samples is between 10 and 30.

Data from the two buildings can be considered as two separate datasets, with no relationship between respective WAP labels and real access points MAC addresses, meaning that two columns with the same WAP name in either dataset may be assigned to different access points.

Similarly to the *UJIIndoorLoc* this dataset has been divided into seven different datasets.

4.1.4 ALCALA2017 Tutorial

This dataset was created during the 2017 Fingerprinting-based Indoor Positioning tutorial held in the School of Engineering of the University of Alcalá. Data was acquired in the same corridor than the *IPIN2016 Tutorial* dataset. The main differences between both datasets are: (1) a thinner grid was used to capture training data; (2) some users made mistakes labeling the training fingerprints. These errors have not been eliminated since it is a situation that can occur in a real scenario. Users should take into account this situation in their methods.

The database consists of 670 training/reference records and 405 test ones. The 154 attributes contain the Wi-Fi fingerprint (152 WAPs) and the coordinates where it was taken.

4.2 AmbiLoc Dataset

The *AmbiLoc* dataset (Popleteev, 2017) is a collection of ambient radio fingerprints, collected in multiple predefined locations across several testbeds. Instead of Wi-Fi signals, the *AmbiLoc* datasets deals with ambient signals of opportunity, such as those from broadcasting TV and FM radio stations or GSM networks, that are almost always present on most indoor locations. This dataset has been collected in multiple testbeds, including large-scale and multifloor buildings, over the course of one year.

The platform provides some basic information of this dataset and a link to the original source of the data.[11]

4.3 magPIE Dataset

Magnetic field-based indoor positioning (Li et al., 2012, 2013; Montoliu et al., 2016) is an infrastructure-less approach which is based on the uniqueness of the disturbances in the magnetic field produced by the structural elements present in a scenario. The uniqueness of the disturbances can be used as a fingerprint, since it is stable over time.

MagPIE is a publicly available dataset for the evaluation of indoor positioning algorithms that use magnetic anomalies (Hanley et al., 2017). This dataset contains IMU and magnetometer measurements along with ground truth position measurements with

[11] http://ambiloc.org/.

centimeter-level accuracy. To produce this dataset, the authors collected over 13 hours of data from three different buildings, with sensors both handheld and mounted on a wheeled robot, in environments with and without changes in the placement of objects that affect magnetometer measurements.

The platform provides some basic information of this dataset and a link to the original source of the data.[12]

5 Methods Included in the Platform

Two different approaches are considered here for the fingerprinting-based location process: a deterministic, or nonparametric method; and a statistical, or parametric method. In the first, no statistical behavior is assumed, and the location problem is solved according to a set of observations whose positions are known; while the second method makes explicit use of distributions and statistical parameters of the data stored in the radio map to optimize the probabilities in the assignment of the estimated position.

5.1 Deterministic-Based Approach

The deterministic approach (Bahl and Padmanabhan, 2000; Marques et al., 2012; Torres-Sospedra et al., 2015a) relies on the well known k-Nearest Neighbors algorithm (*knn*) (Cover and Hart, 1967) to, given an RSSI vector, select the k more similar training examples from the radio map. The similarity between the RSSI value vectors can be determined, for example, as the *Euclidean* distance between them, but other distance functions can be used instead (Torres-Sospedra et al., 2015a). Once the k neighbors are selected, the method estimates the location of the user by calculating the weighted average of the neighbor's positions.

Fig. 7 shows a possible R source code of this method (with $k = 3$). The R dataframe *training.set* contains the RSSI values and the localization of the training points. The last two columns are the longitude (column name LONG) and latitude (column name LAT) of those points. *The validation.set* dataframe has the same structure. The complete description can be found at: http://indoorlocplatform.uji.es/methods/knn/.

5.2 Probabilistic-Based Approach

Given the limitations of sensors accuracy and the complex character of signal propagation, the RSSI vector stored for a particular position cannot have completely reliable and accurate information about the emitters signal strength. This uncertainty has been usually modeled by a normal distribution (Haeberlen et al., 2004), therefore many readings of the signals at the same position are needed to obtain a representative set of statistical parameters to model each RSSI present at that position. The more measurements for a particular location, the more reliable will be their inferred statistical parameters.

[12]http://bretl.csl.illinois.edu/magpie/.

```
k               <- 3
n_observations  <- nrow(training.set)
n_features      <- ncol(training.set) - 2
distances       <- matrix(0, 1, n_observations)
for (i in 1:n_observations) {
  distances[1, i] <- sqrt(sum((training_set[i, 1:n_features]
      - validation_set[1, 1:n_features])^2))
}
nearest        <- order(distances)[1:k]
weights        <- 1 / distances[nearest]
weights        <- weights / sum(weights)
est.longitude  <- sum(training.set$LONG[nearest] * weights)
est.latitude   <- sum(training.set$LAT[nearest]  * weights)
```

FIG. 7 A possible R source code for the deterministic method.

```
library(dplyr)
k           <- 3
delta       <- 10
data.means  <- training.set         %>%
               group_by(LONG, LAT) %>%
               summarise_each(funs(mean))
data.sds    <- training.set         %>%
               group_by(LONG, LAT) %>%
               summarise_each(funs(sd))
n_max       <- nrow(data.means)
n_waps      <- ncol(data.means) - 2
p           <- matrix(0, n_max, n_waps)
for (n in 1:n_max) {
    n_means  <- as.numeric(data.means[n, 3:(n_waps + 2)])
    n_sds    <- as.numeric(data.sds[n, 3:(n_waps + 2)])

    for (j in 1:n_waps) {
      o1       <- validation.set[1, j] - delta
      o2       <- validation.set[1, j] + delta
      p1       <- pnorm(o1, mean=n_means[j], sd=n_sds[j],
          lower.tail=FALSE)
      p2       <- pnorm(o2, mean=n_means[j], sd=n_sds[j],
          lower.tail=FALSE)
      p[n, j] <- p1 - p2
    }
}
similarity     <- rowSums(p)
similar        <- order(similarity, decreasing=TRUE)[1:k]
weights        <- similarity[similar]/sum(similarity[similar])
est.longitude <- sum(data.means$LONG[similar] * weights)
est.latitude  <- sum(data.means$LAT[similar] * weights)
```

FIG. 8 A possible R source code for the probabilistic-based method.

In the probabilistic-based approach (Haeberlen et al., 2004; Youssef and Agrawala, 2005; Madigan et al., 2005), the initial collection of RSSI observations associated with a particular point is transformed into a pair of vectors containing the means and the standard deviations of the RSSI for each beacon, and then the complete training data is stored as a set of statistical parameters. Then, given a test fingerprint, for each beacon, it is possible to estimate a probability value that expresses the similarity between the observation measurement at this beacon and the training data for a particular location. An evaluation of the total similarity for every location can be computed as a function of these individual probabilities.

The algorithm selects the k training samples with higher probability and, similarly to the deterministic method, it estimates the location of the user by calculating the weighted average of the selected samples' positions.

Fig. 8 shows a possible R source code of this method (with $k = 3$). Input data has the same structure than in the deterministic method. The complete description can be found at:[13]

6 Experiments

The two methods explained in Section 5 have been tested with the four Wi-Fi-based datasets described in Section 4, using the tools included in the *Dashboard* section of the platform. Therefore, they are easily reproducible. All possible combination of the parameters has been tested. Only the combination of tuning parameters obtaining the best result is showed. In all cases, the test dataset has been used to assess the performance of the algorithms. The figure of merit used to provide an estimation of the performance of the methods is the mean localization error between the estimated position and the real one (internally known by the platform) of all test samples.

Table 5 shows the results obtained using the *UJIIndoorLoc* dataset. Note that there is no dataset for the building 0, floor 3 and for the building 2, floor 0, since there are no samples for these floors in the test set.

Table 6 shows the results on the *Tampere* dataset. In this case, only the results obtained with the deterministic approach are showed, since the probabilistic-based method can only be applied when there are enough samples at each position to calculate the estimation of the statistical parameters needed for the correct operation of this method.

Finally, Tables 7 and 8 show the results on the *IPIN2016 Tutorial* and *ALCALA2017 Tutorial* datasets. In both cases, all the samples are in the same building and floor, therefore it is not necessary to divide the data into subsets.

In the case of the *UJIIndoorLoc* dataset, the deterministic method provides better results than the probabilistic one in almost all the cases. The differences in the results obtained across buildings and floors depend on the quality of the radio map capture at each scenario and also on the structural characteristics of each scenario. According to the mean accuracy, the deterministic-based approach is preferable.

[13] http://indoorlocplatform.uji.es/methods/probabilistic/

Table 5 Mean Positioning Error (in Meters) of Both Methods on the *UJIIndoorLoc* Dataset

Building	Floor	Number of Samples	Deterministic	Probabilistic
0	0	17	4.26	7.83
0	1	17	5.65	6.77
0	2	60	6.06	5.79
1	0	105	9.62	11.26
1	1	147	7.65	20.42
1	2	132	5.40	8.99
1	3	140	8.16	11.0
2	1	20	6.64	10.09
2	2	19	7.96	9.07
2	3	19	3.88	4.57
2	4	22	12.50	21.31
	Mean	704	7.18	10.64

Table 6 Mean Positioning Error (in Meters) of Both Methods on the *Tampere* Dataset

Building	Floor	Number of Samples	Deterministic
1	1	156	9.83
1	2	110	14.21
1	3	118	8.01
1	4	105	13.03
2	1	61	15.87
2	2	77	8.38
2	3	37	6.74
	Mean	664	10.86

Table 7 Mean Positioning Error (in Meters) of Both Methods on the *IPIN2016 Tutorial* Dataset

Number of Samples	Deterministic	Probabilistic
702	4.21	3.55

Table 8 Mean Positioning Error (in Meters) of Both Methods on the *ALCALA2017 Tutorial* Dataset

Number of Samples	Deterministic	Probabilistic
405	5.03	2.53

Table 9 Value of *RMID* Obtained for Each Dataset

Dataset	Building	Floor	*RMID*
Alcalá2017 Tutorial	1	1	0.05
IPIN2016 Tutorial	1	1	0.08
UJIIndoorLoc	0	1	0.22
UJIIndoorLoc	0	0	0.23
UJIIndoorLoc	2	2	0.29
UJIIndoorLoc	1	2	0.33
UJIIndoorLoc	2	3	0.33
UJIIndoorLoc	0	3	0.34
UJIIndoorLoc	2	1	0.35
UJIIndoorLoc	0	2	0.36
UJIIndoorLoc	2	0	0.49
UJIIndoorLoc	1	3	0.65
UJIIndoorLoc	1	1	0.66
Tampere	1	4	0.67
UJIIndoorLoc	1	0	0.67
UJIIndoorLoc	2	4	0.70
Tampere	1	3	0.72
Tampere	1	2	0.77
Tampere	1	1	0.80
Tampere	2	1	0.89
Tampere	2	2	0.90
Tampere	2	3	0.91

The data is ordered by ascending *RMID* value.

There is also a high variability across buildings and floors in the results obtained for the *Tampere* dataset due to the same reasons than in the *UJIIndoorLoc* dataset.

In the case of the *IPIN2016 Tutorial* and *ALCALA2017 Tutorial*, the results are very similar and in both cases the probabilistic approach is preferable. Note that, in the *ALCALA2017 Tutorial* dataset, the difference is quite significant since the probabilistic-based approach can deal with the unintentional mistakes introduced by some of the dataset creators. Results obtained for these datasets are better than the ones obtained for the *UJIIndoorLoc* and the *Tampere* datasets since the scenarios of the Tutorial datasets correspond to small areas and more fingerprints per m^2 than the other two. Therefore position error is lower.

Table 9 shows the Radio Map Inherent Difficulty (RMID) value of each dataset (Sansano et al., 2017). This measure gives an estimation of the inherent difficulty of a radio map to obtain accurate estimates. According to this value, *IPIN2016 Tutorial* and *ALCALA2017 Tutorial* are scenarios where it is easier to obtain accurate results. However, as confirmed by the results showed in Tables 5 and 6, the RMID value of the *UJIIndoorLoc* and *Tampere* datasets shows that both datasets are quite complex and therefore, it is quite difficult to obtain positioning estimation with high accuracy.

Taking into account all the datasets, the mean localization error of the deterministic-based method is 6.79 m without including the results of the *Tampere* dataset, and 8.21 including it. The mean error of the probabilistic-based method is 9.47. Therefore, according to the results, in general, it seems that the deterministic-based method is preferable. However, taking into account the differences of the four scenarios, the deterministic-based approach gets better results in big scenarios with low density of data (*UJIIndoorLoc* and *Tampere* datasets), while the probabilistic based one is preferable in small ones with high density of data (*IPIN2016 Tutorial* and *ALCALA2017 Tutorial* datasets).

Note that the results obtained with the methods included in the platform can be effectively improved using more sophisticated algorithms, and also using modern machine learning techniques. For instance, the ranking of the *UJIIndoorLoc* dataset shows better results than the ones presented in Table 5, since they are the best results obtained in the *Wi-Fi fingerprinting in large environments* (IPIN'15) competition.

7 The Platform in Use

The performance of the platform was tested during the 2017 Fingerprinting-based Indoor Positioning tutorial held in the School of Engineering of the University of Alcalá. As an activity of the course, an indoor localization competition took place using the *ALCALA2017 Tutorial* dataset.

The 15 attendees were invited to use the platform to upload the results of their proposals. Some of them used the *Dashboard* section of the platform to test different parameter configurations of the localization methods included in the platform, and others manually programmed their own method from the source code provided by the course instructors. After a very competitive and exciting competition, the winner team got an error of only 2.14 m using a probabilistic method. This result is even better than the best one that can be directly obtained using the Dashboard included in the platform.

In general, tutorial attendees were able to easily use the platform, mainly the *Dataset* download section, the *Dashboard* section and, obviously, the *Ranking* sections. Almost no queries to the course introduction were produced, showing the effective user-centered design applied to the platform.

8 Conclusions

In this chapter, the *IndoorLoc Platform* has been presented. It is a public repository for comparing and evaluating indoor positioning algorithms. The proposed web platform can be used to download datasets, learn how some well-known algorithms work, study the source code of those algorithms, test the methods, and even upload results of the user's methods to check the accuracy when comparing against the results provided by other methods already included in a ranking, among other functionalities.

To present a real example of the usage of the platform, a comparative study of the accuracy of two well-known fingerprinting-based indoor localization algorithms, using four of the datasets included in the platform, have also been presented. According to the results obtained, the deterministic-based approach gets better results in big scenarios while the probabilistic based one is preferable in small scenarios. These experiments are easily reproducible using the tools included in the platform.

This web platform is an ongoing project, and future versions will implement new algorithms and include more datasets, with the aim to provide an interesting tool for researchers and become a reference web platform for indoor positioning research. For this purpose, researchers are invited to include more methods and datasets in the platform.

Acknowledgments

The authors gratefully acknowledge funding from the Spanish Ministry of Economy and Competitiveness in the projects: "Proyectos I+D Excelencia" TIN2015-70202-P and "Redes de Excelencia" TEC2015-71426-REDT.

References

Bahl, P., Padmanabhan, V.N., 2000. RADAR: an in-building RF-based user location and tracking system. In: Proceedings of the 19th Annual Joint Conference of the IEEE Computer and Communications Societies (INFOCOM'00), pp. 775–784.

Barsocchi, P., Crivello, A., Rosa, D.L., Palumbo, F., 2016. A multisource and multivariate dataset for indoor localization methods based on WLAN and geo-magnetic field fingerprinting. In: Proceedings of the Seventh Conference on Indoor Positioning and Indoor Navigation (IPIN'16).

Bishop, C.M., 2006. Pattern Recognition and Machine Learning (Information Science and Statistics). Springer-Verlag New York, Inc.

Cover, T., Hart, P., 1967. Nearest neighbor pattern classification. IEEE Trans. Inform. Theory 13 (1), 21–27.

Cramariuc, A., Lohan, E.S., 2016. Open-access WiFi measurement data and python-based data analysis. Available from: http://www.cs.tut.fi/tlt/pos/meas.htm.

García, S., Molina, D., Lozano, M., Herrera, M.F., 2009. A study on the use of non-parametric tests for analyzing the evolutionary algorithms' behaviour: a case study on the cec'2005 special session on real parameter optimization. J. Heuristics 15 (6), 617–644.

Goldbloom, A., Hamner, B., Moser, J., Cukierski, M., 2017. Kaggle: your home for data science. Available from: https://www.kaggle.com/.

Haeberlen, A., Flannery, E., Ladd, A.M., Rudys, A., Wallach, D.S., Kavraki, L.E., 2004. Practical robust localization over large-scale 802.11 wireless networks. In: Proceedings of the 10th Annual International Conference on Mobile Computing and Networking (ModiCom'04), pp. 70–84.

Han, D., Jung, S.H., Lee, M., Yoon, G.-W., 2014. Building a practical Wi-Fi-based indoor navigation system. IEEE Pervasive Comput. 13, 72–79.

Hanley, D., Faustino, A.B., Zelman, S.D., Degenhardt, D.A., Bretl, T., 2017. Magpie: a dataset for positioning with magnetic anomalies. In: Proceedings of the 2017 International Conference on Indoor Positioning and Indoor Navigation (IPIN'17).

He, S., Chan, S.G., 2016. Wi-Fi fingerprint-based indoor positioning: recent advances and comparisons. IEEE Commun. Surv. Tutorials 18 (3), 466–490.

Lemic, F., Handziski, V., Wirstrom, N., Van Haute, T., De Poorter, E., Voigt, T., Wolisz, A., 2015. Web-based platform for evaluation of RF-based indoor localization algorithms. In: Proceedings of the 2015 IEEE International Conference on Communication Workshop (ICCW'15).

Li, B., Gallagher, T., Dempster, A.G., Rizos, C., 2012. How feasible is the use of magnetic field alone for indoor positioning? In: 3th International Conference on Indoor Positioning and Indoor Navigation.

Li, B., Gallagher, T., Rizos, C., Dempster, A., 2013. Using geomagnetic field for indoor positioning. In: Proceedings of the International Global Navigation Satellite Systems (IGNSS) Society Symposium.

Lichman, M., 2013. UCI machine learning repository. Available from: http://archive.ics.uci.edu/ml.

Lymberopoulos, D., Choudhury, R.R., Yang, X., Sen, S., 2014. Microsoft indoor localization competition (IPSN'14). Available from: https://www.microsoft.com/en-us/research/event/microsoft-indoor-localization-competition-ipsn-2014.

Lymberopoulos, D., Liu, J., Yang, X., Naguib, A., Rowe, A., Trigoni, N., Moayeri, N., 2015. Microsoft indoor localization competition (IPSN'15). Available from: https://www.microsoft.com/en-us/research/event/microsoft-indoor-localization-competition-ipsn-2015.

Lymberopoulos, D., Liu, J., Zhang, Y., Dutta, P., Yang, X., Rowe, A., 2016. Microsoft indoor localization competition (IPSN 2016). Available from: https://www.microsoft.com/en-us/research/event/microsoft-indoor-localization-competition-ipsn-2016.

Lymberopoulos, D., Liu, J., Bocca, M., Sequeira, V., Trigoni, N., Yang, X., 2017. Microsoft indoor localization competition (IPSN 2017). Available from: https://www.microsoft.com/en-us/research/event/microsoft-indoor-localization-competition-ipsn-2017.

Madigan, D., Einahrawy, E., Martin, R.P., Ju, W.-H., Krishnan, P., Krishnakumar, A.S., 2005. Bayesian indoor positioning systems. In: Proceedings of the 24th Annual Joint Conference of the IEEE Computer and Communications Societies (INFOCOM'05), pp. 1217–1227.

Marques, N., Meneses, F., Moreira, A., 2012. Combining similarity functions and majority rules for multi-building, multi-floor, WiFi positioning. In: Proceedings of the 2012 International Conference on Indoor Positioning and Indoor Navigation (IPIN'12), pp. 1–9.

Moayeri, N., Ergin, O., Lemic, F., Handziski, V., Wolisz, A., 2016. Perfloc: an extensive data repository for development and a web-based capability for performance evaluation of smartphone indoor localization apps. In: Proceedings of the 27th IEEE International Symposium on Personal, Indoor and Mobile Radio Communications (PIMRC'16).

Montoliu, R., Torres-Sospedra, J., Belmonte, O., 2016. Magnetic field based indoor positioning using the bag of words paradigm. In: Proceedings of the International Conference on Indoor Positioning and Indoor Navigation (IPIN'16).

Montoliu, R., Sansano, E., Torres-Sospedra, J., Belmonte, O., 2017. IndoorLoc platform: a public repository for comparing and evaluating indoor positioning systems. In: Proceedings of the 2017 International Conference on Indoor Positioning and Indoor Navigation (IPIN'17).

Nahrstedt, K., Vu, L., 2012. CRAWDAD dataset uiuc/uim (v. 2012-01-24). Available from: http://crawdad.org/uiuc/uim/20120124.

Popleteev, A., 2017. AmbiLoc: a year-long dataset of FM, TV and GSM fingerprints for ambient indoor localization. In: Proceedings of the 2017 International Conference on Indoor Positioning and Indoor Navigation (IPIN'17).

Potortì, F., Barsocchi, P., Girolami, M., Torres-Sospedra, J., Montoliu, R., 2015. Evaluating indoor localization solutions in large environments through competitive benchmarking: the EvAAL-ETRI competition. In: Proceedings of the Sixth Conference on Indoor Positioning and Indoor Navigation (IPIN'15).

Sansano, E., Montoliu, R., Torres-Sospedra, J., 2017. A novel methodology to estimate a measurement of the inherent difficulty of an indoor localization radio map. In: Proceedings of the 2017 International Conference on Indoor Positioning and Indoor Navigation (IPIN'17).

Talvitie, J., Lohan, E.S., Renfors, M., 2014. The effect of coverage gaps and measurement inaccuracies in fingerprinting based indoor localization. In: Proceedings of International Conference on Localization and GNSS 2014 (ICL-GNSS'14).

Torres-Sospedra, J., Montoliu, R., Usó, A.M., Avariento, J.P., Arnau, T.J., Benedito-Bordonau, M., Huerta, J., 2014. UJIIndoorLoc: a new multi-building and multi-floor database for WLAN fingerprint-based indoor localization problems. In: Proceedings of the 2014 International Conference on Indoor Positioning and Indoor Navigation (IPIN'14), pp. 261–270.

Torres-Sospedra, J., Montoliu, R., Trilles, S., Belmonte, O., Huerta, J., 2015a. Comprehensive analysis of distance and similarity measures for Wi-Fi fingerprinting indoor positioning systems. Expert Syst. Appl. 42 (23), 9263–9278.

Torres-Sospedra, J., Rambla, D., Montoliu, R., Belmonte, O., Huerta, J., 2015b. UJIIndoorLoc-Mag: a new database for magnetic field-based localization problems. In: Proceedings of the Sixth Conference on Indoor Positioning and Indoor Navigation (IPIN'15).

Torres-Sospedra, J., Jiménez, A.R., Knauth, S., Moreira, A., Beer, Y., Fetzer, T., Ta, V.-C., Montoliu, R., Seco, F., Mendoza-Silva, G.M., Belmonte, O., Koukofikis, A., Nicolau, M.J., Costa, A., Meneses, F., Ebner, F., Deinzer, F., Vaufreydaz, D., Dao, T.-K., Castelli, E., 2017a. The smartphone-based offline indoor location competition at IPIN 2016: analysis and future work. Sensors 17 (3).

Torres-Sospedra, J., Moreira, A., Knauth, S., Berkvens, R., Montoliu-Cols, R., Belmonte-Fernndez, O., Trilles, S., Nicolau, M.J., Meneses, F., Costa, A., Koukofikis, A., Weyn, M., Peremans, H., 2017b. Realistic evaluation of indoor positioning systems based on Wi-Fi fingerprinting: the 2015 EvAAL-ETRI competition. J. Ambient Intell. Smart Environ. 9, 263–279.

Wu, C., Yang, Z., Liu, Y., Xi, W., 2013. WILL: Wireless indoor localization without site survey. IEEE Trans. Parallel Distrib. Syst. 24 (4), 839–848.

Youssef, M., Agrawala, A., 2005. The Horus WLAN location determination system. In: Proceedings of the 3rd International Conference on Mobile Systems, Applications, and Services (MobiSys'05), pp. 205–218.

13

Challenges and Solutions in Received Signal Strength-Based Seamless Positioning

Pedro Figueiredo e Silva, Philipp Richter, Jukka Talvitie, Elina Laitinen, Elena Simona Lohan

LABORATORY OF ELECTRONICS AND COMMUNICATIONS ENGINEERING, TAMPERE UNIVERSITY OF TECHNOLOGY, TAMPERE, FINLAND

1 Introduction and Definitions

According to Webster dictionary, seamless means "having no awkward transitions, interruptions, or indications of disparity" and it is often used synonymously with "smooth" and "continuous." One of the first papers to discuss in detail which are some possible mechanisms to attain a seamless positioning on mobile devices has been published in Sun et al. (2005). The authors in Sun et al. (2005) identified five main directions of research to help in creating seamless positioning platforms, namely: (i) data fusion approaches, (ii) single-step localization solutions, which nowadays can be translated into Simultaneous Localization and Mapping (SLAM) solutions (Taylor et al., 2011), (iii) use of smart antennas and Angle of Arrival (AOA) techniques, (iv) designing the network topologies with positioning in mind (unlike the classical approach where the network topology is designed fully according to communication targets), and (v) interoperability at positioning level, such as systems (cellular, WLAN, etc.) designed to include interoperable positioning features that can be then smoothly and easily accessed and combined.

Seamless positioning targets often go hand in hand with the targets of low-power consumption at the mobile side and the need of low-cost solutions at the network side. Interoperability of different positioning systems is also an important design criterion. Low-cost, low-power, and interoperable solutions can be usually achieved when there is no need of additional infrastructure or hardware updates and the localization is done based on available signals and infrastructures. This is why, the Received Signal Strength (RSS) solutions are one of the gaining techniques in the realm of low-cost, low-power, seamless, and interoperable positioning, and they form the core of this chapter.

Geographical and Fingerprinting Data for Positioning and Navigation Systems. https://doi.org/10.1016/B978-0-12-813189-3.00013-7

The RSS-based approaches can be basically used with any radio frequency (RF) communication systems whose standard supports RSS measurements or reporting. For example, cellular, WLAN, Bluetooth Low Energy (BLE), Radio Frequency Identification (RFID) are all systems which currently support RSS measurements or enable some form of RSS indication, and thus can be used for RSS-based positioning. The RSS-based positioning typically relies on fingerprinting techniques, involving two stages:

1. A training stage, where data about the environment, such as RSS measurements, are collected with different labels or position tags and stored into a training database available at the Location Service Provider (LSP) side, and
2. An estimation stage, where the mobile gets access to the full or partial information stored into the training database by the LSP and computes its position by matching the real-time RSS measurements with the measurements stored in the training database. This matching (or fingerprinting) process outputs a location label or a position, according to what was initially stored in the database.

One-stage approaches are also possible with RSS-based positioning, but they are much less accurate than the two-stage fingerprinting approaches. In the one-stage approaches, the receiver has to assume that a certain statistical path-loss model is valid in the location area. With an underlying path-loss model, and assuming that the transmitter location is known, an user receiving RSS measurements from three or more transmitters (or Access Nodes) in range can compute its most likely position.

The different types of RSS-based positioning via fingerprinting are described in Section 2.

2 Overview of Fingerprinting Methods

Following the discussion in the previous section, we can classify the RSS-based fingerprinting methods as shown in Fig. 1. The two-stage approaches are described in more detail in what follows. The path-loss (PL) probabilistic techniques can be used in both two-stage and one-stage approaches. The SLAM techniques are not covered in this chapter. For good surveys of SLAM techniques, the readers are referred to Agarwal et al. (2014) and Cadena et al. (2016).

2.1 Methods With Full Training Databases

The classical fingerprinting with a full training database refers to the two-stage approach where RSS measurements are first collected at different locations in a building and stored into a training database under the following quadruplet per access point $(P_{a,i}, x_i, y_i, z_i)$, where $P_{a,i}$ is the RSS value measured from the ath Access Point (AP) in the ith location in the building, and x_i, y_i, z_i are the 3D coordinates at the ith location. If an access point is not heard at a certain location, its RSS is set according to a convention. For example, in a Matlab-based analysis, we can use the convention that $P_{a,i}$ = NaN (Not A Number) if the

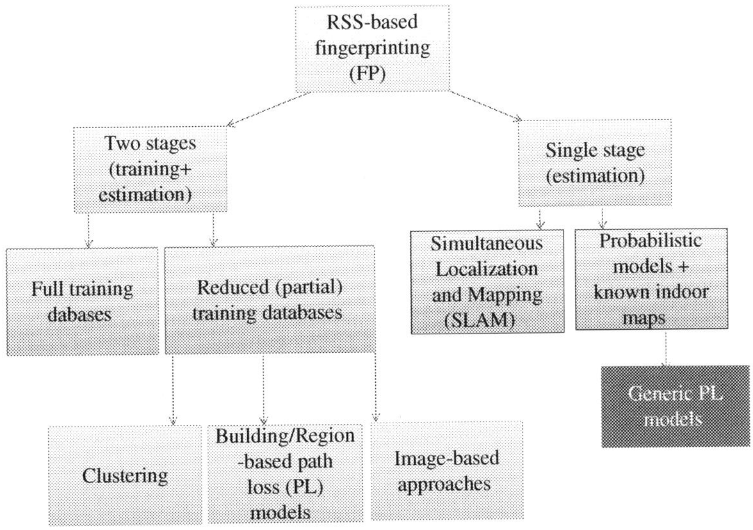

FIG. 1 Classification of fingerprinting approaches.

ath access point is not heard in that particular point. Alternatively, one can set the missing $P_{a,i}$ values to something very small, such as −200 dB, or simply ignore it.

Let us denote by N_{AP} the number of AP in the building and by N_m the number of measurement points or locations. In the estimation phase, the mobile measures the RSS $O_a, a = 1, \ldots, N_{AP}$ from the hearable access points in its range and compares the observed O_a RSS values with the $P_{a,i}$ values stored in the fingerprint database. We use the convention that O_a = NaN if the ath access point is not heard by the mobile. The comparisons are done in terms of some metrics M_i, shown for example, in Table 1 and computed in each of the N_m measured locations. In Table 1, $\mathbf{O} = [O_1, O_2, \ldots, O_{N_{AP}}]^T$, $\mathbf{P}_i = [P_{1,i}, P_{2,i}, \ldots, P_{N_{AP},i}]^T$, and $S = \text{diag}(\sigma_1^2, \sigma_2^2, \ldots, \sigma_{N_{AP}}^2)$, where σ_a^2 is the noise variance per measured RSS in the ath AP. In each row of Table 1, only the commonly heard AP between the training database and the mobile's measurement are considered, according to the mentioned conventions. The choice of how to encode that information depends on these metrics, because they process that information differently (Torres-Sospedra et al., 2015) and not all combinations are feasible.

2.2 Methods With Reduced Training Databases

In order to store a lower number of parameters in the training database instead of all measured RSS values from all heard AP, several methods are available, such as clustering or classification (Cramariuc et al., 2016; Abdou et al., 2016), statistical path-loss modeling, compression or image-based (Talvitie et al., 2016), compressed sensing (Feng et al., 2012), feature extraction (Lin and Chen, 2016), principal component analysis (Mo et al., 2015),

Table 1 Examples of RSS Distance Metrics Used in Fingerprinting

Method	Metric M_i	Comparison Criterion
Euclidean	$M_i = \left(\sum_{a=1}^{N_{AP}} \|\|O_a - P_{a,i}\|\|^2\right)^{1/2}$	$\min_i(M_i)$
Square of Euclidean (Spearman distance)	$M_i = \sum_{a=1}^{N_{AP}} \|\|O_a - P_{a,i}\|\|^2$	$\min_i(M_i)$
Manhattan distance	$M_i = \sum_{a=1}^{N_{AP}} \|\|O_a - P_{a,i}\|\|$	$\min_i(M_i)$
Mahalanobis distance	$M_i = \sqrt{(\mathbf{O} - \mathbf{P}_i)^T S^{-1} (\mathbf{O} - \mathbf{P}_i)}$	$\min_i(M_i)$
Logarithmic Gaussian likelihood	$M_i = \sum_{a=1}^{N_{AP}} \log\left(\frac{1}{\sqrt{2\pi\sigma_a^2}} e^{\left(-\frac{(O_a - P_{a,i})^2}{2\sigma_a^2}\right)}\right)$	$\max_i(M_i)$

etc. The main approaches to reduce a training database are explained in the following subsections.

2.2.1 Clustering Methods

Fig. 2 shows a classification of clustering approaches which can be used to decrease the size of the training database. Basically, the clustering or database reduction can be done either in coordinates domain (2D or 3D modeling), or in RSS or AP domain, but combined or hierarchical methods are also possible, for example, those based on Support Vector Machines (SVM) as reported in Khullar and Dong (2017). For example, the 3D coordinates clustering was compared with RSS clustering in Cramariuc et al. (2016) and it has been shown that the 3D coordinate clustering gives typically better positioning accuracy than the RSS clustering, at the expense of a higher computational complexity. Affinity propagation clustering combined with SVM has been studied, for example, in Abdou et al. (2016) for the purpose of keeping only the relevant AP in the location estimation and to achieve a low complexity algorithm implementation. This problem is also related with the AP reduction problem, addressed in Section 3.2.2.

2.2.2 Path-Loss Approaches

The path-loss (PL) approaches are another set of methods which can be used to reduce the size of the training databases. They rely on the assumption that each AP in a network obeys a certain, intrinsic path-loss propagation model, characterized by a set of few parameters, which can be modeled statistically. The most encountered PL model is the one-slope PL model, where the measured RSS in dB scale $P_{a,i}$ is given by:

$$P_{a,i} = P_{T_a} - 10 n_a \log(d_{a,i}) + \eta_{a,i},$$

where P_{T_a} is an "apparent" transmit power (in dB) of the ath AP in the network (typically measured at 1 m away from the AP), n_a is a path-loss (constant) coefficient characterizing

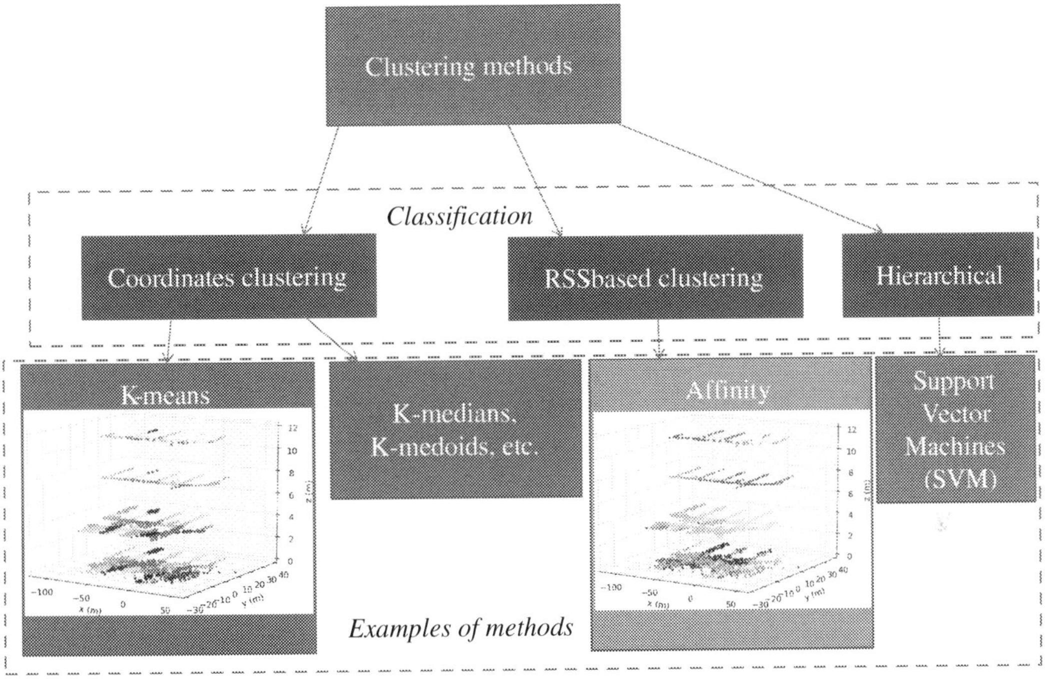

FIG. 2 Classification of clustering approaches with few examples.

the *a*th AP and the $\eta_{a,i}$ is a noise term, typically modeled via a Gaussian zero-mean process with variance σ_a^2, which incorporates the shadowing effects and the measurement errors. With this model, each AP is characterized by only three parameters: P_{T_a}, n_a, σ_a^2. Thus, instead of storing all the measurements at various training locations, with this PL approach it is enough to store these three parameters and the AP location x_{AP}, y_{AP}, z_{AP} per AP.

A major drawback with PL approaches is the additional inaccuracy created by the model misfit to the real data. More advanced PL models, for example, including floor losses and two-slope models, have been studied, for example, by Shrestha et al. (2013), where it was shown that the positioning error differences between the different PL models are very low, and that we usually loose between 20% and 30% in the positioning accuracy in PL models versus the full-training fingerprinting, but we gain up to 11 times in database reduction.

Multislope PL models have also been studied in Karttunen et al. (2016) and Munir et al. (2017), in a different context than wireless positioning. It is still an open research topic whether the multislope PL models can benefit the indoor wireless positioning and which PL models are best suitable for high-accuracy fingerprinting with reduced training database.

2.2.3 Image-Based Approaches

When considering global-scale positioning systems, one attractive approach for reducing the size of fingerprint databases is the spectral compression method presented in Talvitie et al. (2016). There, the fundamental idea has been to conduct the compression in frequency domain by applying an appropriate frequency transformation, such as Discrete Fourier Transform (DFT) or Discrete Cosine Transform (DCT).

In the spectral compression approach, the RSS values in the fingerprint database are processed individually for each network AP in the form of an RSS image, which represents a map of RSS values in a rectangular grid of geographical coordinates. Due to the fundamental characteristics of radio wave propagation, such as path-loss and shadowing effects, the nearby RSS values in the RSS image are correlated. Now, by considering the DFT, the Power Spectral Density (PSD) function of the RSS image can be defined as the DFT of the corresponding autocorrelation function.

Thus, based on the well-known time/frequency-scaling property of the Fourier transform, the PSD of the RSS image gets narrower, as the autocorrelation function gets wider. Physically, this means that the energy of the DFT-transformed RSS image is concentrated more tightly compared to the original RSS image. Moreover, the compression can be achieved by storing only the most significant frequency domain components which should include the majority of the useful energy of the original RSS image. However, since in practice some of the energy of the original RSS image is always left outside of the stored DFT coefficients, the described spectral compression approach is fundamentally a lossy compression technique.

It is convenient to represent the RSS image, that is, the observed RSS values for a single AP, as a matrix $\mathbf{G}_{\mathrm{RSS}} \in \mathbb{R}^{M \times N}$, whose elements denote the RSS values (in dBm) in a rectangular grid. Furthermore, each column $n = 0, \ldots, N-1$ and each row $m = 0, \ldots, M-1$ determine a specific x-coordinate and y-coordinate for the corresponding RSS value. The total number of measurements is $N_m = N \times M$. Quite often some of the elements of $\mathbf{G}_{\mathrm{RSS}}$ are missing, as there are no RSS measurements taken from those locations. Therefore, in order to complete the matrix, appropriate interpolation and extrapolation methods can be exploited as studied, for example, in Talvitie et al. (2015). In addition to interpolation and extrapolation methods, nonuniformly sampled frequency transformations are also feasible, but in these cases the computational complexity might grow unnecessarily large with respect to the gained advantage.

Typically the average observed RSS value of a RSS image, pointing out the zero-frequency element in the frequency domain, is significantly larger compared to the other frequency domain elements. As a result, the average RSS value would always be included in the compressed data, and therefore we can remove it from the RSS image and store it beforehand. By this way, a zero mean RSS image can be obtained as $\mathbf{G}^0_{\mathrm{RSS}} = \mathbf{G}_{\mathrm{RSS}} - \mu_{\mathbf{G}}$, where $\mu_{\mathbf{G}} = 1/(MN) \sum_{i=0}^{MN-1} \mathbf{G}_{\mathrm{RSS}}(i)$, and where $\mathbf{G}_{\mathrm{RSS}}(i)$ denotes the ith element of the RSS image matrix by using column-wise indexing. Now, in order to achieve the frequency domain representation without the need of using complex-valued quantities, a DCT is employed in the following example. Consequently, the frequency transformed RSS image $\mathbf{H}_{\text{RSS}}$ (of the same size with $\mathbf{G}_{\text{RSS}}$) can be given as

$$\mathbf{H}_{\text{RSS}}(k,l) = w_k w_l \sum_{m=0}^{M-1} \sum_{n=0}^{N-1} \mathbf{G}_{\text{RSS}}^{(0)}(m,n) \cos\left(\frac{\pi(2m+1)k}{2M}\right) \cos\left(\frac{\pi(2n+1)l}{2N}\right), \text{ where}$$

$$w_k = \begin{cases} 1/\sqrt{M}, & k=0 \\ \sqrt{2/M}, & 1 \le k \le M \end{cases} \quad \text{and} \quad w_l = \begin{cases} 1/\sqrt{N}, & l=0, \\ \sqrt{2/N}, & 1 \le l \le N, \end{cases}$$

where $\mathbf{H}_{\text{RSS}}(k,l)$ denotes the element of $\mathbf{H}_{\text{RSS}}$ in the kth row and lth column. As show in Fig. 3, the energy of the frequency transformed image $\mathbf{H}_{\text{RSS}}$ is heavily concentrated on one corner of the image. The original RSS image (on top of Fig. 3) is constructed based on linear interpolation and minimum method extrapolation, as given in Talvitie et al. (2015), by using the known RSS values in locations marked with $\times$. From the frequency-domain representation of the RSS image (on left), only the K largest energy components marked with $\circ$ are used to construct the compressed RSS image (on right).

Thus, by storing only the K largest elements from $\mathbf{H}_{\text{RSS}}$ it is possible to generally reduce the amount of information required to represent the original image with adequate accuracy. It is not a straightforward task to choose an appropriate global value for the number of stored elements K, since the image size and the information content varies over separate APs. Hence, by considering each image separately, and choosing the K per image basis, will typically improve the performance of the compression process. Nonetheless, by assuming that the stored frequency domain elements are selected based on the order of element energy $|\mathbf{H}_{\text{RSS}}(i)|^2$, the stored frequency components can be given as

$$\left\{\mathbf{H}_{\text{RSS}}\left(q_s\right) \; : \; s = 0, \ldots, K-1\right\},$$

where q_s indicates the K elements with the highest energy. Besides the actual frequency component value $\mathbf{H}_{\text{RSS}}(q_s)$, also the index q_s has to be stored into the database in order to recover the compressed RSS image. In addition, other parameters required to be saved are the image size parameters M and N, the physical image location, for example (latitude, longitude, altitude)-coordinate of one corner point, and the AP identity. It is worth noticing that the required bit resolution and the number of bits required to represent separate parameters might differ considerably, according to the environment. For instance, as the stored index values are concentrated on a single corner of the matrix, a reasonably designed entropy coder can easily achieve high compression ratios.

The reconstruction of the RSS images from the stored frequency domain components is rather straightforward. First, the compressed frequency domain image $\hat{\mathbf{H}}_{\text{RSS}}$ is initialized with zero elements and the stored frequency domain components $\mathbf{H}_{\text{RSS}}(q_s)$ are inserted in the corresponding element indices as

$$\hat{\mathbf{H}}_{\text{RSS}}(q) = \begin{cases} \mathbf{H}_{\text{RSS}}(q_s), & \text{when } q = q_s \\ 0, & \text{otherwise.} \end{cases}$$

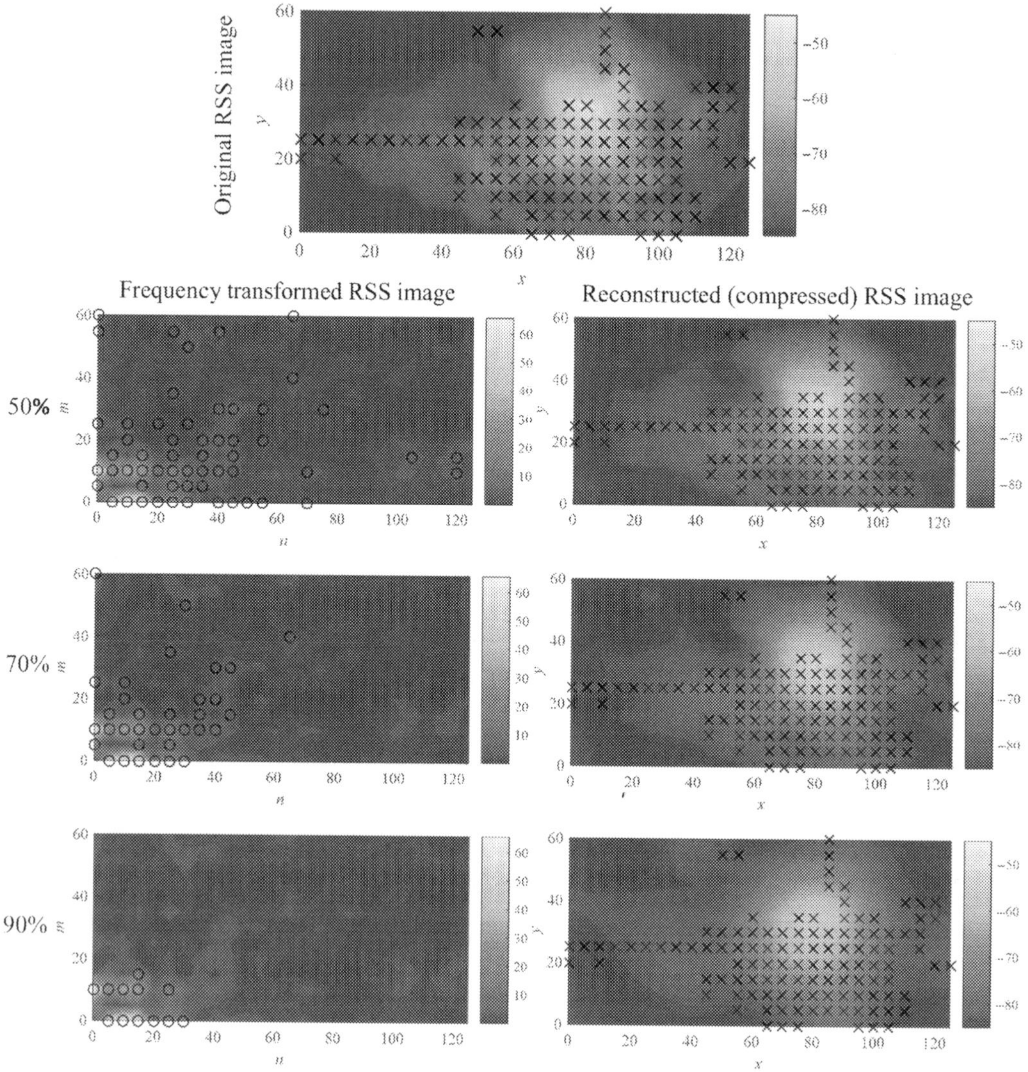

FIG. 3 Example of an image compression process with 50%, 70%, and 90% compression ratios. The original uncompressed RSS image (on the top) has been obtained by using the linear interpolation and minimum method extrapolation as given in Talvitie et al. (2015). The locations of the original RSS measurements used in the interpolation and extrapolation are marked with ×.

After this, the compressed zero mean RSS image is obtained by taking the inverse frequency transformation, which in this case is the inverse DCT, given as

$$\hat{\mathbf{G}}_{\mathrm{RSS}}^{(0)}(m,n)=\sum_{k=0}^{M-1}\sum_{l=0}^{N-1}w_k w_l \hat{\mathbf{H}}_{\mathrm{RSS}}(k,l)\cos\left(\frac{\pi(2m+1)k}{2M}\right)\cos\left(\frac{\pi(2n+1)l}{2N}\right).$$

Finally, the compressed RSS image is obtained by adding the stored mean value as $\hat{\mathbf{G}}_{\mathrm{RSS}} = \hat{\mathbf{G}}^{(0)}_{\mathrm{RSS}} + \mu_{\mathbf{G}}$. In Fig. 3, the above described compression process has been illustrated for one indoor real-life RSS image of a WLAN AP, by using 50%, 70%, and 90% compression ratios.

In practice, when appropriately designed, the lossy nature of the spectral compression approach can actually benefit the positioning system. This is because of the built-in noise filtering property, in which during the compression, low energy spectral coefficients with poor signal-to-noise ratio (SNR) are automatically discarded from the database. It is worth noticing that by assuming uncorrelated noise between the RSS values in the original database, the noise energy spectrum is flat. Hence, the energy of the noise is distributed evenly on top of all elements of the frequency transformed RSS image.

One challenge is to find a good balance between the compression ratio and the noise filtering volume. A large compression ratio leads to good SNR of the database, but also to poor representation of the RSS data, and vice versa. Nonetheless, with appropriate parametrization, the positioning performance by using spectrally compressed database can reach, and even exceed, the performance of conventional fingerprinting with an uncompressed database. This has been illustrated by the results in Talvitie et al. (2016), as shown in Fig. 4, where the average positioning error has been provided by considering WLAN measurements over five separate buildings. Here, the positioning performance of the spectral compression is compared with the conventional fingerprinting and PL models as a function of compression ratio. It can be seen that the spectral compression can provide comparable performance with the fingerprinting method for compression ratios up to 70%–80%.

2.2.4 Other Approaches

A variety of other approaches have been proposed to deal with reduced-database fingerprinting, in various contexts, including wireless localization. For example, compressive sensing approaches (Milioris et al., 2011; Feng et al., 2012) take advantage of the inherent sparse structure of the RSS measurements inside a building, due to the fact that an AP is heard only in parts of a building. The feature extraction (Lin and Chen, 2016) or principal

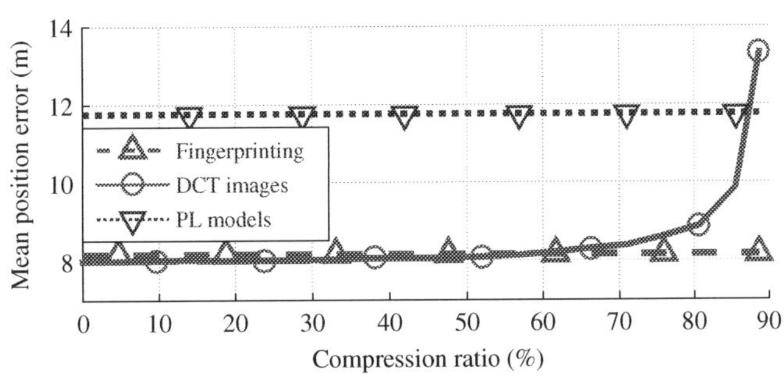

FIG. 4 The mean positioning error as a function of compression ratio of the spectral compression for conventional fingerprinting, path-loss models, and the RSS image approach.

component analysis (Mo et al., 2015) methods extract some relevant features from the set of the training data and use them in the estimation process. Multiclassifiers, extremely randomized trees and other machine learning approaches have also been studied, for example, in Uddin and Islam (2015) and Zou et al. (2016).

3 Challenges and Solutions in Fingerprinting

Table 2 summarizes the main challenges in RSS-based fingerprinting. Some of them have been already briefly addressed in the previous section, and some others are explained in more details below. Some others are only mentioned in this table with references for further reading. Table 2 also discusses different solutions to address these challenges and how some of these solutions support the objective of seamless positioning.

Several different phenomena can cause RSS fluctuations or an offset in the RSS values. Naturally, the positioning accuracy may decrease if the RSS in the database is clearly different than the RSS observed by the user in the estimation phase. RSS offsets can be divided roughly into constant, random, or localized bias (Laitinen, 2017). If an offset is caused by different equipment type, it is most typically modeled as a constant positive or negative bias. Temporal propagation dynamics such as user orientation or body losses in crowded periods during the training phase compared with the estimation phase can result in RSS fluctuations, i.e., random bias.

One of the main challenges in the two-phase localization methods is that the device used in the data collection process may be different than the device used in the estimation phase. As different devices with different chipsets scale the measured RSS values differently, the observed RSS at the same physical location may vary between different devices. According to Della Rosa et al. (2010a,b) and Kaemarungsi (2006), even 25–30 dB differences in RSSs for different devices have been noticed. In some cases the observed RSS values can be linear, as was noticed by Haeberlen et al. (2004) and Kjaergaard (2006), but due to chipset sensitivity, antenna spacings, antenna gains, and operating systems, linearity is not a guarantee (Della Rosa et al., 2010a; Wang and Wong, 2014). Fig. 5 illustrates the relation between RSS values in dB and the RSS indicator (RSSI) for three different chipsets (Atheros, Symbol, and Cisco). It can be seen that both the RSS scale and the steps differ between the chipset manufacturers remarkably.

Besides possible RSS offsets, another challenge in fingerprinting is the huge amount of data to deal with. Nowadays, the transmitter configuration can be very dense, meaning that the number of AP in an office building with several floors and plenty of offices is typically hundreds. High transmitter density naturally means that a lot of transmitters are heard in each measurement (i.e., in the training phase), leading into a huge amount of data to be handled. As also the number of fingerprints can be significant, the memory requirements for the database in large areas or with many buildings may become overwhelming, and data transmission may become a bottleneck for the positioning system, especially with fingerprinting.

Table 2 Overview of Challenges and Solutions in Fingerprinting

Challenge	Description	Possible Solutions and Seamless Support
Calibration issues	Difference or offsets in RSSs reported by different devices	Calibration methods.
Database sizes	Large training databases	Compressing the databases, e.g., spectral compression (Talvitie et al., 2016), AP reduction, sparse fingerprints, and interpolation.
Database gaps	Parts of the buildings are not accessible	Interpolation and extrapolation, e.g., in Talvitie et al. (2015) and Richter and Toledano-Ayala (2015). It may increase the positioning accuracy and the coverage area of the positioning system that integrates fingerprinting.
Height estimation	Height dimensions is often difficult to estimate accurately, especially in buildings with open spaces between floors	Use of 3D modeling and sensor aiding.
Channel modeling for single-stage methods	Finding adequate path-loss and shadowing models for the probabilistic approaches	Solutions found, e.g., in Shrestha et al. (2013), Karttunen et al. (2016), and Munir et al. (2017). May extend RSS-based seamless positioning to much larger areas.
Unknown location of emitters	For probabilistic approaches, knowledge of the AP location is typically important	Estimating the emitter location based on training data, e.g., as in Shrestha et al. (2013) and Varzandian et al. (2013).
Training data collection and updating	As wireless environments are highly dynamic, data need to be continuously collected, and updated or stored	Crowdsourced solutions, e.g., Chen and Wang (2015) and Wang et al. (2016) or automatization. Improves the accuracy of fingerprinting and, thus, the performance of the overall positioning system.
Server-mobile protocols and signaling	Location-related data needs to be transferred securely and robustly between server and mobile	Protocol design and secure communications solutions. Important for many positioning systems that rely on a client-server set-up. Standardized protocols facilitate the integration of those systems.
Low-cost positioning algorithms	This refers to the choice of the positioning signals, positioning algorithm, and data fusion approaches	Various studies can be found in Laitinen (2017), Kasebzadeh et al. (2014), and Liu et al. (2017), etc. Improved positioning and fusion algorithms improve the seamless positioning experience, e.g., Richter and Toledano-Ayala (2017).
Heterogeneity of training databases	Diversity of database structure, content and access to databases	Harmonization of fingerprint databases through, e.g., open (quasi) standards; hybridization with heterogeneous data. It facilitates the integration of different systems into one positioning system.

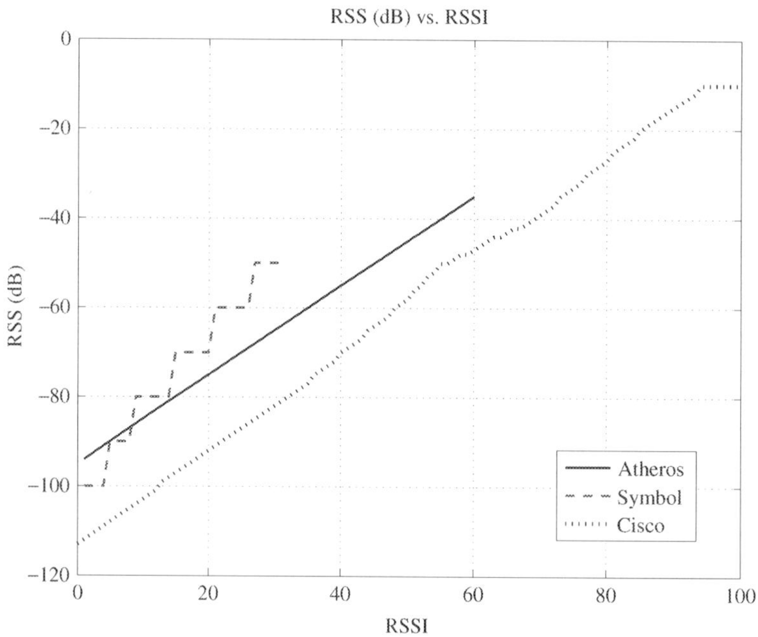

FIG. 5 The relation between RSS values in dB and RSSI for three different chipset manufacturers.

3.1 Calibration Issues

3.1.1 The Effect of RSS Offsets

RSS offsets and their impact on the positioning accuracy were studied, for example, in Chen et al. (2005), Vaupel et al. (2010), Laitinen et al. (2015), and Laitinen (2017). Chen et al. (2005) concentrated on the effect of three dynamic factors, namely relative humidity level, people presence and movements, and open/closed doors, on the positioning accuracy. Vaupel et al. (2010) studied the positioning performance of different devices in the same environment. More extensive research of the effect of an offset between the RSS values in the training and estimation phases for WLAN-based positioning was presented in Laitinen (2017) and Laitinen et al. (2015).

In Laitinen (2017), the effect of an offset between the RSS values in the training and estimation phases was studied by adding artificially a bias, b, to the original measured RSS. All results presented by Laitinen (2017) and Laitinen et al. (2015) are based on real RSS measurements in several multifloor buildings. Different offset types have been considered, and the bias can be chosen to be either a constant one, a random one or a localized constant bias, where a RSS offset occurs just for a certain part of the fingerprints.

Based on the research in Laitinen (2017), Fig. 6 presents two examples of the Cumulative Distribution Function (CDF) of absolute distance error for two different measurement set-ups, for different bias types. For constant biases, the artificial bias is chosen to be $b = \pm 10$ dB, $b = \pm 20$ dB, and $b = \pm 30$ dB. In the case of random bias, the amount of a bias

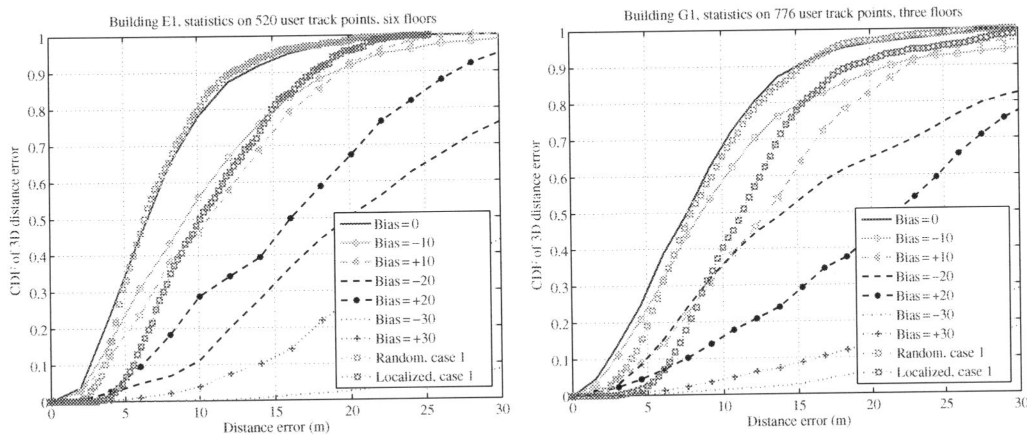

FIG. 6 CDF of absolute distance error for two measurement set-ups: six-floor shopping mall *(left)* and three-floor shopping mall *(right)*.

is varied randomly between −10 dB and +10 dB according to uniform distribution. In the case of localized bias, −10 dB artificial bias is set up for 50% of the FPs in the database, and +10 dB bias for the remaining 50% of the FPs.

As seen in Fig. 6, a random bias between −10 dB and +10 dB affects the results very little. Thus, small RSS fluctuations can be easily neglected. Constant RSS offsets, including the localized case, clearly affect more. A negative constant bias of −10 dB increases the mean distance error less than 20% in general, whereas a positive constant bias of +10 dB is more severe, it increases the averaged mean distance error more than 50% (Laitinen, 2017; Laitinen et al., 2015).

3.1.2 Possible Calibration Methods

If the RSS offset between different equipments is wide, some calibration is needed in order to compensate the possible deterioration in the positioning accuracy. Various calibration techniques have been studied in the literature, for example, Haeberlen et al. (2004), Vaupel et al. (2010), Machaj et al. (2011), Kjaergaard (2006), Kjaergaard and Munk (2008), Cheng et al. (2013), and Wang and Wong (2014). Calibration methods can be performed in offline (training) or online (estimation) phase. Offline calibration means premapping between different equipment in the training phase. It requires a lot of measurements and large datasets in order to be able to calculate models for appropriate mapping. If the dataset is collected with one device only, it is also possible to use so-called learning period before the localization phase with another device (Haeberlen et al., 2004; Kjaergaard, 2006). In online calibration, no mapping models are needed, but the adaptation is obtained using, for example, RSS differences (Hossain et al., 2013) or RSS ratios (Kjaergaard and Munk, 2008; Cheng et al., 2013) instead of absolute RSS values in the training database.

The smallest enclosing circle-based fingerprint clustering from Liu et al. (2016) is another method proved to help in calibration and to improve the position accuracy up to 29%. An interesting alternative to online adaptation is to use so-called test rank based method proposed in Machaj et al. (2011), where the detected transmitters in the measurement are sorted according to their RSS values. The order of the transmitters forms the so-called ranking vector, and in the positioning phase, only these ranking vectors of the FP and the mobile are compared. A similar approach is also presented in Yedavalli et al. (2005). However, also these approaches, assume that the order of the sorted RSS values is the same between different devices. More discussion on different calibration schemes can be found in Jung et al. (2017), for example.

3.2 Database-Size Reduction

Another challenge in fingerprinting is when we deal with mobile-centric positioning and need to transfer a large amount of data from the location server side to the mobile side in order to enable the mobile device to compute its own position. In such situations, it is important to reduce the size of the databases transferred from the server side to the mobile side, not only to increase the speed of the position computation, but also to lower the mobile's battery consumption and to decrease the data consumption at the mobile side. Some of the methods to achieve a reduction in the training databases transferred to the mobile are discussed in the following subsections.

An obvious way to reduce the amount of transmitted data is to decrease the number of fingerprints. Increasing the average distance between fingerprint locations of the training database is one option to achieve that, for example, by increasing the fingerprint grid size and average the neighboring RSS, by simply subsample the fingerprints, or by collecting the fingerprints sparsely, in the first place, and then interpolate RSS spatially. Naturally, there is a trade-off between the positioning accuracy and the density of fingerprints.

3.2.1 Compression and Clustering

Compression and clustering methods for training databases are a further option to decrease the amount of data to be transmitted from the network or location server to the mobile side. At this point, the reader is referred to the discussion in Section 2.

3.2.2 Access Point Number Reduction

The primary target for the WLAN network configuration is to guarantee a good coverage and to serve plenty of simultaneous users as efficiently as possible. The AP infrastructure may therefore include several transmitters located very close to each others, or transmitters that are part of a virtual WLAN, using multiple MAC addresses and Basic Service Set Identifiers (BSSID), which will be seen as several transmitters at the same location. From a positioning point of view, the data transmitted by those APs is likely to be highly correlated and therefore redundant. Thus, not all of the heard APs add information to the positioning, that is, the position accuracy is retained if that information were removed. This data only increases the storage demands and the complexity of the localization process unnecessarily and a reduction is commonly desired.

AP selection has been widely studied (Fang and Lin, 2012; Zou et al., 2015; Zhou et al., 2013; Miao et al., 2014; Liang et al., 2015; Youssef et al., 2003; Chen et al., 2006; Kushki et al., 2007). In addition, AP selection together with the grid interval and with several positioning approaches has been addressed in Laitinen (2017) and Laitinen and Lohan (2016). AP selection can be done in the online positioning phase (Kushki et al., 2007; Zou et al., 2015), in the offline training phase (Youssef et al., 2003; Chen et al., 2006; Zhou et al., 2013), or by taking into account both phases (Laitinen et al., 2012; Miao et al., 2014). However, the biggest advantages are achieved when the selection is performed already in the training or offline phase, in order to reduce the amount of stored and transferred data.

The most used AP selection methods are max-mean proposed in Youssef et al. (2003) and InfoGain (also called Entropy) proposed in Chen et al. (2006). The best selection criteria for both fingerprinting and path-loss-based positioning approaches was shown in Laitinen (2017) to be the maxRSS criterion, where 40% or more of the available APs are chosen to form the AP subset, based on the RSS values sorted from maximum to minimum. Another interesting criterion is called multiple BSSID selection, where the idea is to choose only one AP among the closely located ones (i.e., within one meter), in order to remove the correlated data (Laitinen, 2017). According to Laitinen (2017) and Laitinen and Lohan (2016), up to 60% of the APs can be removed without deteriorating significantly the localization performance. However, with this criterion, the number of closely located AP or those supporting multiple BSSID is dependent on the building, and therefore the percentage of removed APs may vary between only few percent and up to 60%. Thus the gain in the database reduction can be smaller than in the case when maxRSS selection criterion is used.

3.3 Measurement Gaps

For a global localization system, it is extremely challenging to achieve a complete fingerprint measurement set with no geographical measurement gaps. Even if one invested in particularly extensive measurement campaigns, there would always be areas with restricted access, which basically would decline the opportunity for achieving full coverage databases. In addition, maintaining and updating a global fingerprint database are enormous tasks, which can by themselves lead to coverage gaps. For instance, this could happen when the measurement data from specific area is declared outdated before new measurements become available.

In order to mitigate the negative effects of the coverage gaps, the missing fingerprints need to be estimated. This is fundamentally an interpolation and/or extrapolation task. In this context, the term interpolation refers to the process where RSS values are estimated for measurement gaps between available fingerprint coordinates, that is, inside the convex hull of the fingerprint coordinates. Moreover, the term extrapolation refers to the estimation of RSS values outside the convex hull of the fingerprint coordinates. From these two, the extrapolation task is usually more challenging than the interpolation task due to the poor geometry of the known RSS values and large average distances to the closest known RSS values. In addition, whereas with conventional interpolation techniques the RSS estimates are naturally limited by the known neighboring RSS values, many extrapolation techniques, such as gradient-based extrapolation, might produce RSS estimates which are out of the range from sensible physical values.

When considering a target area with precollected fingerprints, the actual total number of fingerprints is not necessarily the most important parameter, but more likely the distribution of the fingerprint coordinates is the crucial information. Hence, missing a fingerprint here and there does not affect as much as missing a large chunk of fingerprints from a certain location. However, due to the complicated nature of the positioning system, which is highly dependent on the considered environment, it is challenging to approximate how the possible coverage gaps in the database affect the system performance when compared to full coverage database. In addition, it is as interesting to find out how the above-mentioned interpolation and extrapolation techniques can facilitate the mitigation of negative effects caused by the coverage gaps.

One approach in order to study the effect of coverage gaps on the system performance is to first collect a solid set of fingerprints from the target area, and then, to artificially remove fingerprints from the dataset. Now, by comparing the performance between the original full coverage dataset and the dataset with removed fingerprints, the effects of missing fingerprints should become evident. However, as stated before, it should be noticed that removing arbitrary fingerprints here and there does not reflect the case of realistic coverage gaps in the database. As studied in Talvitie et al. (2015); Talvitie et al. (2014), one way to mimic realistic coverage gaps is to remove fingerprints in specific areas, where all fingerprints within a certain distance from a randomly selected location, are removed from the database. Then, after removing the desired number of fingerprint chunks, the resulted database can be used for studying the positioning performance.

An outcome of an arbitrary fingerprint removal process for a database with indoor WLAN measurements is illustrated in Fig. 7, where the original database is compared with the databases from which 30% of the data points have been artificially removed. In the other set, the data points have been removed uniformly, and in the other, the removal process has been done based on the fingerprint chunks. From here it is evident that the chunk-based removal method introduces more substantial coverage gaps in the database compared to the uniform removal. The radius of the fingerprint chunk can be determined based on the physical dimensions of the studied positioning environment, for example, based on the average room size for indoor scenarios, or on the average block size in a street plan for outdoor city environments.

The effect of coverage gaps on the positioning accuracy has been presented in Talvitie et al. (2015), where an indoor database consisting of WLAN APs in a large university building is studied over different database coverage gap scenarios.

In Fig. 8, the effect of coverage gaps on the positioning accuracy is compared between the original database, the databases with coverage gaps, and the databases where the coverage gaps have been filled by interpolation and extrapolation. In this case, the interpolation and extrapolation has been performed based on Inverse Distance Weighting (IDW) method, where an RSS estimate for any location (x, y) is obtained as a weighted average over the known RSS values as

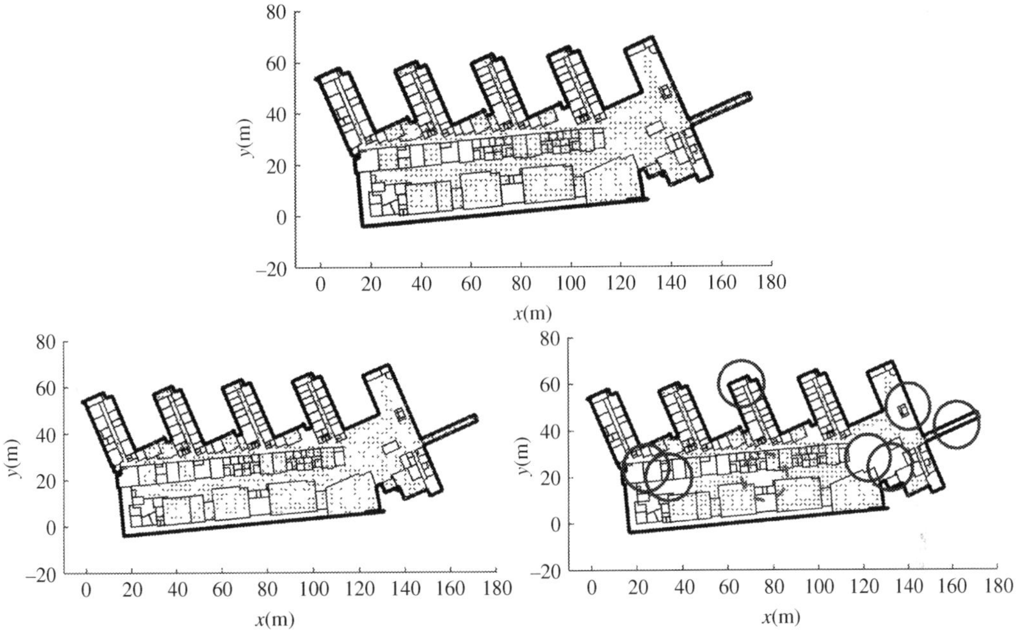

FIG. 7 Example of original fingerprint database (on *top*), and the corresponding databases after randomly removing 30% of the data points, based on uniform removal (on *left*) and chunk-based removal (on *right*). The *continuous circles* indicate the locations of the removed chunks, and the *dashed circle* is the last removed area, where part of the fingerprints has been retrieved in order to achieve the exact desired database reduction percent.

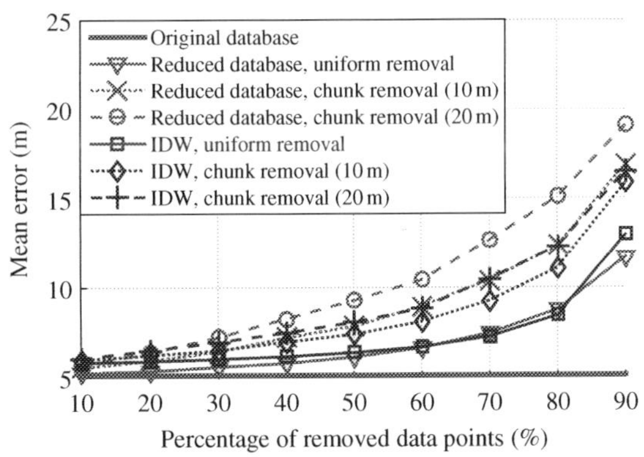

FIG. 8 Mean positioning error as a function of data point removal percentage for the original database and the reduced databases, including both the uniform removal approach and the chunk removal approach (with 10 m and 20 m chunk radii), and for the IDW-based interpolated/extrapolated database.

$$\hat{P}_{RSS}(x, y) = \sum_{i=0}^{L-1} w_i P_i(x_i, y_i), \text{ with } w_i = \frac{d_i^{-u}}{\sum_{j=0}^{L-1} d_j^{-u}} \quad \text{and} \quad d_i = \sqrt{(x_i - x)^2 + (y_i - y)^2},$$

where w_i and $P_i(x_i, y_i)$ are the ith weight and the corresponding known RSS value at location (x, y). The AP subscript has been dropped here for clarity, as such approaches are applied for each AP. Moreover, d_i is the distance between the estimate location and the location of the ith known RSS value, and L is the number of known RSS values in total. Fig. 8 shows that when the number of coverage gaps increases, that is, the number of fingerprints in the database is reduced, the performance offered by the interpolation/extrapolation improves. In addition, it can be observed that the performance degradation caused by uniformly removed fingerprints is at considerably lower level compared with the chunk-based removal process.

3.4 Height or Floor Estimation

A further challenge in indoor fingerprinting is encountered in multifloor buildings with open spaces, such as malls or automation halls. The floor identification or the exact height estimation in such scenarios is very challenging. There are two ways of looking at this issue: On one hand, the fingerprints can be collected and stored per floor, with a floor index $f = 1, 2, \ldots$ saved as the height dimension, no matter of the height of the measurement device used to collect the measurements. On the other hand, the exact height of the fingerprint can be inferred from additional sensor measurements such as barometer or accelerometer readings, then the z-coordinate would reflect the mobile's height. At the present time, there are very few articles investigating which of these two approaches is better. For example, Xiao et al. (2017) shows that vertical accuracy can be improved with a 3D model where the fingerprints height is perfectly known. Floor detection studies can also be found in Jia et al. (2016), Moreira et al. (2016), and Razavi et al. (2015).

4 Integration of WLAN With Other Signals of Opportunity

While there are already solutions capable of achieving submeter level performance in some indoor environments (Kotaru et al., 2015), not a single system can be seen as the ultimate solution for indoor scenarios or for seamless positioning. Especially one that can accommodate cost, battery performance, and different degrees of positioning accuracy (Liu et al., 2007).

The plenitude of man-made and natural signals offer a significant advantage indoors compared to outdoor scenarios, where cellular network and satellite signals are often more predominant. Hence, in indoor scenarios there are many man-made signals, from different spectra such as radio and light, or sound, and with their own very specific purposes. For example, WLAN (or WiFi) networks have the purpose of exchanging data between network peers, visible light installations aim to illuminate the building areas, while others, such as pressure and magnetic fields, are a naturally occurrence due to our planet's inner workings.

However, all of these have their own characteristics and traits which can be exploited for other purposes, including positioning. This section aims to cover the use of some of these sources, called signals of opportunity—signals not originally created with the positioning purpose in mind—and their combination to tackle the problem of indoor positioning.

4.1 Signals of Opportunity and Their Characteristics

As mentioned before, a signal of opportunity (SoO) is any wireless signal which can be used for positioning purpose, but which has been designed with other objective in mind, such as communication, or illumination, etc. When it comes to identifying signals of opportunity, the main goal is to build on existing systems and their own characteristics (Coluccia et al., 2014). In indoor scenarios, WLAN networks have become one of the prime systems that have been exploited for positioning purposes, especially due to their ubiquitousness and integration with mostly all mobile devices. In addition to WLAN, several Bluetooth-complaint technologies are making their way into the majority of user electronics. Nowadays, BLE, for example, is widely encountered on many wireless mobile devices. RFID is another technology that provides SoO and which is more and more used in various applications such as logistics, inventory tracking, conference attendee tracking, or interactive management.

Considering ubiquitousness of a technology provides a good way to identify a SoO. A second concern is the exploitation of the most appropriate system characteristic. In this regard, one has to take into account the principle of operation of such systems and look for specific parameters, such as clock accuracy, channel bandwidth, signal modulation, propagation, among others. In addition to these, one also has to take into account how to aggregate and process these measurements in order to provide a position solution.

With this in mind, Table 3 provides a qualitative overview on certain SoO measurements and their advantages and disadvantages. The measurements in the time and angle domain typically require dedicated hardware. For that reason, RSS is one of the most interesting feature to build upon for an opportunistic positioning.

Table 3 A Qualitative View Regarding Derivatives From Potential Signals of Opportunity

Measurement	Advantages	Disadvantages
Received Signal Strength	Inherent characteristic Low cost and burden	Hardware dependency Sensitive to multipath
Time of Arrival	Lax clock requirements for the network	Requires precise clock in the device Sensitive to small clock errors Usually requires estimation of time component
Time Difference of Arrival	Lax clock requirements for the device	Requires precise clock in the network
Angle of Arrival	Builds on signal's phase	Requires antenna arrays or specialized antennas Accuracy decreases with distance to target

Thus, the next two sections offer two case studies on the combination of RSS from several systems, such as WLAN, BLE, and RFID. While not the focus of our study, geomagnetic fields can also be used as a signal of opportunity as pointed out in Pasku et al. (2017), Shu et al. (2015), and Zhang et al. (2015).

4.2 WLAN and BLE

This section highlights the use of WLAN and standard BLE RSS measurements. This case study uses data from two measurement campaigns carried at the Tampere University of Technology.

BLE has risen in popularity over the last years, especially as a connectivity standard for smaller gadgets, such as smart watches and TVs (DeCuir, 2014). However, its low power has also enabled several location-based services, through the use of its beacon functionality. In such mode, the devices advertise, at a time, in one of three advertisement channels. As such advertisement is caught by scanning devices, such as mobile phones or other dedicated hardware, the RSS indicates the proximity to such tag. Hence, this system is finding its way in the indoor positioning market, where it is often used as an advertisement means to push people into buying a specific product.

The predominance of both BLE and WLAN signals in indoor environments make them an interesting subject of study, allowing one to delve into the benefits of combining RSS measurements from such systems. Hence, this study combines data from two separate measurement campaigns, each used to capture either WLAN or BLE signals, by merging them in postprocessing. Merging these datasets is possible as both datasets are georeferenced in the same local frame.

Overall, our case study shows a small increase in accuracy when combining both technologies, as shown in Fig. 9. In terms of RMSE the combination of both technologies leads to 6.07 m versus, 7.81 m for BLE and 6.71 m for WLAN. While there is an improvement in accuracy when merging both technology, it is not significantly higher than the WLAN-only positioning. One might conclude that the reasons behind this are twofold: on one hand, both technologies operate in the same frequency band, which leads to similar propagation conditions for both technologies, and thus to correlated measurements, and on the other hand, as the WLAN network in the studied building is very dense (more than 500 unique SSIDs within four floors and a horizontal area of about 150 $m \times 100$ m), adding additional BLE transmitters brings in little additional benefit. An open question here is how such a WLAN-BLE fusion would behave in buildings with very low density of APs.

4.3 WLAN and RFID

In this section, we present a case study of combined WLAN and RFID measurements as the inputs for a fingerprinting algorithm with full training database. This study is based on simulations of RSS in WLAN and RFID, followed by field experiments conducted in several studies (especially Hasani et al., 2015 and Lohan et al., 2014) at the Tampere University of Technology.

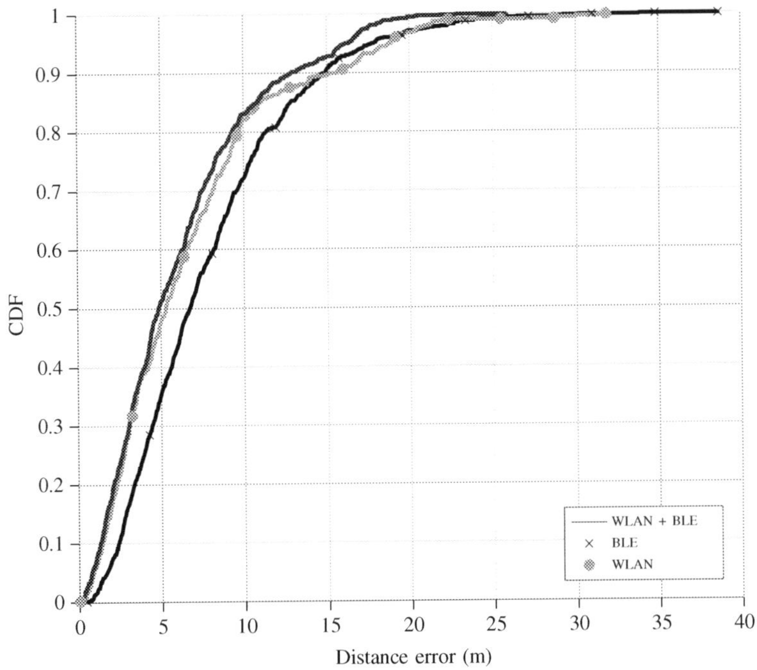

FIG. 9 Positioning performance with WLAN, BLE, and both technologies combined.

RFID has been a leading solution in the field of security for many years and seen as a contender to solve the indoor positioning dilemma. While the cost of its tags is rather negligible, its infrastructure costs is demanding, not only monetarily, but also power consumption wise. Nevertheless, despite its shortcomings, it is a feasible contender in the indoor positioning challenge. The goal of this study was to investigate the benefits that RFID could bring when matched together with WLAN.

In the case study shown here, we made the assumption that four RFID readers were placed in the center of a 25 by 25 m area in two distinct environments: (i) one where WLAN signals are highly available and (ii) another where WLAN signals are scarce. In more detail, for such environments, we assume approximately four APs per 100 m^2 (high availability) and about two APs per 100 m^2. The results are shown in Fig. 10A and B.

Fig. 10 shows first a benefit in both cases of a combined approach compared with the RFID-only and the WLAN-only approaches. Second, it can be seen that the gain in low density WLAN environments is much higher than in a high density WLAN environment, as intuitively expected. In Fig. 10B, the performance gain of the joint positioning is approximately 20% at 4 m error compared with the single-system positioning.

Finally, the figures also point out that position estimates based on RFID are only achievable 40% of the times. This means that the remaining 60% of times, a position estimate is unavailable as the signal strength is below the receiver sensitivity. In other words, the short range of RFID is a liability for a positioning system focused solely on this technology.

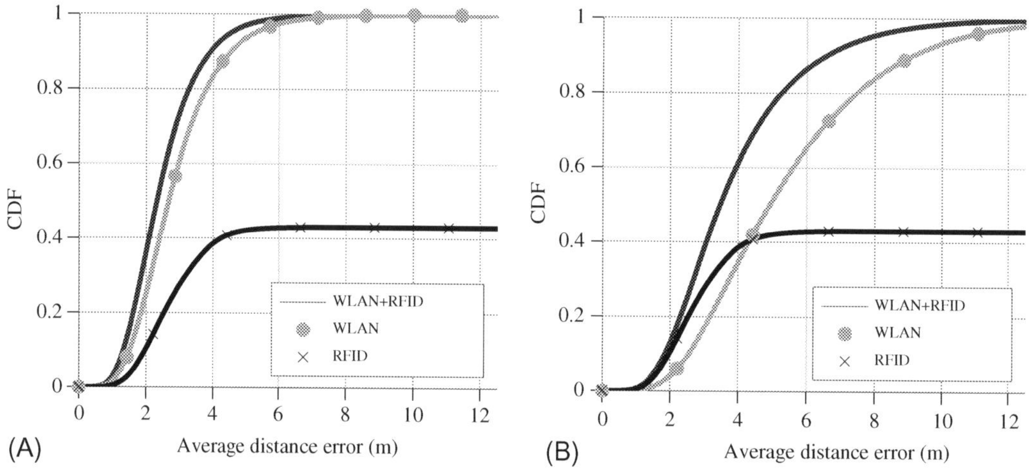

FIG. 10 Positioning performance with WLAN, RFID, and both technologies combined in a simulated scenario. (A) High availability of WLAN; (B) Low availability of WLAN.

5 Integration of WLAN With GNSS

As pointed out in the previous sections, a positioning system that provides seamless accurate location information in indoor and outdoor environments is still lacking. An open issue is particularly the position accuracy in environments that are not indoors, but where GNSS positioning is heavily degraded due to the lack of line-of-sight (LOS) to the satellites. Examples for such environments are the transition zones between indoors and outdoors, semiopen spaces with light or only partial roofs, and also densely constructed city centers.

A common strategy to provide accurate positioning in these areas is to combine GNSS with indoor location systems, such as WLAN-based positioning. The range of WLAN signals and their ability to penetrate structures enable seamless positioning. This section focuses on solutions to seamless indoor/outdoor positioning by merging GNSS and WLAN.

Several different approaches for the integration of GNSS-based positioning and WLAN-based positioning have been published. Depending on their level of integration or type of measurement they can be categorized in three groups:

1. methods that *switch* between GNSS and WLAN derived position estimates,
2. methods that *average* the position estimates from both systems, and
3. methods that *fuse* information from each of the systems and then compute a joint position estimate.

Most of the systems that belong to the first category simply use the availability of GNSS to choose which system to use. Another criterion to decide for one or the other system is the energy consumption, or an additionally derived accuracy measure for the position estimate of the subsystems.

Methods of the second category combine the position estimates from GNSS and WLAN by either computing the weighted average, or by sequential Bayesian filtering. To perform reliably, these approaches need to determine adequate weights (in a Kalman filter, e.g., the weights are determined by the measurement covariance); this is crucial and usually the biggest challenge.

The third category of approaches yields a position estimate by fusing GNSS pseudoranges with either ranges or RSS deduced from WLAN signals. Such methods can be used even when less than four satellites are in view. RSS from just a few AP can be beneficial, so that the robustness compared with the stand-alone solutions is improved, together with accuracy and precision.

In particular the methods that belong to the latter group, that integrate raw-data to a common solution, are able to yield accurate seamless positioning and are therefore detailed in the following subsections. For further information about the other approaches we refer to Richter et al. (2014) and the references therein.

5.1 Fusing GNSS Pseudoranges With WLAN Ranges

The integration of GNSS pseudoranges and WLAN ranges is conceptually straightforward. GNSS pseudoranges and WLAN ranges are joined in a common set of multilateration equations which can be solved by a least-squares method or a closed form solution such as Bancroft's algorithm. The works by Li and O'Keefe (2013) and Nur et al. (2013) are notable advances in that topic. They also showed that, based on the current hardware, timing, synchronization, and NLOS condition still pose a challenge for ranging with WLAN signals. The LOS requirement is especially hard to fulfill and hinders these approaches in achieving accurate seamless positioning. WLAN hardware supporting the Fine Timing Measurement procedure and eventually the new generation WLAN standard (IEEE 802.11az) are expected to push the accuracy of these fusion approaches further.

5.2 Fusing GNSS Pseudoranges With WLAN RSS

Fusing GNSS pseudoranges and WLAN RSS means to combine different physical quantities. This is achieved by encoding the information of the different measurements in a universal manner. The existing approaches (Hejc et al., 2014; Richter and Toledano-Ayala, 2017) compute a joint likelihood function from the observed pseudoranges and RSS. To obtain a position estimate from the joint likelihood function one may compute the maximum likelihood estimate. The Bayesian point estimates are also a common choice.

Due to the use of fingerprinting on the WLAN side, the overall system shares the challenges of fingerprinting methods, most prominently, those referring to the construction and maintenance of the training database. An additional issue when it comes to fusing pseudoranges and RSS are the different state spaces. The state space of pseudoranges is continuous, whereas the fingerprints are defined on a spatially discrete state space. Recently, Gaussian process (GP) regression, a machine learning technique, was proposed to alleviate both issues (Richter and Toledano-Ayala, 2017).

Gaussian processes (Rasmussen and Williams, 2006) are a generalization of multivariate normal distributions: instead of a mean vector and a covariance matrix, they are specified by a mean and covariance function. In the context of WLAN fingerprinting, the Gaussian processes exploit the spatial correlation of the RSS samples (fingerprints) to establish a model that permits to regress RSS spatially. Gaussian process regression is also known as "kriging." After the model is learned, fitted to the RSS fingerprints of the training database, RSS can be interpolated at arbitrary positions. The effort of constructing the radio map is alleviated, because fewer fingerprints are required to achieve comparable localization performance as using dense fingerprints and no interpolation (Schwaighofer et al., 2004). The different state spaces are addressed, because the Gaussian process model provides a continuous functional representation of the spatial RSS distribution.

The method reported in Richter and Toledano-Ayala (2017) is illustrated in Fig. 11. It inherits the training and estimation stage from fingerprinting and incorporates Gaussian process regression in a particle filter (PF) in the estimation stage.

5.2.1 Training Stage

During the training stage, or offline phase, the algorithm proposed by Richter and Toledano-Ayala (2017) creates the radio map and computes a Gaussian process model for each access point. To find an appropriate Gaussian process model for one access point a mean and a covariance function has to be specified a priori. In a second step the parameters of these functions, so called hyperparameters, have to be found by fitting the

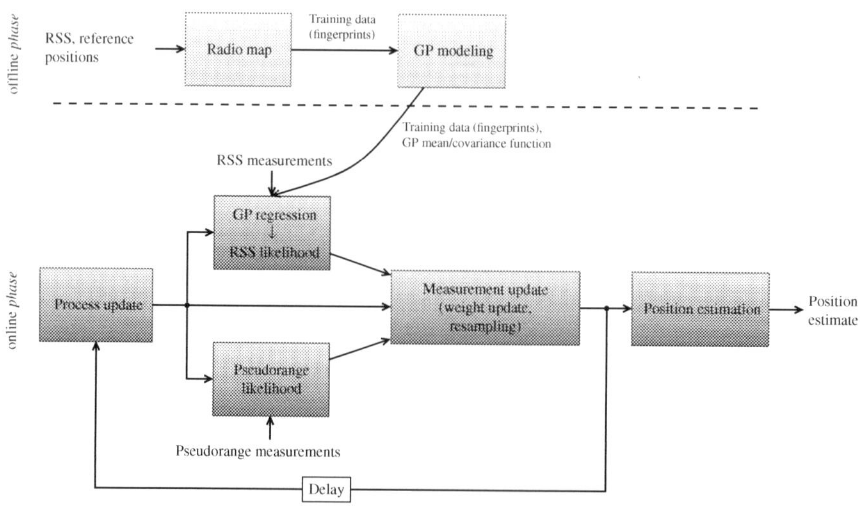

FIG. 11 Flow chart hybrid positioning system based on Bayesian filter fusing GPS pseudoranges and WLAN RSS.

mean and covariance function to the fingerprints (They define the mean and covariance functions' specific form.). Once the model is trained, that is, the optimal hyperparameters for the chosen mean and covariance function are found, the resulting Gaussian process model provides a continuous functional description of RSS over space. Such a model has to be obtained for each access point.

From the mean function of the trained Gaussian process RSSs can be predicted, and the covariance function gives corresponding variance estimates for these RSSs. The variance stems from the spatial correlation captured in the model. The correlation between neighboring RSS declines the larger their distant. Predicted RSS farther away from fingerprints have larger variance and, thus, areas lacking training data have larger variance than areas in which fingerprints are available.

5.2.2 Estimation Stage

In the estimation stage, or online phase, Richter and Toledano-Ayala (2017)'s algorithm employs a generic particle filter (Ristic et al., 2004) to track the mobile. A particle filter approximates numerically the recursive Bayesian estimator consisting of the process update and the measurement update. The process update models the motion of the mobile. The outcome of the process update is a distribution that predicts the position of the mobile at the next time step based on the current position and using the motion model. In Richter and Toledano-Ayala (2017) a random walk velocity process is used as process model.

The measurement update corrects the predictions made before in the process update. This correction incorporates the information of the measurements in form of a likelihood function and provides the posterior distribution. For each set of observables—a set of pseudoranges, one for each satellite, and a set of RSSs (averages), one for each access point—a likelihood function is required. Richter and Toledano-Ayala (2017) model both, pseudoranges and RSS, as normally distributed random variables.

The pseudoranges are modeled as sum of the geometric distance between receiver and satellite, plus the clock offset and normally distributed noise. A variance for each pseudorange is estimated by the GPS receiver and is based on the User Range Accuracy.

To model the RSS average, the Gaussian process model from the training stage is used. The noise is again assumed to be normally distributed and a Gaussian likelihood function is established. Its mean and variance are computed by using the mean and covariance function of the Gaussian process model.

The key idea of particle filtering is to sample the functions involved in the process and measurement update of the Bayesian filter, so that integrals are replaced by sums which then can be solved numerically. During one cycle of the recursion (see online phase in Fig. 11) the process update predicts the mobile's location based on the motion model and a previous estimate of the location, and the measurement update corrects this prediction by incorporating the observations in form of the mentioned likelihood functions. The outcome of the measurement update is the posterior distribution of the mobile's location.

This posterior distribution describes the location estimate and its uncertainty and is also input for the following cycle. A position estimate can simply be the maximum of the posterior distribution or the conditional mean.

Noteworthy is the choice of the variances in Richter and Toledano-Ayala (2017). In the case of pseudoranges they are obtained from the GPS receiver at each epoch. The variance of a RSS observation is provided by the Gaussian process model, thus, encodes spatial information instead of temporal information.

5.2.3 Performance of Pseudorange and RSS Fusion Filter

The performance of the above presented fusion scheme has been investigated in an indoor/outdoor environment harsh for GNSS and WLAN fingerprinting (Richter and Toledano-Ayala, 2017): Large parts of the test tracks are on semiopen, roofed passageways along the buildings, several trees cover the open area of the testbed; moreover, fingerprints have been established not only on the tracks, but as well on the open spaces between the buildings. Thus, this test environment is rich in multipath and ambiguous fingerprints and presents a typical case for seamless positioning.

The positioning results for two test tracks (Track-1 and Track-2) of the proposed PF are compared with that of an extended Kalman filter (EKF). Both filter employ the same process update. The EKF processes pseudoranges, as typical in GNSS receivers, and integrates a position estimate from the fingerprinting-based system.

The general performance of this approach to seamless positioning is depicted in Fig. 12. Track-1 has more indoor sections, it starts in the left building and enters the right building toward the end of the path (cf. solid path with arrows). Track-2 consists of more outdoor sections as it starts outdoors on the large open space, takes a rectangular path on the lawn area toward the right building. The track's only indoor section is in the northern part of the right building, from where it continues to the left building, where it circles the small lawn

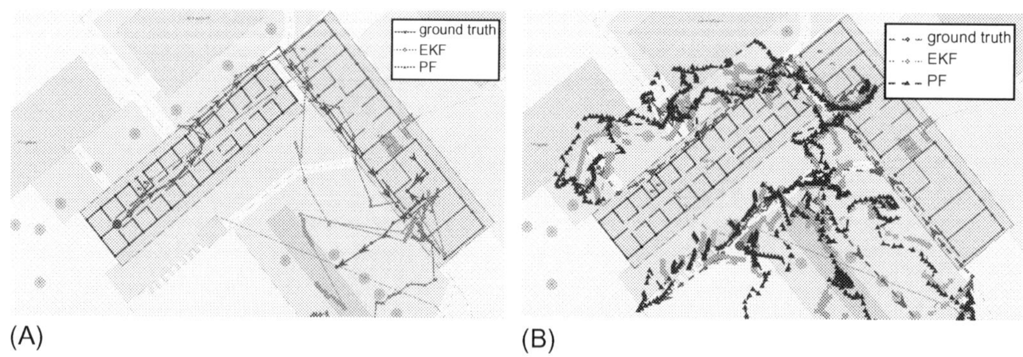

FIG. 12 Testbed, ground truth, and estimates of Track-1 and Track-2 of seamless indoor/outdoor positioning experiment using GPS pseudoranges and WLAN signal strength. (A) Track-1; (B) Track-2 (© OpenStreetMap contributors).

area between the buildings and then returns to the large open space area where it started (cf. red, solid path with arrows).

The PF's estimates of Track-1 (cyan, triangle, solid) follow the ground truth quite well. Larger deviations can be observed in the last third of the track in the indoor part of the right building and in the following outdoor section. The EKF performance on that track (magenta, square, solid) is rather poor. For Track-2, both filter perform similarly. Both filter estimate well the indoor section. During the section outside of the left building the accuracy of the EKF appears higher than that of the PF.

The empirical cumulative distribution function of the two filters is shown in Fig. 13. The PF has a median accuracy of 5 m for both tracks, whereas the EKF yields about 10 m for Track-1 and 6 m for Track-2. Excluding for a moment the EKF for Track-1, the error for fusing pseudoranges with RSS is below 10 m in 90% of time. The large maximum error of the PF for Track-2 is due to the poor initial estimate.

Table 4 summarizes the above results in terms of the mean absolute error (MAE). It compares additionally the PF hybrid solution with particle filter configurations in which either only GPS or only WLAN is used. The combination of both sources of information

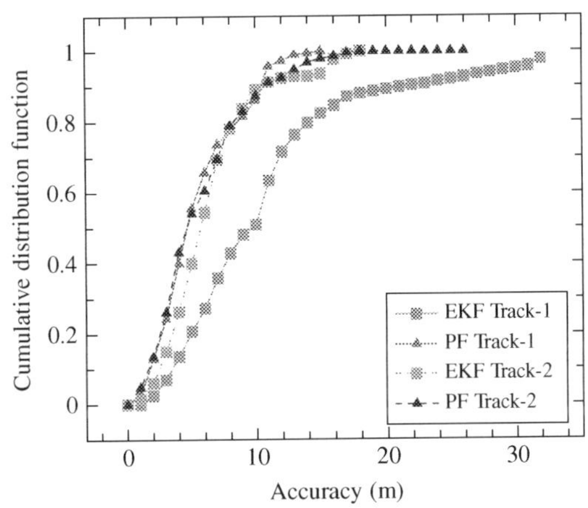

FIG. 13 Cumulative distribution function.

Table 4 Mean Absolute Error for Track-1 and Track-2 Using the PF in GPS+WLAN, GPS-Only, and WLAN-Only Configuration and the EKF

	PF			EKF
	GPS+WLAN	**GPS-Only**	**WLAN-Only**	**GPS+WLAN**
MAE Track 1 (m)	5.36	32.37	6.04	10.72
MAE Track 2 (m)	5.58	6.73	10.46	5.21

clearly outperforms the GPS-only and WLAN-only solutions. The GPS-only solutions fails in estimating Track-1 due the lack of GPS signals. In the case of Track-2, the accuracy of the GPS-only configuration achieves almost the accuracy of the PF filter using pseudoranges and RSS; this suggests that the GPS receiver uses robust methods to estimate the pseudoranges and that the WLAN data does not contribute much to GPS+WLAN solution. The WLAN-only solution performs better on Track-1 than on Track-2, which is reasonable as Track-1 is largely indoors and Track-2 consists of more outdoor sections.

The poor position accuracy of the EKF on Track-1 is due to the unavailability of GPS fixes and because the fingerprint estimates are not trusted sufficiently compared to the pseudoranges. Thus, the weighting between pseudoranges and WLAN position is deficient. Nevertheless, it works fine for Track-2. The PF in contrast performs consistently on both tracks with an average error of 5.5 m. The mechanism to weight pseudoranges and RSS is able to cope with the challenging localization scenario, including a constantly changing sky-view, multipath propagation, and ambiguous fingerprints.

6 Integration of WLAN With Other Data

If WLAN enables the seamless feature, the challenge is the integration with other sensor data in order to yield accurate localization in all environments.

Furthermore, suitable for seamless positioning are all those methods relying on spatially unique sensory data indoors and outdoors such as camera images or RF signals, so that fingerprinting can be applied; or those methods providing uninterrupted reception of sensory data between indoors and outdoors (e.g., source-to-receiver path not blocked by building structures), such as inertial data or magnetic field. A selection of these methods is briefly overviewed in the following subsections.

6.1 Inertial Data

With the development of small and affordable inertial measurement units (IMU) and their proliferation into consumer devices, transforming mobile phones to multisensor platforms, a lot of progress has been made toward seamless positioning. IMUs provide acceleration and angular rate of the device they are attached to and they describe the relative motion of the device through velocity and orientation. These data are independent of the environment and therefore well suited for seamless localization. To integrate inertial data with RSS, recursive Bayesian filters, such as Kalman filters, are applied, where the inertial data are fed into the motion update step. Many schemes to fuse IMU data exist, as it is possible to compute a position relative to the previous position directly from the IMU data. However, such details are out of the scope of this chapter. IMU-assisted WLAN positioning has been addressed, for example, in Jin et al. (2014).

6.2 Vision Navigation

Vision navigation solutions refer to the solutions which rely on images or videos captured by fixed or mobile cameras. A building or a street can be identified also by visual patterns and benchmarks (e.g., a door shape, a tree, etc.), in a similar way with RF-based fingerprinting. The vision navigation relies on comparing visual features collected in an offline training phase with visual captures during the online estimation phase. Visual markers (e.g., paper markers) could be additionally placed in a building to increase the accuracy of the estimates, as done, for example, in Mulloni et al. (2009).

6.3 Visible Light Positioning

Visible light positioning (VLP) is another low-cost positioning method, which relies on line-of-sight (LOS) measurements from light emitting diodes (LED). VLP typically relies on Time of Arrival or Angle of Arrival information and can achieve submeter accuracy. A good overview of various VLP solutions can be found, for example, in Fang et al. (2017).

6.4 Magnetic Field Navigation

Magnetic field navigation is based on the observation that magnetic patterns in various parts of a building have certain characteristic features, which enable a mobile device to recognize its location based on pattern matching or fingerprinting with these magnetic field patterns. Many mobile devices already support magneto-resistive sensors and thus have the capability of doing magnetic field measurements (Nurmi et al., 2017).

6.5 Positioning With Sounds or Ultrasonic Waves

Sound-based positioning systems have been investigated in Hazas and Hopper (2006). By difference with the RF waves, which move at light speed, the sound speed is only about 346 m/s. The sound-based solutions typically rely on Time of Arrival and Angle of Arrival measurements and they are able to offer an increased location privacy to the user. The results reported in Hazas and Hopper (2006) showed cm-level achievable accuracy with an ultrasonic positioning system. The main disadvantages are the need for additional hardware and large and bulky sensors needed at both transmitter and receiver side. Sound-based positioning systems are at the moment rather little investigated or analyzed.

Fig. 14 summarizes how the different systems addressed above can be used with fingerprinting and hybridization. As seen here, most of the systems provide some form of signal strength or intensity which enables fingerprinting for positioning. Afterward any combination of two or more such systems can be in principle hybridized, either at measurement level or at position-estimate level. Examples of hybridization algorithms are enumerated in the last column of Fig. 14.

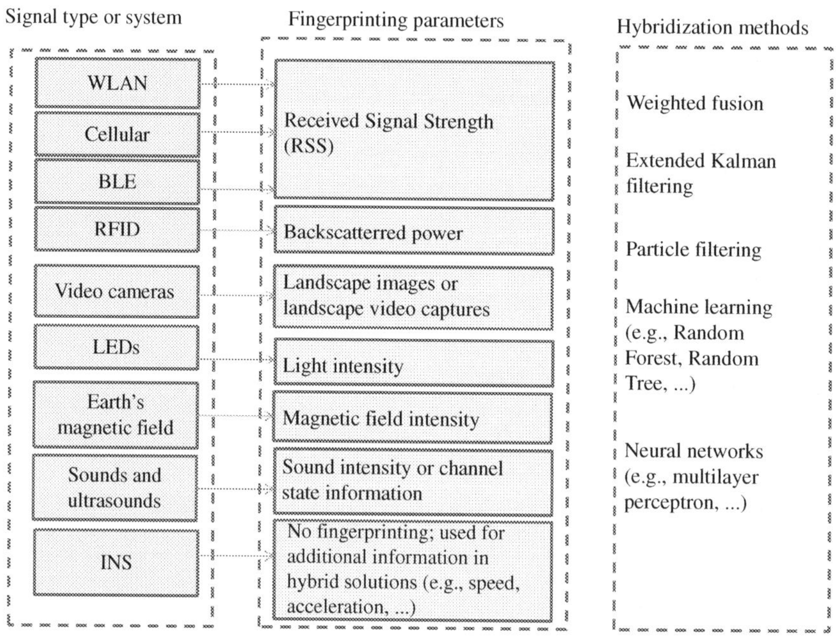

FIG. 14 Summary of hybridization methods based on fingerprinting approaches.

6.6 Multimodal Positioning

From a user perspective, the seamless positioning is not only desired between different environments but between different types of motion as well. Multimodal transportation, meaning that various type of transport is used, such as bike, bus, car, tram, etc., is a typical scenario in urban environments. An example might be a user taking a bus to the train station to travel per train to the next city where he walks or takes a taxi to his destination. In this example the user employs different vehicles, which have all different types of motion. To achieve accurate seamless positioning, these modes of motion need to be appropriately modeled and then detected by the positioning algorithm. The integration of such models with sensory (e.g., RSS) and/or location data is usually achieved by Kalman filter type methods. The interacting multiple model (IMM) algorithm allows further to detect different motions and weights between the outcomes of parallel running Kalman filters which each employ a different motion model (Bar-Shalom et al., 2001). The integration of inertial data into the motion models is particularly beneficial for this approach.

6.7 Cloud Architectures

In order to decrease the computational resources at the mobile side, and thus to increase the mobile battery life, cloud-based solutions such as the cloud GNSS concept

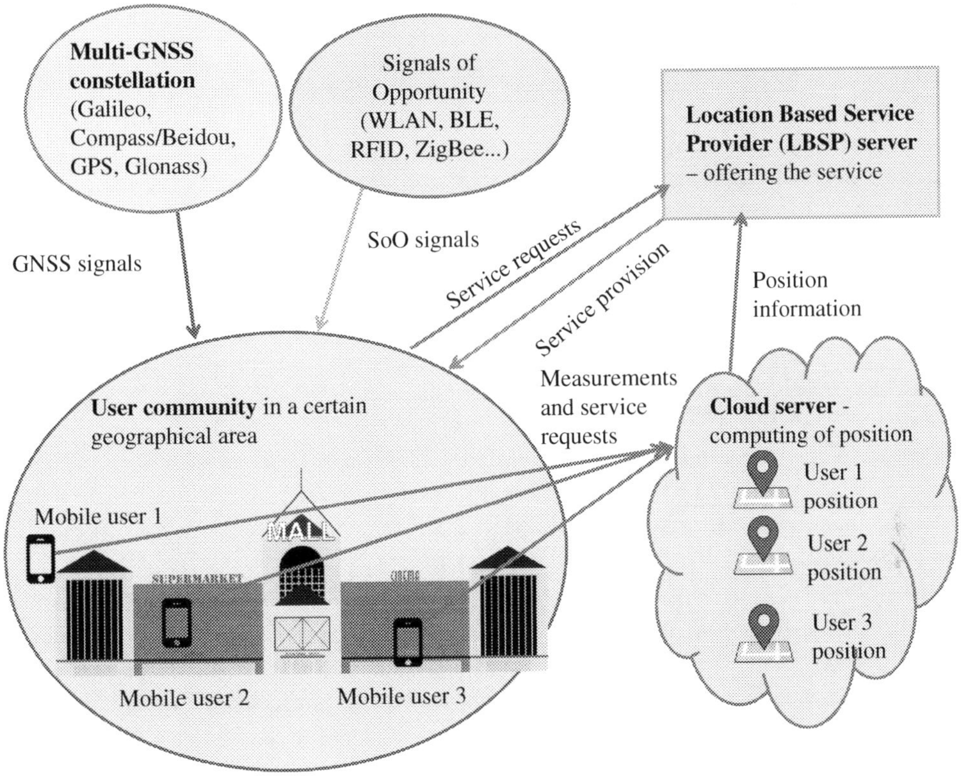

FIG. 15 Cloud positioning concept.

(Lucas-Sabola et al., 2016) are emerging. The concept is illustrated in Fig. 15 and it relies on the idea that various measurements provided by users in a certain geographical areas, such as GNSS pseudoranges, WLAN RSS, etc., are sent to a cloud server. The cloud server uses the information provided by all the users in the area and estimates jointly the user positions. In such a way, a user with clear GNSS signal (e.g., placed outdoors near a shopping mall) can help via his/her measurements the position estimate of the users nearby, who are indoors. The other users' location is not visible to the end users; only the cloud server has access at the full information. Such architectures of course may raise issues of privacy and security of the localization solutions, which are outside the scope of this chapter. The readers interested in more details on the security of location can found a good survey in chapter 3 of Zekavat and Buehrer (2011).

The readers interested in more details about existing navigation methods on smartphones, a very good survey has been recently published in Davidson and Piché (2017).

7 Open Issues and Conclusions

Fingerprinting solutions, typically based on RSS measurements are among the most spread low-cost localization solutions nowadays in indoor and urban areas. In order to achieve seamless localization, one key point is the seamless integration of different fingerprint databases and other navigation solutions, such as GNSS, sensors, etc. The emerging communication systems such as cellular 5G or 802.11az WLAN solutions are moving away from fingerprinting toward ranging solutions, relying more and more heavily on angle and time measurements and on the presence of the line-of-sight. Indeed, future wireless communication networks are expected to be very dense, especially in urban areas where the user location needs are stringent, and to operate in wider bands, thus providing inherently a better location accuracy (Koivisto et al., 2017).

This chapter started with an overview of RSS-based fingerprinting with WLAN signals (see Section 2), which currently are the most widespread indoor and urban wireless channels. The main motivation came from the fact that WLAN signals can be used for positioning with no additional hardware or infrastructures and with only minor software updates at network or mobile side. Typical fingerprinting methods, that can be divided into methods relying on full training database, and methods with reduced training database, such as clustering, path-loss, or image-based approaches, were described.

In the following Section 3, the focus was on various challenges in fingerprinting. The main challenges were summarized (see Table 2) and some of these challenges, such as the effect of RSS offsets, database reduction through different mechanisms and dealing with coverage gaps, were explained in more detail together with their possible solutions.

In Section 4, two studies of combining WLAN signals with BLE and RFID signals were described, respectively in order to take better advantage of existing signals of opportunity. A very simple combining mechanism based on fingerprinting data was employed and it was shown that under some assumptions, such as scarce AP distribution of some systems, adding a second system for the positioning is indeed beneficial.

In addition, seamless positioning solutions that rely on GNSS and WLAN were presented in Section 5. Both systems are available world-wide, infrastructure as well as inexpensive end-user devices that provide pseudoranges (Humphreys et al., 2016), so that more accurate and even ubiquitous localization is achievable with that combination. New developments, such as ultra-dense IoT networks, are supposed to deliver submeter accuracy in a seamless fashion and may provide in combination with GNSS true ubiquitous localization.

Fusing GNSS pseudoranges and RSS signatures in particular (see Section 5.2), can overcome the drawbacks that GNSS presents as a ranging-based system. Because WLAN fingerprinting is to some extent complementary to GNSS and it can enhance information of GNSS more than WLAN ranging-based methods. The different error mechanisms of GNSS and WLAN fingerprinting also invite to develop GNSS multipath detection and mitigation strategies. However, ubiquitous localization with that approach requires ubiquity of fingerprint databases with a harmonized access and datasets.

Finally, Section 6 discussed more at large the different hybridization options existing nowadays for seamless positioning and what are other potential systems which can be used for an increased accuracy and/or availability of the positioning estimates.

To conclude, it is the authors' belief that seamless positioning can be achieved only by adequate combining or hybridization of different existing solutions, rather than inventing a new system equally good for indoor and outdoor solutions. With the advent of cheaper and lower complexity GNSS solutions, such as cloud GNSS solutions, it is very likely that future seamless localization systems will have a cloud component, partly relying on GNSS data, and a user-centric component, relying on low-cost easily accessible measurements such as various signal strengths.

References

Abdou, A.S., Aziem, M.A., Aboshosha, A., 2016. An efficient indoor localization system based on affinity propagation and support vector regression. In: 2016 Sixth International Conference on Digital Information Processing and Communications (ICDIPC).

Agarwal, P., Burgard, W., Stachniss, C., 2014. Survey of geodetic mapping methods: geodetic approaches to mapping and the relationship to graph-based SLAM. IEEE Robot. Autom. Mag. 21 (3), 63–80. https://doi.org/10.1109/MRA.2014.2322282.

Bar-Shalom, Y., Li, X.-R., Kirubarajan, T., 2001. Estimation with Applications to Tracking and Navigation. John Wiley & Sons, Inc..

Cadena, C., Carlone, L., Carrillo, H., Latif, Y., Scaramuzza, D., Neira, J., Reid, I., Leonard, J.J., 2016. Past, present, and future of simultaneous localization and mapping: toward the robust-perception age. IEEE Trans. Robot. 32 (6), 1309–1332. https://doi.org/10.1109/TRO.2016.2624754.

Chen, Q., Wang, B., 2015. FinCCM: fingerprint crowdsourcing, clustering and matching for indoor subarea localization. IEEE Wireless Commun. Lett. 4 (6), 677–680. https://doi.org/10.1109/LWC.2015.\break{}2482971.

Chen, Y.-C., Chiang, J.-R., Chu, H.-H., Huang, P., Tsui, A.W., 2005. Sensor-assisted WiFi indoor location system for adapting to environmental dynamics. In: Proceedings of 8th ACM International Symposium on Modeling, Analysis and Simulation of Wireless and Mobile Systems, pp. 118–125.

Chen, W., Chang, Q., Hou, H.-T., Wang, W.-P., 2006. Power-efficient access-point selection for indoor location estimation. IEEE Trans. Knowl. Data Eng. 18 (7), 877–888.

Cheng, W., Tan, K., Omwando, V., Zhu, J., Mohapatra, P., 2013. RSS-ratio for enhancing performance of RSS-based applications. In: Proceedings of IEEE INFOCOM, pp. 3075–3083.

Coluccia, A., Ricciato, F., Ricci, G., 2014. Positioning based on signals of opportunity. IEEE Commun. Lett. 18 (2), 356–359. https://doi.org/10.1109/LCOMM.2013.123013.132297.

Cramariuc, A., Huttunen, H., Lohan, E.S., 2016. Clustering benefits in mobile-centric WiFi positioning in multi-floor buildings. In: 2016 International Conference on Localization and GNSS (ICL-GNSS).

Davidson, P., Piché, R., 2017. A survey of selected indoor positioning methods for smartphones. IEEE Commun. Surv. Tutorials 19 (2), 1347–1370. https://doi.org/10.1109/COMST.2016.2637663.

DeCuir, J., 2014. Introducing Bluetooth Smart: Part I: A look at both classic and new technologies. IEEE Consumer Electron. Mag. 3 (1), 12–18. https://doi.org/10.1109/MCE.2013.2284932.

Della Rosa, F., Leppäkoski, H., Biancullo, S., Nurmi, J., 2010a. Ad-hoc networks aiding indoor calibrations of heterogeneous devices for fingerprinting applications. In: Proceedings of International Conference on Indoor Positioning and Indoor Navigation (IPIN).

Della Rosa, F., Paakki, T., Leppäkoski, H., Nurmi, J., 2010b. A cooperative framework for path loss calibration and indoor mobile positioning. In: Proceedings of 7th Workshop on Positioning Navigation and Communication (WPNC), Dresden, Germany.

Fang, S.-H., Lin, T.-N., 2012. Principal component localization in indoor WLAN environments. IEEE Trans. Mob. Comput. 11 (1), 100–110.

Fang, J., Yang, Z., Long, S., Wu, Z., Zhao, X., Liang, F., Jiang, Z.L., Chen, Z., 2017. High-speed indoor navigation system based on visible light and mobile phone. IEEE Photonics J. 9 (2), 1–11. https://doi.org/10.1109/JPHOT.2017.2687947.

Feng, C., Au, W.S.A., Valaee, S., Tan, Z., 2012. Received-signal-strength-based indoor positioning using compressive sensing. IEEE Trans. Mob. Comput. 11 (12), 1983–1993. https://doi.org/10.1109/TMC.2011.216.

Haeberlen, A., Rudys, A., Flannery, E., Wallach, D.S., Ladd, A.M., Kavraki, L.E., 2004. Practical robust localization over large-scale 802.11 wireless networks. In: Proceedings of 10th Annual International Conference on Mobile Computing and Networking (MOBICOM).

Hasani, M., Talvitie, J., Sydänheimo, L., Lohan, E.S., Ukkonen, L., 2015. Hybrid WLAN-RFID indoor localization solution utilizing textile tag. IEEE Antennas Wirel. Propag. Lett. 14, 1358–1361. https://doi.org/10.1109/LAWP.2015.2406951.

Hazas, M., Hopper, A., 2006. Broadband ultrasonic location systems for improved indoor positioning. IEEE Trans. Mob. Comput. 5 (5), 536–547. https://doi.org/10.1109/TMC.2006.57.

Hejc, G., Seitz, J., Vaupel, T., 2014. Bayesian sensor fusion of Wi-Fi signal strengths and GNSS code and carrier phases for positioning in urban environments. In: IEEE/ION Position, Location and Navigation Symposium—PLANS 2014, pp. 1026–1032.

Hossain, A.K.M.M., Jin, Y., Soh, W.-S., Van, H.N., 2013. SSD: a robust RF location fingerprint addressing mobile devices' heterogeneity. IEEE Trans. Mob. Comput. 12 (1), 65–77.

Humphreys, T.E., Murrian, M., van Diggelen, F., Podshivalov, S., Pesyna, K.M., 2016. On the feasibility of cm-accurate positioning via a smartphone's antenna and GNSS chip. In: 2016 IEEE/ION Position, Location and Navigation Symposium (PLANS), pp. 232–242.

Jia, M., Yang, Y., Kuang, L., Xu, W., Chu, T., Song, H., 2016. An indoor and outdoor seamless positioning system based on android platform. In: 2016 IEEE Trustcom/BigDataSE/ISPA, pp. 1114–1120.

Jin, M., Koo, B., Lee, S., Park, C., Lee, M.J., Kim, S., 2014. IMU-assisted nearest neighbor selection for real-time WiFi fingerprinting positioning. In: 2014 International Conference on Indoor Positioning and Indoor Navigation (IPIN), pp. 745–748.

Jung, S.H., Moon, B.C., Han, D., 2017. Performance evaluation of radio map construction methods for Wi-Fi positioning systems. IEEE Trans. Intell. Transp. Syst. 18 (4), 880–889. https://doi.org/10.1109/TITS.2016.2594479.

Kaemarungsi, K., 2006. Distribution of WLAN received signal strength indication for indoor location determination. In: Proceedings of 1st International Symposium on Wireless Pervasive Computing.

Karttunen, A., Gustafson, C., Molisch, A.F., Wang, R., Hur, S., Zhang, J., Park, J., 2016. Path loss models with distance-dependent weighted fitting and estimation of censored path loss data. IET Microwaves Antennas Propag. 10 (14), 1467–1474. https://doi.org/10.1049/iet-map.2016.0042.

Kasebzadeh, P., Granados, G.S., Lohan, E.S., 2014. Indoor localization via WLAN path-loss models and Dempster-Shafer combining. In: International Conference on Localization and GNSS 2014 (ICL-GNSS 2014).

Khullar, R., Dong, Z., 2017. Indoor localization framework with WiFi fingerprinting. In: 2017 26th Wireless and Optical Communication Conference (WOCC).

Kjaergaard, M.B., 2006. Automatic mitigation of sensor variations for signal strength based location systems. In: Proceedings of Second International Workshop on Location and Context Awareness.

Kjaergaard, M.B., Munk, C.V., 2008. Hyperbolic location fingerprinting: a calibration-free solution for handling differences in signal strength. In: Proceedings of Sixth Annual International Conference on Pervasive Computing and Communications, pp. 110–116.

Koivisto, M., Costa, M., Werner, J., Heiska, K., Talvitie, J., Leppänen, K., Koivunen, V., Valkama, M., 2017. Joint device positioning and clock synchronization in 5G ultra-dense networks. IEEE Trans. Wirel. Commun. 16 (5), 2866–2881. https://doi.org/10.1109/TWC.2017.2669963.

Kotaru, M., Joshi, K., Bharadia, D., Katti, S., 2015. SpotFi: Decimeter level localization using WiFi. SIGCOMM Comput. Commun. Rev. 45 (4), 269–282. https://doi.org/10.1145/2829988.2787487.

Kushki, A., Plataniotis, K.N., Venetsanopoulos, A.N., 2007. Kernel-based positioning in wireless local area networks. IEEE Trans. Mob. Comput. 6 (6), 689–705.

Laitinen, E., 2017. Physical layer challenges and solutions in seamless positioning via GNSS, cellular and WLAN systems. Ph.D. thesis, Tampere University of Technology, Tampere, Finland.

Laitinen, E., Lohan, E.S., 2016. On the choice of access point selection criterion and other position estimation characteristics for WLAN-based indoor positioning. MDPI Sens. 16 (5).

Laitinen, E., Lohan, E.S., Talvitie, J., Shrestha, S., 2012. Access point significance measures in WLAN-based location. In: Proceedings of 9th Workshop on Positioning Navigation and Communication (WPNC), Dresden, Germany, pp. 24–29.

Laitinen, E., Talvitie, J., Lohan, E.S., 2015. On the RSS biases in WLAN-based indoor positioning. In: Proceedings of IEEE International Communication Workshop (ICCW), London, UK, pp. 797–802.

Li, B., O'Keefe, K., 2013. WLAN TOA ranging with GNSS hybrid system for indoor navigation. In: Proceedings of the 26th International Technical Meeting of the Satellite Division of the Institute of Navigation, pp. 416–425.

Liang, D., Zhang, Z., Peng, M., 2015. Access point reselection and adaptive cluster splitting-based indoor localization in wireless local area networks. IEEE Internet Things J. 2 (6), 573–585.

Lin, H., Chen, L., 2016. An optimized fingerprint positioning algorithm for underground garage environment. In: 2016 International Conference on Information Networking (ICOIN), pp. 291–296.

Liu, H., Darabi, H., Banerjee, P., Liu, J., 2007. Survey of wireless indoor positioning techniques and systems. IEEE Trans. Syst. Man Cybern. 37 (6), 1067–1080.

Liu, W., Fu, X., Deng, Z., Xu, L., Jiao, J., 2016. Smallest enclosing circle-based fingerprint clustering and modified-WKNN matching algorithm for indoor positioning. In: 2016 International Conference on Indoor Positioning and Indoor Navigation (IPIN).

Liu, S., Si, P., Xu, M., He, Y., Zhang, Y., 2017. Edge big data-enabled low-cost indoor localization based on Bayesian analysis of RSS. In: 2017 IEEE Wireless Communications and Networking Conference (WCNC).

Lohan, E.S., Koski, K., Talvitie, J., Ukkonen, L., 2014. WLAN and RFID Propagation channels for hybrid indoor positioning. In: International Conference on Localization GNSS 2014 (ICL-GNSS 2014). IEEE.

Lucas-Sabola, V., Seco-Granados, G., López-Salcedo, J.A., García-Molina, J.A., Crisci, M., 2016. Cloud GNSS receivers: new advanced applications made possible. In: 2016 International Conference on Localization and GNSS (ICL-GNSS).

Machaj, J., Brida, P., Piché, R., 2011. Rank based fingerprinting algorithm for indoor positioning. In: Proceedings of International Conference on Indoor Positioning and Indoor Navigation (IPIN).

Miao, H., Wang, Z., Wang, J., Zhang, L., Zhengfeng, L., 2014. A novel access point selection strategy for indoor location with Wi-Fi. In: Proceedings of Control and Decision Conference (CCDC), Changsha, China, pp. 5260–5265.

Milioris, D., Tzagkarakis, G., Jacquet, P., Tsakalides, P., 2011. Indoor positioning in wireless LANS using compressive sensing signal-strength fingerprints. In: 2011 19th European Signal Processing Conference, pp. 1776–1780.

Mo, Y., Zhang, Z., Meng, W., Agha, G., 2015. Space division and dimensional reduction methods for indoor positioning system. In: 2015 IEEE International Conference on Communications (ICC), pp. 3263–3268.

Moreira, A., Nicolau, M.J., Costa, A., Meneses, F., 2016. Indoor tracking from multidimensional sensor data. In: 2016 International Conference on Indoor Positioning and Indoor Navigation (IPIN).

Mulloni, A., Wagner, D., Barakonyi, I., Schmalstieg, D., 2009. Indoor positioning and navigation with camera phones. IEEE Pervasive Comput. 8 (2), 22–31. https://doi.org/10.1109/MPRV.2009.30.

Munir, H., Hassan, S.A., Pervaiz, H., Ni, Q., Musavian, L., 2017. Resource optimization in multi-tier HetNets exploiting multi-slope path loss model. IEEE Access 5, 8714–8726. https://doi.org/10.1109/ACCESS.2017.2699941.

Nur, K., Feng, S., Ling, C., Ochieng, W., 2013. Integration of GPS with a WiFi high accuracy ranging functionality. Geo-spatial Inform. Sci. 16 (3), 155–168. https://doi.org/10.1080/10095020.2013.817106.

Nurmi, J., Lohan, E.S., Wymeersch, H., Seco-Granados, G., Nykanen, O. (Eds.), 2017. Multi-Technology Positioning. Springer.

Pasku, V., Angelis, A.D., Angelis, G.D., Arumugam, D.D., Dionigi, M., Carbone, P., Moschitta, A., Ricketts, D.S., 2017. Magnetic field-based positioning systems. IEEE Commun. Surv. Tutorials 19 (3), 2003–2017. https://doi.org/10.1109/COMST.2017.2684087.

Rasmussen, C.E., Williams, C.K.I., 2006. Gaussian Processes for Machine Learning. MIT Press.

Razavi, A., Valkama, M., Lohan, E.S., 2015. K-means fingerprint clustering for low-complexity floor estimation in indoor mobile localization. In: 2015 IEEE Globecom Workshops (GC Wkshps).

Richter, P., Toledano-Ayala, M., 2015. Revisiting Gaussian process regression modeling for localization in wireless sensor networks. Sensors 15 (9), 22587–22615. https://doi.org/10.3390/s150922587.

Richter, P., Toledano-Ayala, M., 2017. Ubiquitous and seamless localization: fusing GNSS pseudoranges and WLAN signal strengths. Mob. Inf. Syst. 2017, 16, Article ID 8260746. https://doi.org/10.1155/2017/8260746.

Richter, P., Toledano-Ayala, M., Soto-Zarazúa, G.M., Rivas-Araiza, E.A., 2014. A survey of hybridisation methods of GNSS and wireless LAN based positioning system. J. Ambient Intell. Smart Environ. 6 (6), 723–738. https://doi.org/10.3233/AIS-140289.

Ristic, B., Arulampalam, S., Gordon, N., 2004. Beyond the Kalman Filter: Particle Filters for Tracking Applications. Artech House Publishers.

Schwaighofer, A., Grigoras, M., Tresp, V., Hoffmann, C., 2004. GPPS: a Gaussian process positioning system for cellular networks. In: Thrun, S., Saul, L.K., Schölkopf, B.P. (Eds.) Advances in Neural Information Processing Systems 16. MIT Press, Cambridge, MA, pp. 579–586.

Shrestha, S., Talvitie, J., Lohan, E.S., 2013. Deconvolution-based indoor localization with WLAN signals and unknown access point locations. In: 2013 International Conference on Localization and GNSS (ICL-GNSS).

Shu, Y., Bo, C., Shen, G., Zhao, C., Li, L., Zhao, F., 2015. Magicol: indoor localization using pervasive magnetic field and opportunistic WiFi sensing. IEEE J. Sel. Areas Commun. 33 (7), 1443–1457. https://doi.org/10.1109/JSAC.2015.2430274.

Sun, G., Chen, J., Guo, W., Liu, K.J.R., 2005. Signal processing techniques in network-aided positioning: a survey of state-of-the-art positioning designs. IEEE Signal Process. Mag. 22 (4), 12–23. https://doi.org/10.1109/MSP.2005.1458273.

Talvitie, J., Lohan, E.S., Renfors, M., 2014. The effect of coverage gaps and measurement inaccuracies in fingerprinting based indoor localization. In: International Conference on Localization and GNSS 2014 (ICL-GNSS 2014).

Talvitie, J., Renfors, M., Lohan, E.S., 2015. Distance-based interpolation and extrapolation methods for RSS-based localization with indoor wireless signals. IEEE Trans. Veh. Technol. 64 (4), 1340–1353. https://doi.org/10.1109/TVT.2015.2397598.

Talvitie, J., Renfors, M., Lohan, E.S., 2016. Novel indoor positioning mechanism via spectral compression. IEEE Commun. Lett. 20 (2), 352–355. https://doi.org/10.1109/LCOMM.2015.2504097.

Taylor, C.N., Veth, M.J., Raquet, J.F., Miller, M.M., 2011. Comparison of two image and inertial sensor fusion techniques for navigation in unmapped environments. IEEE Trans. Aerosp. Electron. Syst. 47 (2), 946–958. https://doi.org/10.1109/TAES.2011.5751236.

Torres-Sospedra, J., Montoliu, R., Trilles, S., Belmonte, O., Huerta, J., 2015. Comprehensive analysis of distance and similarity measures for Wi-Fi fingerprinting indoor positioning systems. Expert Syst. Appl. 42 (23), 9263–9278. https://doi.org/10.1016/j.eswa.2015.08.013.

Uddin, M.T., Islam, M.M., 2015. Extremely randomized trees for Wi-Fi fingerprint-based indoor positioning. In: 2015 18th International Conference on Computer and Information Technology (ICCIT), pp. 105–110.

Varzandian, S., Zakeri, H., Ozgoli, S., 2013. Locating WiFi access points in indoor environments using non-monotonic signal propagation model. In: 2013 9th Asian Control Conference (ASCC).

Vaupel, T., Seitz, J., Kiefer, F., Haimerl, S., Thielecke, J., 2010. Wi-Fi positioning: system considerations and device calibration. In: Proceedings of Indoor Positioning and Indoor Navigation (IPIN).

Wang, L., Wong, W.-C., 2014. A novel rss model and power-bias mitigation algorithm in fingerprinting-based indoor localization in wireless local area networks. In: Proceedings of 20th European Wireless Conference, Barcelona, Spain.

Wang, B., Chen, Q., Yang, L.T., Chao, H.C., 2016. Indoor smartphone localization via fingerprint crowdsourcing: challenges and approaches. IEEE Wirel. Commun. 23 (3), 82–89. https://doi.org/10.1109/MWC.2016.7498078.

Xiao, C., Yang, D., Chen, Z., Tan, G., 2017. 3-D BLE indoor localization based on denoising autoencoder. IEEE Access 5, 12751–12760. https://doi.org/10.1109/ACCESS.2017.2720164.

Yedavalli, K., Krishnamachari, B., Ravula, S., Srinivasan, B., 2005. Ecolocation: a sequence based technique for RF localization in wireless sensor networks. In: IPSN 2005. Fourth International Symposium on Information Processing in Sensor Networks, 2005, pp. 285–292.

Youssef, M.A., Agrawala, A., Shankar, A.U., 2003. WLAN location determination via clustering and probability distributions. In: Proceedings of Pervasive Computing and Communications, Fort Worth, TX, USA, pp. 143–150.

Zekavat, R., Buehrer, R.M. (Eds.), 2011. Handbook of Position Location: Theory, Practice and Advances. Wiley.

Zhang, C., Subbu, K.P., Luo, J., Wu, J., 2015. GROPING: Geomagnetism and cROwdsensing Powered Indoor Navigation. IEEE Trans. Mob. Comput. 14 (2), 387–400. https://doi.org/10.1109/TMC.2014.2319824.

Zhou, Y., Chen, X., Zeng, S., Liu, J., Liang, D.A., 2013. AP selection algorithm in WLAN indoor localization. Inform. Technol. J. 12, 3773–3776.

Zou, H., Luo, Y., Lu, X., Jiang, H., Xie, L., 2015. A mutual information based online access point selection strategy for WiFi indoor localization. In: Proceedings of International Conference on Automation Science and Engineering (CASE), Gothenburg, Sweden, pp. 180–185.

Zou, H., Huang, B., Lu, X., Jiang, H., Xie, L., 2016. A robust indoor positioning system based on the procrustes analysis and weighted extreme learning machine. IEEE Trans. Wirel. Commun. 15 (2), 1252–1266. https://doi.org/10.1109/TWC.2015.2487963.

14

Deployment of a Passive Localization System for Occupancy Services in a Lecture Building

Pedro E. Lopez-de-Teruel, Felix J. Garcia, Oscar Canovas
FACULTY OF COMPUTER SCIENCE, UNIVERSITY OF MURCIA, MURCIA, SPAIN

1 Introduction

Nowadays, building energy and environmental quality management is an important aspect which requires solutions and strategies that can be carried out in response to real-time changes due to the mainly dynamic nature of the presence of occupants, as referred by Yang et al. (2016). This implies a direct impact on, for example, heating, ventilation, and air conditioning (HVAC) systems. In the scenario that we present in this work, the main goal was to deploy a sensing system that would be connected to the HVAC system in order to take into account the information about occupancy in the different lecture rooms of a lecture building.

Based on our previous background and experience described in Lopez-de Teruel et al. (2017b), we developed an indoor localization system based on fingerprinting that was designed taking into account three main requirements: (a) there is no specific software component running on the mobile devices in order to scan and probably send signal observations to a particular localization server since, generally speaking, it is well known that users are reluctant to install apps that are battery consuming; (b) there are many different mobile devices to be tracked since they are personal devices belonging to university members and, therefore, they tend to generate signals with very different Received Signal Strength (RSS) and temporal patterns; and (c) training should be as quick and less time consuming as possible (Yang et al., 2015) since the proposal has to be scalable to a whole campus level when required.

Consequently, our system is characterized by some design decisions that address these requirements and that will be outlined in this chapter. First, our proposal is able to track unmodified mobile devices using monitoring equipment in the areas of interest—i.e., it performs *passive* localization—and therefore does not require the explicit collaboration

Geographical and Fingerprinting Data for Positioning and Navigation Systems. https://doi.org/10.1016/B978-0-12-813189-3.00014-9

of the users. Taking advantage of the fact that mobile devices periodically scan 802.11 channels for access points, which involve the transmission of probe messages, or send data frames—if they are already connected to some existing wireless network—we can, in both cases, capture the corresponding radio signals generated in order to perform the localization. In the scenario presented here, this did not necessarily imply the deployment of new elements, since we made use of existing computers in order to add the monitoring functionality. Second, our proposal is able to cope with the device heterogeneity problem identified by Park et al. (2011) by using different data representation methods which are mainly based on the order relationship information between RSS values, thus discarding the absolute values which require the adoption of calibration methods. Finally, our proposed training stage involves only a lightweight site survey based on the definition of a minimum number of points of interest that must be covered by an operator during a nonexhaustive recording procedure performed by a training app. This lightweight process is suitable and feasible thanks to the adapted representation methods that we will define.

The practical information that we include in this work has been obtained from an already working system running in a lecture building of 6000 square meters with 20 classrooms. During a 18-month operation period we detected more than 200,000 different MAC addresses, though a more detailed temporal analysis determined that the actual number of frequent users was only around 4000 (after eliminating those MACs simply corresponding to sporadic or nearby passing devices). These remaining devices still constitute a rich dataset that is used in this chapter in order to illustrate some practical decisions that may arise when dealing with real scenarios. The main aim of the chapter, therefore, is to get clear insights into all the practical considerations to take into account when deploying a fully operational passive localization system based on wireless signals, suitable to offer indoor occupancy information-based services in a practical and agile way. In our opinion, there are several real life scenarios that could adopt this approach to infer occupancy information and to make use of that information for higher level services. Therefore, we hope that the information provided here would be helpful for designers and developers with similar requirements.

2 Overview of the Localization System

As we previously mentioned, our system is based on passive indoor positioning. The following set of features intrinsically characterizes this kind of systems, in contrast with active ones:

- There is no need for special software installed on the mobile devices to be tracked.
- They require the deployment of special purpose devices, usually called *monitors*, whose number and specific positions are decided by the system designers.
- Estimations and other calculations are performed by an external element, instead of in the mobile devices themselves.

- Mobile devices are assumed to be heterogeneous, so a noncalibrated approach is required.
- Traffic patterns are unpredictable since there is no (or minimal) control over the mobile devices.
- The granularity of the localization tends to be coarser, based on zone classification, rather than exact position regression.

2.1 Deployment Cycle

Fig. 1 shows an overview of the deployment cycle of our passive localization system. First of all, the administrators decide a set of interest zones Z_i in the target scenario (1), and the system designers determine the number of monitors M_j to deploy and their respective initial positions (2). In a broad sense, a monitor is any hardware element running software able to capture wireless (802.11) traffic and export the relevant information of this captured data to a central server. In this particular scenario, a monitor is a standard PC, usually employed for teaching purposes, with a dedicated WiFi adapter configured to capture wireless traffic in promiscuous mode. Monitors rely only on capturing the frames transmitted by the mobile devices as part of their usual connections or active scanning

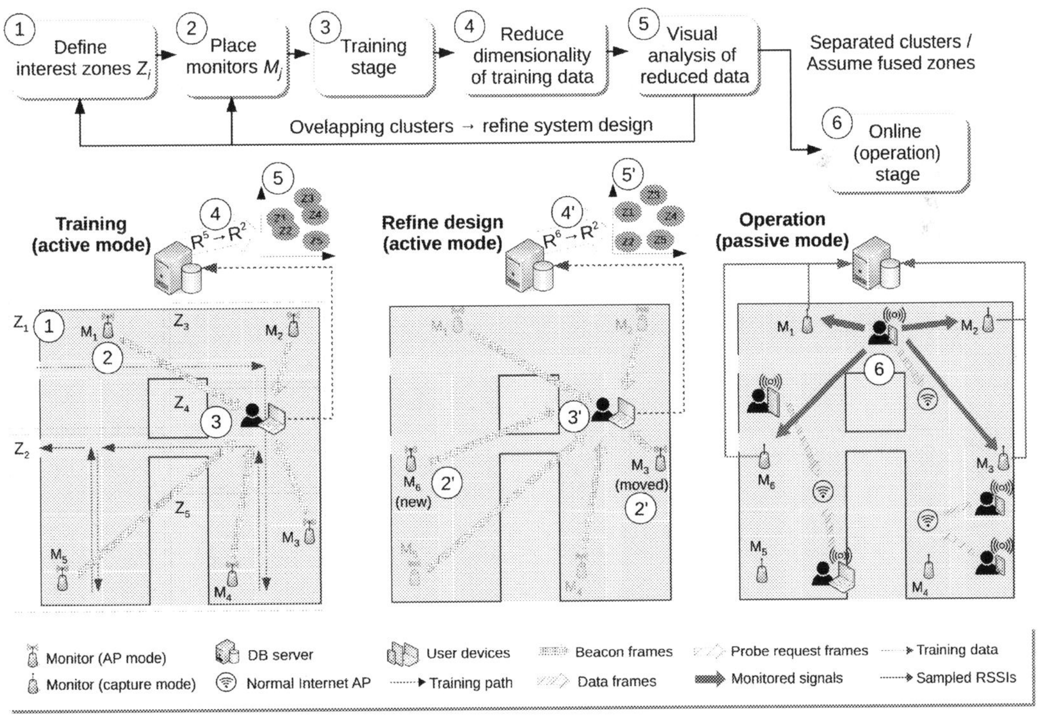

FIG. 1 Overview of the refinement process.

periods. In this scenario we do not make use of prompting techniques like Musa and Eriksson (2012) to increase the number of packets received from them.

Then, as in any fingerprinting-based technique, a training phase which involves a manual site survey process is performed (3), where an operator carrying a device running specific software follows a given walking path. The specific details of this training software are provided in Section 2.2. During this phase monitors, instead of capturing traffic, are configured to broadcast beacon frames (i.e., we set them in *access point mode*, for 802.11). The RSSIs observed by the training device for those frames are then recorded by the software, and used to save the corresponding geotagged observations. Each of these observations is thus formed by a vector of RSSI measures of dimension equal to the number of monitors, plus a ground truth (x, y) position in a coordinate system locally defined in the scenario. Using this position, the specific zone of each vector can also be easily determined.

In this particular scenario we also make use of dimensionality reduction techniques to know whether the current position of the monitors allows for correct classification of users location according to the desired zone partition. For additional information about this stage, the reader can refer to our work (Lopez-de Teruel et al., 2017a).

Once system designers have decided that the current monitor positions and distribution of zones are adequate, and using the corresponding training data as the final fingerprint map, monitors can be switched to *capture mode*, and the system can start its operation (6). We have verified that the user mobile devices (smartphones, tablets, laptops, etc.) being monitored show a wide variety of hardware, WiFi interfaces, antennas, operating systems, and the like. Moreover, some of them are connected to local available Internet access points, and thus generating normal data frames traffic, while others are not currently connected, and thus just generating sporadic probe frames in order to request information from available access points. Consequently, they produce signals with very different strength and temporal patterns. Note also that training vectors are based on RSSI of frames emitted from our monitors and captured by the training device, while in the operational phase just the opposite occurs, that is, RSSIs are obtained for frames emitted by user devices and measured by the monitoring elements. Though the respective RSSIs will be clearly correlated, as determined by the device-monitor distance, in general these measures will not have to be exactly the same. This asymmetry adds another source of heterogeneity to the related RSSIs fingerprints. For all these reasons, heterogeneity had to be addressed in our system, by means of some design decisions regarding data representation that will be explained later.

2.2 Training Approach

Our training phase involves a site survey process to build the corresponding geopositioned fingerprinting database. This process is traditionally assumed to be time consuming, but we have used a different approach to make it faster and more straightforward. Occupancy is mainly based on a per-zone classification problem, rather than exact position regression, which alleviates the need for an otherwise typically exhaustive sampling procedure.

Training observations are tagged (x, y) using a locally defined coordinate system, and those approximate geopositions are obtained as the operator follows the indications of the training app, which provides continuous visual feedback about the required walking path for the site survey. As the example in Fig. 2 shows, a set of connected waypoints forms the path to be followed by the operator. The app is also responsible for collecting the 802.11 fingerprints that will be geotagged using the coordinates where the operator has to be physically at that moment. All the operator has to do is to follow the path shown as accurately as possible and remain still at the designated waypoints (smaller points in the figure) for the required scanning time. Given a particular scenario, our application provides the mechanisms to define these waypoint-based paths, as well as how much scan time is required for each waypoint. The walking speed of the operator can also be configured. In the particular scenario of the building being described here, the required time for training was around 1 hour for the whole building (approximately 4000 square meters of the total 6000 square meters were mapped). The training was divided in five different walking paths of approximately 10 minutes each, one of them shown as example in Fig. 2.

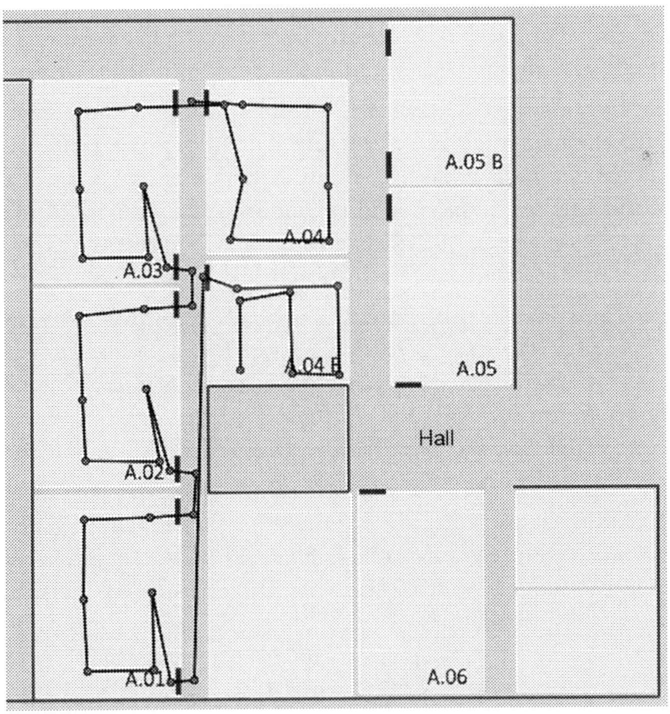

FIG. 2 User interface of the training application based on waypoints.

We are aware that this method can generate some noisy observations, since the current position of the operator does not always match the exact coordinates shown by the application. The set of obtained samples also tends to be relatively sparse, due to the relatively short sampling periods. However, for occupancy purposes, the resulting fingerprinting maps offer an excellent trade-off between the accuracy subsequently obtained in the online phase and the invested training times. As we will see, this light survey technique does not significantly affect the correct classification rate, and it makes the training stage clearly feasible even for relatively large scale scenarios.

2.3 Data Representation

As we have already stated, our monitors are in charge of collecting the 802.11 frames emitted by the user devices, and then they send the relevant information to a central server to be processed. Using a sufficiently long sampling period Δ (typically 1–3 minutes), we build *raw vectors* $\mathbf{r} = (r_1, \ldots, r_M) \in R^M$ for every captured device during that sampling period, where r_i refers to the maximum RSSI value (in dBms) observed by monitor i for the different frames transmitted by that particular device in the corresponding Δ-length time interval. We use the maximum value in order to attenuate fading and multipath effects that might affect the RSSI received, and also to minimize the impact of those values obtained when the monitors were capturing in channels which are not the central frequency used by the device to transmit the frames. If any $\mathbf{r}_i$ value is unavailable (because the corresponding monitor did not capture any frame from the corresponding device), a minimum value of -100 dBm is assigned to it, in order to get a completely defined vector.

These raw measures are then transformed into two alternative representation methods, which we call *order vectors* and *ternary vectors*. The purpose of these alternative representations is to build a vector that is well-fitted to apply different distance metrics in the k-NN (Nearest Neighbor) classifier, while still being suitable for heterogeneous devices.

The idea behind *order vectors* is to represent just the magnitude relationship between the RSSI measurements of a raw vector, thus discarding the specific r_i values, which might not be very useful, due to the already discussed issue of device heterogeneity. In this case, the output vector obtained represents the relative positions of the RSSI signals perceived by each monitor when the input raw vector components are sorted into a descending order. This way, fluctuations in the RSSI values will not alter the resulting vectors as long as the relative order of the signal strengths for the different monitors is maintained.

Ternary vectors are just another alternative to avoid using absolute RSSI values, while still keeping the relevant magnitude order relationships among every pair of monitors. This time the ternary vector is built using all the $\binom{M}{2} = \frac{M*(M-1)}{2}$ combinations of monitors by pairs. Additional information about these representation methods and their related distance metrics can be found in our work (Lopez-de Teruel et al., 2017b).

3 Real Scenario: Occupancy for a Lecture Building

3.1 Overview

Our system is deployed in a lecture building of 6000 square meters with 20 classrooms whose floor plan is shown in Fig. 3. Every classroom, except one, is equipped with a teaching computer connected to the university intranet (represented as circles in Fig. 3). We use these computers to install monitoring software able to capture 802.11 traffic and transmit the relevant information via Gigabit Ethernet to a central server. The use of teaching computers as monitors has the twofold advantage of avoiding the ad hoc deployment of new equipment and saving costs. The only additional hardware needed was an inexpensive off-the-shelf dual-band WiFi card installed in each of those computers to perform the monitoring. According to our study based on dimensionality reduction techniques published in Lopez-de Teruel et al. (2017a), one monitor in each classroom is enough for our purposes.

Several zones of interest (represented by dotted rectangles in Fig. 3) were defined, which refer to the different classrooms and the main hall. Our occupancy sensing system provides a characterization of the different passing users (i.e., students and professors) and their usual behavior.

In order to perform its scanning, each of our monitors simply scans the different 802.11 channels (both 2.4 GHz and 5 GHz) periodically, following a plain round-robin schedule. Parameters of this continuous process, such as the scan time for each channel, the set of channels to scan, or the maximum amount of time before a monitor transmits the collected information to the server, are fully configurable. For each captured packet the

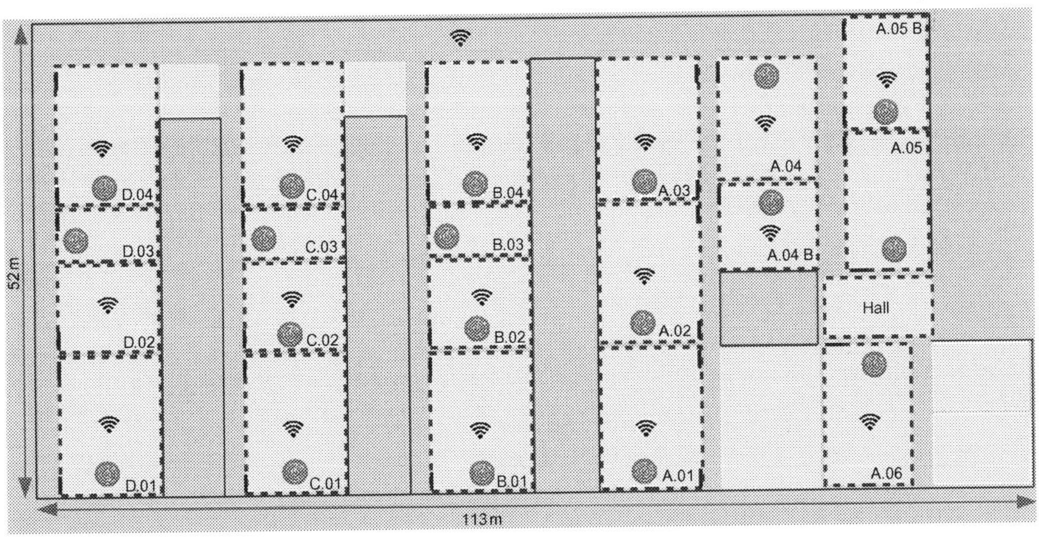

FIG. 3 Floor plan of the lecture room building.

only information which is used is the MAC address of the emitting mobile device—which is key-hashed for privacy reasons—the RSSI value and the corresponding time stamp.

A central server is in charge of hosting the localization engine itself. Another central element of this engine is a database containing both the fingerprinting data collected during the training phase and the continuously updated information of the captures sent to it by the monitors. The server is also in charge of running the localization software responsible for calculating occupancy and positioning information when required. The server provides a RESTful API which is used by higher-level occupancy services.

3.2 Characterization of the Passive Sensing

As has been made apparent, we rely on the data frames sent to the APs pertaining to the network infrastructure (represented as black WiFi icons in Fig. 3) or on the probe requests transmitted by the user devices. One well known issue in this kind of passive systems is that, due to both temporal and spatial sparsity of observations, it is not possible to guarantee a tracking performance similar to that of active systems. Therefore, and in order to characterize our particular environment, we conducted a statistical analysis to aid us in determining some important deployment and validation parameters which clearly distinguish us from typical active systems.

One of the most important deployment parameters of a passive system is the *time window* Δ. It should be noted that we do not have any control of the exact time when each device emits a frame to be captured by our monitors. Moreover, the monitors themselves could also be desynchronized when scanning the different channels. So, monitors just capture a set of individual raw RSSI samples per (*device*, *monitor*) pair for irregularly sampled timestamps. In order to collect useful monitoring data for classification, the central server groups these individual samples by time intervals to obtain vectors including RSSIs for several monitors. Of course, there will be a clear dependence of the number of active (i.e., capturing) monitors for each vector on this Δ value. Fig. 4 shows different probability distributions of the number of monitors capturing signals from a device depending on this time window value, as obtained in our scenario. Of course, the greater the time window, the more likely a given device will be captured by more monitors, thus getting more informative vectors. On the downside, the greater the time window, the less precise will be our system for tracking moving devices. Nevertheless, as we have verified, people in a lecture room building tend to stay relatively static for long periods of time and Δ values of up to 2 or 3 minutes are assumable. We also observe that for values of $\Delta > 180$ seconds the number of active monitors per aggregated vector tends to stay stable.

We have also verified the influence of the Δ parameter in the accuracy of the location estimations. Using data representations based on order and a 5-NN (Nearest Neighbors) classifier, we analyzed the classification accuracy on a test set composed by samples obtained with six different devices when varying the passive time window interval. We clearly observed how for too small values of Δ, the accuracy clearly degrades, while $\Delta = 90$ seconds offers a good compromise between accuracy and time granularity of

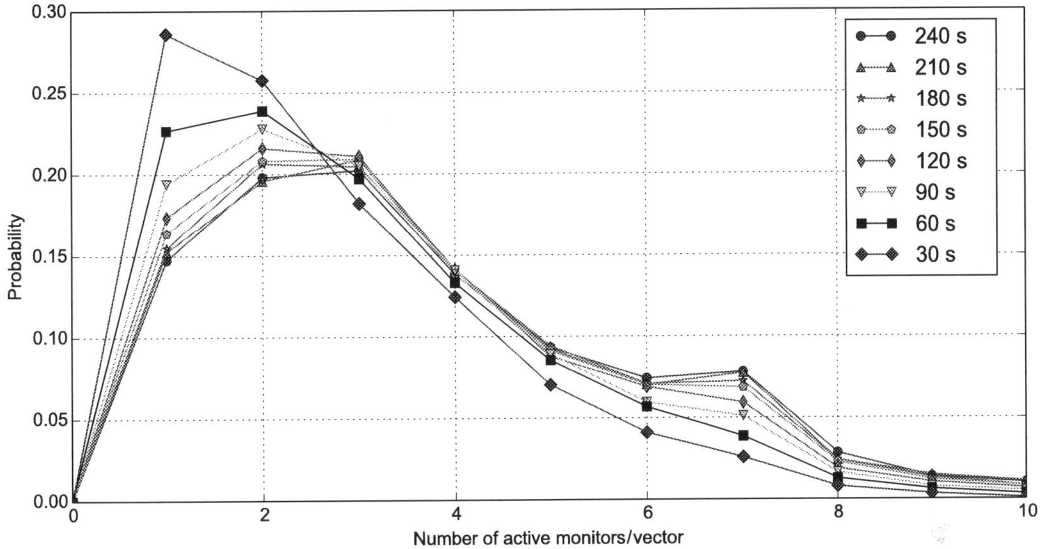

FIG. 4 Probability distributions of the number of monitors capturing signals from a device depending on the time window Δ.

the resulting passive classification system. Accuracy level is around 80%. These are very good results, taking into account that they are obtained on a challenging test dataset of six different devices, with none of them executing any specifically dedicated software.

The spatial coverage of each monitor is another important value to take into account when designing passive localization systems. In our deployment, every monitor covers device locations up to 35–40 m from its corresponding position, or even up to 55 m in some cases (see Fig. 5, which is based on the dataset we obtained during the training phase). Given our spatial distribution of monitors, the system has a minimum coverage of 5–6 monitors on every position of the building (and up to 12–13 on some specific, centered positions). Given an adequate Δ sampling time interval, this was enough to obtain meaningful raw measurements vectors.

3.3 Considerations About Accuracy

Fig. 6 shows classification results in relation to the test set, this time disaggregated by device model. We also show the accuracy when we consider a location estimation wrongly assigned to an adjacent zone as approximately correct (note, of course, that this could be more or less adequate depending on the specific application of the occupancy sensing system). We observe that results are then well above ranges of 90%–95% accuracy, with slight variations depending on the specific devices. In general, we also observed that the laptop seems to be slightly better located than smartphones and tablets, and that some mobile devices (Samsung S3 and Galaxy smartphones) are better located than others (i.e., Galaxy Tab2 tablet), although in fact this could be just an artifact caused by the testing dataset.

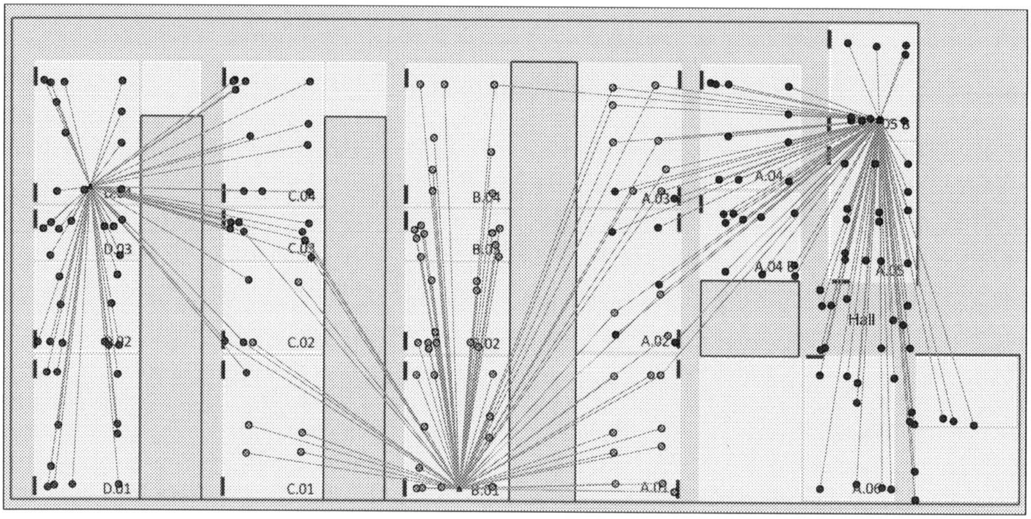

FIG. 5 Spatial sampling coverage for three arbitrary monitors (using training set).

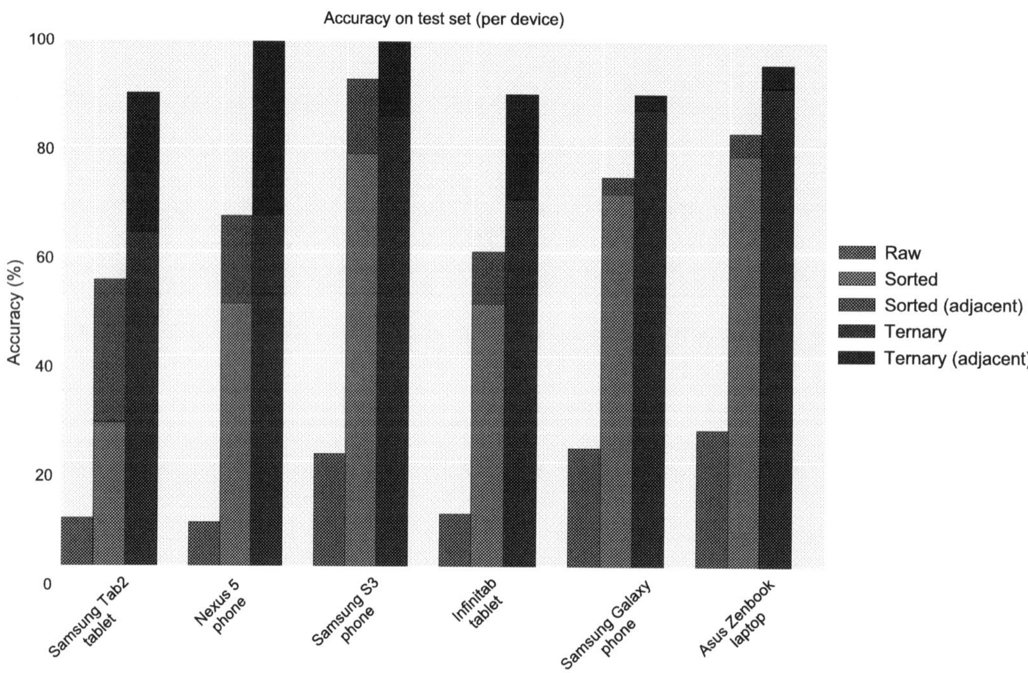

FIG. 6 Accuracy results per device in the test set. Note that overall accuracy gets slightly increased if we consider adjacent zones' errors as also correct, which could be acceptable or not depending on the final application of the occupancy system.

Finally, and given the special characteristics of our environment, where practically every relevant zone (mostly lecture rooms) has its own dedicated monitor, the reader might be wondering how a simple "zone with monitor with strongest RSSI" classification technique would perform. Our analysis determined that, for some specific values of Δ, we can obtain even slightly better individual classification results (up to 90%). However, not only that type of classification would not be adequate for many other types of less structured environments, but also the resulting passive classification systems would be much less robust to sporadic monitor failures. We have tested the resilience of our system to such events, by removing a varying number of monitors when classifying the test set. Using ternary vectors, it is possible to obtain up to 70% classification accuracy even after removing three monitors.

3.4 Occupancy Services

As we have already commented, the localization system provides a RESTful API that can be used to develop specific location-based services. In this particular scenario there are several services for occupancy purposes. We present two examples that could be meaningful for illustration purposes. On the one hand, as Fig. 7 shows, it is possible to obtain occupancy heat maps, which are useful to show real-time information, to analyze occupancy during a given period of time, and, in our case, to provide information to the HVAC system. In the example shown, the system provides the number of devices in every

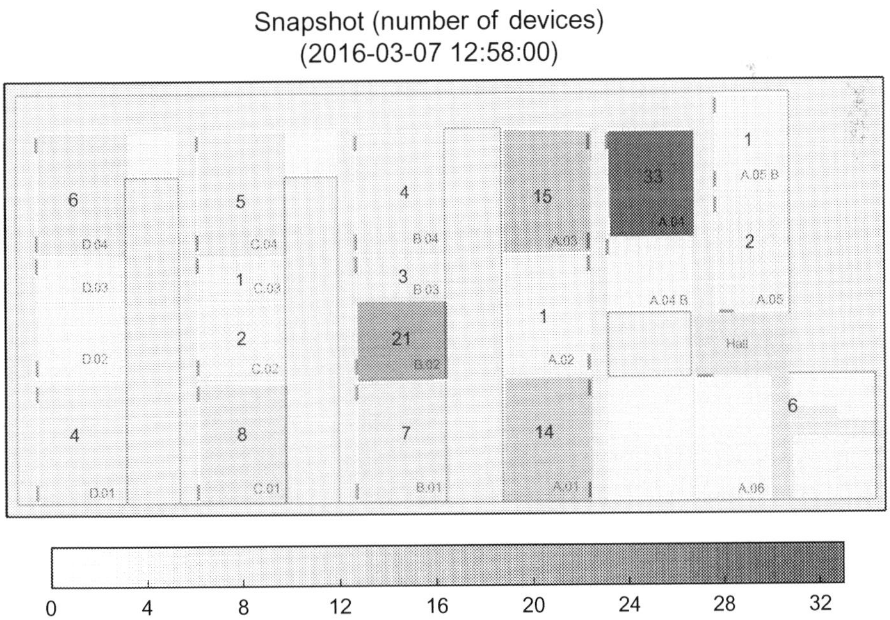

FIG. 7 Example heatmap obtained during the online operation of the system.

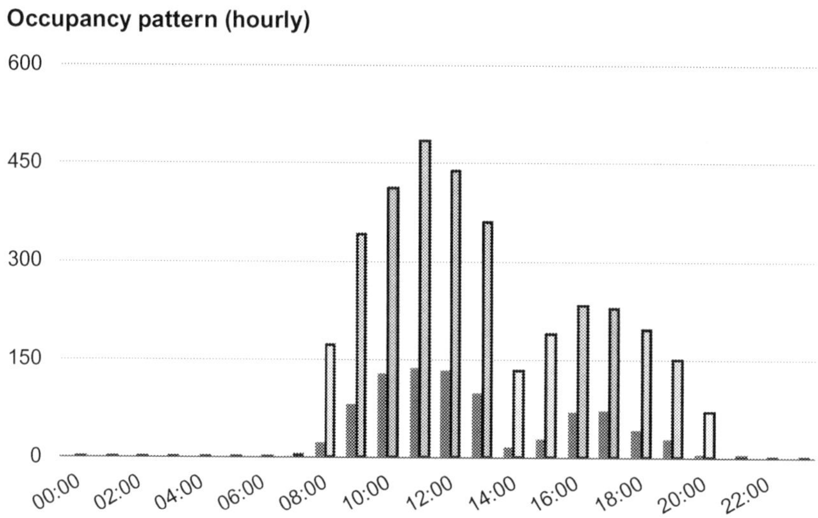

FIG. 8 Temporal occupancy analysis for a given period (one day in the example).

zone of the building (this is related with the potential number of occupants if we assume that every user has a single device). On the other hand, we can also analyze the temporal occupancy pattern of a building (or just a particular zone) over a given period. For example, as Fig. 8 shows, we can determine the number of devices that were present in the building for every hour of a particular day. It is worth noting that since our localization engine is able to provide not only the location of the devices, but also the amount of time they spend at each location, we can provide two kinds of data. For every time slot we display two different bars. Right bars in the figure refer to the total number of different devices that were present at each particular hour in the zone of interest. But, additionally, we also display a finer-grained information, shown in the corresponding left bars, which is directly related to the amount of time that those devices remained in that zone. This information is measured in *devices* × *hour* units, and is calculated taking into account the total time spent by each device in the targeted zone. In terms of this measure, a mobile device which remained in the zone for a whole hour would contribute with 1 *device* × *hour* unit, while another that only stayed there for, say, 6 minutes, would contribute with 0.1. This constitutes a very useful indicator to distinguish passing areas from other zones where users tend to stay for longer periods of time.

4 Conclusions

Occupancy sensing systems have become the subject of much attention recently due to the increasing number of sensors and devices with wireless connectivity. As Christensen et al. (2014) showed, many scenarios follow a similar approach to the one we present here, that is, to infer information from existing infrastructure elements.

As we have shown in this work, our passive localization system based on wireless signals is suitable to offer indoor occupancy information-based services and it addresses the requirements imposed for the deployment. First, our lightweight training procedure, based on waypoints and real-time feedback, is a key stone to reduce the time required to deploy such kind of systems in a practical way. Second, we use data representations based on relative signal strength order (rather than raw measures) to cope robustly with the existing device heterogeneity in an environment like the one present here. Finally, we have verified several parameters that influence the estimation accuracy and we have presented some existing location-based services already running for occupancy purposes.

As a statement of direction, we have already initiated a research line focused on a more detailed characterization of the environment and we are applying clustering techniques to identify sets of users and behaviors that might provide additional information for occupancy systems. More information is available in Lopez-de Teruel et al. (2017c).

References

Christensen, K., Melfi, R., Nordman, B., Rosenblum, B., Viera, R., 2014. Using existing network infrastructure to estimate building occupancy and control plugged-in devices in user workspaces. Int. J. Commun. Netw. Distrib. Syst. 12 (1), 4–29.

Lopez-de Teruel, P.E., Canovas, O., Garcia, F.J., 2017a. Using dimensionality reduction techniques for refining passive indoor positioning systems based on radio fingerprinting. Sensors 17 (4), 871–895.

Lopez-de Teruel, P.E., Garcia, F.J., Canovas, O., 2017b. Practical passive localization system based on wireless signals for fast deployment of occupancy services. Futur. Gener. Comput. Syst. https://doi.org/10.1016/j.future.2017.09.022.

Lopez-de Teruel, P.E., Garcia, F.J., Canovas, O., Gonzalez, R., Carrasco, J.A., 2017c. Human behavior monitoring using a passive indoor positioning system: a case study in a SME. Proc. Comput. Sci. 110, 182–189. 14th International Conference on Mobile Systems and Pervasive Computing (MobiSPC 2017)/12th International Conference on Future Networks and Communications (FNC 2017)/Affiliated Workshops.

Musa, A., Eriksson, J., 2012. Tracking unmodified smartphones using Wi-Fi monitors. In: Proceedings of the 10th ACM Conference on Embedded Network Sensor Systems, pp. 281–294.

Park, J., Curtis, D., Teller, S., Ledlie, J., 2011. Implications of device diversity for organic localization. In: Proceedings of the IEEE International Conference on Computer Communications (INFOCOM).

Yang, Z., Wu, C., Zhou, Z., Zhang, X., Wang, X., Liu, Y., 2015. Mobility increases localizability: a survey on wireless indoor localization using inertial sensors. ACM Comput. Surv. 47 (3), 54.

Yang, J., Santamouris, M., Lee, S.E., 2016. Review of occupancy sensing systems and occupancy modeling methodologies for the application in institutional buildings. Energy Buildings 121, 344–349.

15

Remote Monitoring for Safety of Workers in Industrial Plants: Learned Lessons Beyond Technical Issues

Jose María Cabero Lopez
TECNALIA RESEARCH & INNOVATION, DERIO, SPAIN

1 Motivation

All tasks in a refinery can be classified in two main groups:

- Turnarounds for maintenance activities. They are characterized mainly by the high quantity of external workers from different contractors collaborating together to carry out the maintenance tasks as soon as possible. Maintenance activities comprise piecework tasks all along the refinery, from which a high percentage is inside confined spaces. A typical turnaround comprises more than 1500 workers from many contractors (normally more than 50), working 24/7 during and average period around two months.
- Daily tasks, comprising operational and maintenance activities. These tasks do not include activities inside confined spaces, which are close during regular season.

The former ones are especially risky because they require staff who is not used to specific company safety policies, in addition to exhaustive long shifts piecework. In 2015, we were contacted by a prestigious refinery worried about workers' safety during turnarounds. Half a year later they had planned a turnaround and they desired to check new technologies to let them increase workers' safety inside confined spaces and their localization all over the plant. Refinery's requirements focused mainly on remote monitoring, long autonomy and possibility of interaction between workers and supervisors to communicate aid-demanding situations (from workers to supervisors) and evacuation situations (from supervisors to workers).

Our design to satisfy their needs became a remote monitoring system for the safety of workers in industrial plants. A brief description of the system is described in Section 2.

In addition to technical performance, such a system must consider a plethora of issues related to sociological and logistic aspects difficult to estimate a priori, such as privacy and

Geographical and Fingerprinting Data for Positioning and Navigation Systems. https://doi.org/10.1016/B978-0-12-813189-3.00015-0

system maintenance. Thus, in order to consider these collateral but very important issues, the system development has evolved in parallel to several turnarounds (one per year since 2015), which has allowed us to understand real needs and facts that otherwise would have been missed. This chapter describes briefly the remote monitoring system and the lessons learned in three turnarounds so far.

2 Remote Monitoring System for Safety of Workers in Refineries

The remote monitoring system has been designed in the form of an open platform to monitor the location of workers in a refinery, both inside and outside confined spaces, and any emergency situation that may require external help, such as accidents and fainting situations.

2.1 The Architecture

This system is based on a three-level architecture consisting of: wearable devices, a fixed communication infrastructure deployed all over the monitored site and a control center in the Cloud. Next, the main functionalities of every part are explained.

2.1.1 Wearable Devices: The Wristband

The monitoring system is based on a set of wearable devices to capture information related to people that wear them. So far, the wearable device used in refineries is a customized antiexplosive wristband that provides two main functionalities: localization of workers and detection of aid-demanding situations, with the following approaches:

a. Localization at two levels:
    i. Presence information: through the usage of a NFC tag that is used to record the exit and entry of the corresponding worker in a confined space.
    ii. Localization information: the wristband provides periodical communications which are used by the control center to estimate their localization in the refinery. These communications are based on Bluetooth Low Energy (BLE) for proximity-based localization (based on received signal strength) and UltraWideBand (UWB) for accurate localization (based on time of flight of the received radio signal). The control center runs a localization engine that is able to determine their position based on this periodical information.
b. The wristband includes two mechanisms to detect aid-demanding situations:
    - No motion: a reactive mechanism, based on an automatic motion detection algorithm that uses the information collected by a built-in accelerometer.
    - Panic button: a proactive mechanism, based on a built-in alarm button that must be pressed by the worker in case of need.
c. Additionally the wristband has a led and vibrator to interact with the worker.

The wristband is battery operated and it has an autonomy of more than one year. It can operate in explosive environments, which is critical in refineries (antiexplosive (ATEX) certified). Next figure shows a detail of the wristband.

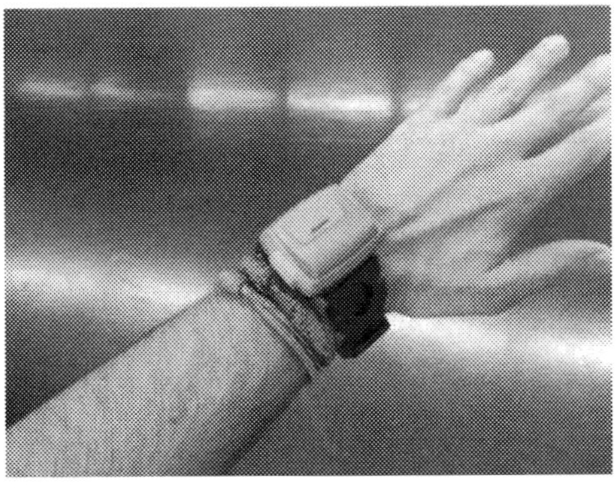

The aim of the system is to monitor people rather than devices; so it is important to detect those situations where the device is not with the person. The system considers three possibilities: wristband forgotten in the plant, outside the plant, or broken. Based on wristbands activity and mobility, these situations are automatically detected and reported to the control center in the form of an alert that will be sent to the supervisors of the refinery.

With respect to the flexibility of the system, the monitoring platform is designed to allow the connection of any commercial wearable devices or sensors with WiFi or Bluetooth Low energy communication capabilities. This is possible through the capabilities of the communication infrastructure explained next.

2.1.2 Communication Infrastructure

The monitoring platform is based on a wireless communication infrastructure that is deployed all over the zone of interest. This infrastructure consists of a set of nodes called anchors that act as communication bridges between wearable devices and the control center in the Cloud. This communication is bidirectional, to collect data from wearable devices (upstream), and also to carry out remote configuration tasks in anchors and wearable devices (downstream).

An anchor consists of the following modules:

- **Communication module:** the anchor is designed to be highly flexible and reliable in terms of robustness. From the communication perspective, this means the possibility to deploy the communication structure in any type of environment, with and without previous communication infrastructure in the specific site. The anchor consists of WiFi and 3G for direct connection between anchors and the control center in the Internet. Both wireless interfaces switch according to the dynamic conditions of the site, that is, if WiFi access points were available in the deployment site, every anchor could connect directly to this WiFi infrastructure to reach the control center. If WiFi infrastructure failed, every anchor would switch automatically to communicate through 3G network. The approach would be similar in case of 3G network as primary network and WiFi as back-up resource.

- **Visualization module:** it consists of a set of alphanumeric displays to show relevant information at the place where they are deployed. It consists also of a buzzer that beeps to alert the workers around of specific events that require attention.
- **Localization module:** it consists of a NFC reader for detection of workers in confined spaces, and two wireless communication modules for the communication with wearable devices in the surroundings. These modules are BLE for proximity-based localization (based on received signal strength) and UWB for accurate localization (based on time of flight of the received radio signal).

Whereas wearables are designed to be intrinsically safe from the ATEX perspective, in the case of anchors, the electronic design is being encapsulated inside ATEX metallic enclosures. Next figure shows an anchor with an antideflagration ATEX enclosure.

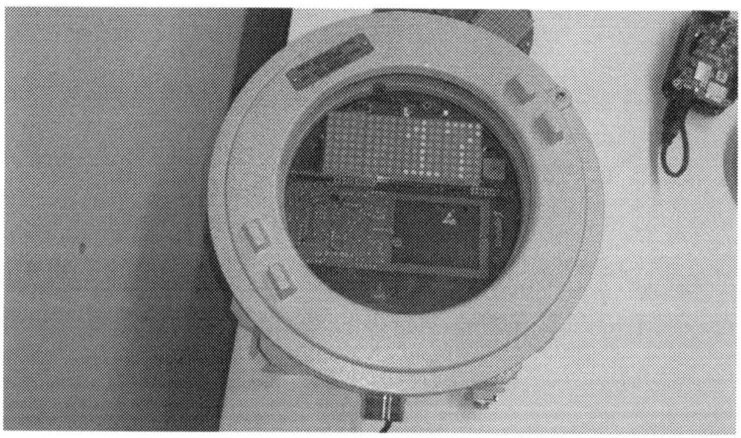

2.1.3 Control Center

It stores and processes the collected data coming from the communication infrastructure in the refinery. It shows permanently the current status of the refinery (localization of workers, active alarms...). Additionally it alerts the supervisors of any aid-demanding situation.

The control center is designed to deal with thousands of simultaneous connections. Additionally, it is completely backed-up to assure continuous service in case of failure. The control center runs in the Cloud, being physically located in different machines, also located in different countries, which assures an effective load balancing process in case of need. On the other hand, its location in the Cloud provides as many resources as necessary in case of exponential growth of system deployments.

The control center consists of the following modules: the database, the processing module, the remote configuration manager, the alert manager, and the graphical user interface, which are explained next:

- **The database:** it stores collected data coming from anchors in the site. Currently this information comes originally from wristbands, but it could come from any sensor connected to the anchor infrastructure through WiFi or Bluetooth Low Energy, or even any other sensor connected to the Internet (through 3G for instance).
- **The processing module:** it runs algorithms to transform raw data into high level information. The most important part of this module is the localization engine, which, based on the incoming information from the wristbands, provides their localization in the area of interest.
- **The remote configuration manager.** This is the part of the system that controls, configures, and updates remotely the infrastructure of anchors and wristbands in the field.
- **The alert manager:** this module receives and generates all alerts in the platform. There are two types of alerts: those generated by the wearable devices to inform supervisors, and those generated from the control center to inform workers.

  When alerts come from workers, this module informs supervisors of the type of alert, where it is coming from, when it happened, and who has generated it.

  So far, there is only one alert that is sent from the control center to the workers, which is an evacuation alert. This situation is broadcasted by the system toward all anchors (it generates a loud and continuous beep from every anchor) and wristbands, making these vibrate.
- **The graphical user interface:** this module shows collected information to end users. It consists of maps of the monitored site, heating maps of activity to show density of workers per area, active alerts, statistics of activity, etc. Next figure shows a specific moment of the activity in a zone of the refinery during the last turnaround, with 136 workers and without active alerts.

2.2 Data Anonymity

Our system collects data from wearable devices, so, linking wearable device ID's and its owner is necessary in most cases if the system must translate between devices and people. Knowing which person wears a specific wearable device offers personalized functionalities that optimize system response to every situation, such as granting or denying access to zones depending on workers' profiles.

The system allows three different strategies from data anonymity point of view:

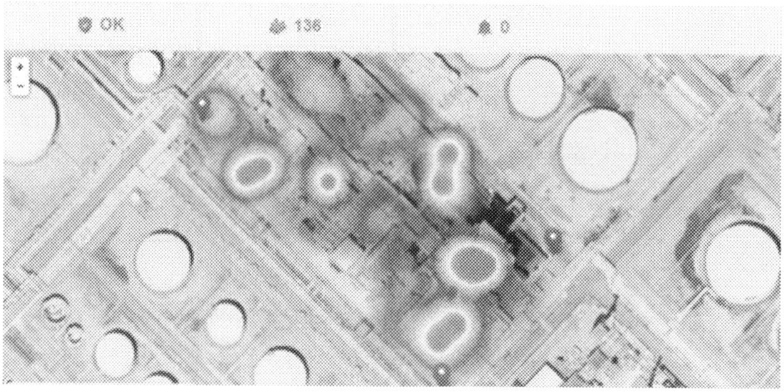

- Complete anonymity: there is no information related to wearable devices and workers. Any rule applies to everyone without exception, for example, access to a zone is granted or denied to everyone no matter who is trying to access.
- Medium anonymity: the system knows who is wearing what device, system user interface does not show any personal identification (name, surname); anyone who is watching the user interface can check only wearable devices identifications (IDs), but system administrator can know who is wearing every device. Rules can be personalized without showing who is whom in the user interface. In the end, if it were necessary, any infraction could be tracked down to the specific worker.
- Low anonymity: the user ID (name, surname, and contractor) is shown in the user interface. This profile implies full availability about personal data to any user watching the system user interface. Full functionality of the system can be applied (personalization).

There are three critical reflections to be done before choosing the level of anonymity. The output of these reflections will determine the success of failure of the system:

- What the regulation says about personal data management: it depends tremendously on the country and region where the system will be deployed, and there is not much to do either technically or socially, but to choose the right level of anonymity.
- What the system is applied for: although the system is originally thought for safety, the system could also be used to monitor people for other purposes: productivity and control. The former aims at an increase of productivity by analyzing data as a whole, without considering specific IDs of the monitored people. The later aims at analyzing who is doing what and where.
- Who monitors information: information can be at disposal of the owner of the plant (all staff or just a set of people), or be extensible for all stakeholders: owner, contractors, and even workers.

3 Learned Lessons in the Field

The system has been developed and is being validated in different turnarounds. Beyond technical evolutions of the system to cope with specific harsh environments in refineries, there are a set of crucial nontechnical aspects that can be extrapolated to other industrial sectors, that are key for the success of these type of services based on monitoring people with wearable devices. Since the first version of the system in 2015, two nontechnical aspects have been our system main threats: person related aspects and logistics. These aspects and potential solutions are explained next.

3.1 Person Related Issues

In the end, workers, as entities to be observed, are key in the success or failure of any monitoring service. They must be convinced about the goodness of the service for their own safety or, at least, be convinced about the system not being a threat. Based on our experience, some aspects are cornerstone to get workers' acceptance:

– *Transparency and privacy:* workers need to understand what the system is used for; it should be a tool open to them; they should be able to watch it running. The objective should be even to have the system operated by them or someone they trust in. In addition, the plant owner should establish a policy to assure and show them that their data are not used for controlling purposes.

  Originally, all industrial deployments had demanded remote monitoring services for safety purposes, although lately, along with the Industry 4.0 movement, more and more industry is demanding remote monitoring services for productivity purposes. So far, all companies have understood that the usage of the system to control workers would make the system fail (be rejected by workers). Our experience shows this aspect as the biggest problem for the system acceptance by workers. The system cannot be considered by workers as a control tool, otherwise it would be condemned to fail.

  Our experience with managers in industrial plants took us to discuss about this aspect many times, sometimes joking, less times seriously, but in the end everyone understood that controlling is out of scope, and the system objective must be only safety and productivity.

  So far, we have deployed the system in industrial plants by applying the medium anonymity approach focused on safety purposes. This way, the system knows who is wearing a wearable device that is being monitored, but the user interface does not show that information to supervisors. Workers are usually shown as numbers (e.g., Worker1, Worker2, etc.).

  Originally, the remote monitoring service was designed to be used by the industrial plan owner. In our experience in industrial plants, many contractors demanded access to its user interface claiming that, this way, any emergency alert related to their workers would be quickly attended by them. This request, combined with privacy-related issues, made us consider the remote system as an open service for all stakeholders in the plant, providing access to the user interface to contractors and workers, being transparent for workers (to disassemble suspicions about privacy) and providing a monitoring tool for contractors as well. Next lines show some of the lessons learnt according to different aspects.

- Ergonomics: wearable devices must be transparent or at least comfortable for them. The remote system is based on a wristband as wearable device for workers. Some complaints have been received about the wristband not being comfortable enough for some tasks, and even seen some workers taking the wristband somewhere else (in their pocket for instance).

  In addition, even for comfortable wearable devices, including a new device to the ones they already wear may not be a good option. Industrial sites where the wristband has been used, clients from now on, are reluctant to include new equipment because of the implicit complexity for workers, and they prefer to expand the functionalities of the existing ones, if it is possible, being totally transparent for workers. To this respect, the concept of smart clothes seems to be the best option for future evolutions of the system.
- Simplicity: ideally, system usage should not require workers actuation. They should keep oblivious to the monitoring system. They already have plenty of procedures to comply with; new ones should be avoided. Our system requires very few instructions for workers: basically panic button usage and confined spaces in/out. Although it is hard to quantify, in the last turnaround supervisors detected many confined spaces in/out protocol infractions, mostly for two reasons: they did not remember the procedure or they did not understand what the system was for. The technical roadmap of the system considers the design of an automatic mechanism to detect confined spaces in/out.
- Technical and procedure robustness: any system deployed in an industrial environment must be adapted to a harsh context. Additionally, in a turnaround, service must be assured without interruptions. If any link of the chain fails, the credibility of the whole system will be put at risk. On the other hand, all relevant procedures, such as installation of the system, maintenance and monitoring must be very clear for all the stakeholders participating in the service. Otherwise, even when the system is technically robust, the service will fail.

  Along the system development, in previous turnarounds, the monitoring system suffered from system technical failures, such as servers down, broken wristbands, partial blackouts, and failures in the procedures, such as problems with logistics and supervisors who did not know the system procedures. The price to pay is that as soon as the system started failing, workers had a good excuse for stop using wristbands.

  After these initiatives and to avoid these problems, it has been decided that any new installation include two initial activities:

  - Intensive training on system functionalities for client supervisors, and integration of the system functionalities on client's safety procedures.

- Incremental installation of the system: start by a pilot to check client's requirements and validate all relevant procedures, and afterward make the system scalable to the whole plant.

An example of the necessity of clear instructions and procedures for those who must run the system is shown in the next subsection.

3.2 Logistics

A remote monitoring system based on wearable devices requires a thorough process of device assignment, replacement and, in case of uninstallation, collection of equipment. When the service is deployed in a turnaround, with plenty of workers performing piece-work tasks, the system must be up and running quick and transparently for workers' daily activities. One very important part of the installation is the assignment, distribution, maintenance, and collection of wearable devices. This task is critical, and usually under-estimated, which can end up in a completely unreliable system.

In a turnaround, most workers participate partially, that is, they fulfill their task, that usually take them several days, and leave to other turnarounds. It is normal for the plant owner to require less wearable devices than workers and reuse them as they come and go. A critical part for the system performance is the distribution, assignment, maintenance, and especially the reassignment of wearable devices.

For these phases it is crucial to have a clear process about how to act and, if possible, just one person and company in charge. This process is especially needed with the scalability of equipment. Our experience in this matter was especially painful in a turnaround with 700 wearable devices; the assignment/reassignment of wearable devices in the control center was carried out by the client, and their distribution to the corresponding workers by a contractor.

After four weeks of turnaround, the client realized that they were watching on the plant map of the user interface, workers who had already left. How was this possible? At the beginning, both assignment and distribution of wristbands were synchronized; after some days and many new workers, the big mistake arose: the contractor had redistributed wearable devices used by workers who had already left to new workers without informing the client, that is, without reassigning the wearable devices to the new workers.

Both client and contractor declare afterward that they did not have clear how the assignment/distribution process worked, which was our very big mistake: be confident about something we considered clear enough and technically easy. We underestimated the necessity of a procedure for these tasks.

4 Conclusions

Monitoring the activity of workers in a refinery can be done with localization techniques and communication technologies. Beyond technical issues, this chapter summarizes our experiences in turnarounds in refineries; it focuses on two aspects that are critical from our point of view: person-related issues and logistics.

Beyond specific technical features, privacy, simplicity, ergonomics, transparency, training, and integration with client safety procedures are presented as factors for the success of any service for human monitoring. Some tips and approaches to these aspects are described based on our experience.

16

A Review of Indoor Localization Methods Based on Inertial Sensors

Estefania Munoz Diaz, Dina Bousdar Ahmed, Susanna Kaiser

INSTITUTE OF COMMUNICATIONS AND NAVIGATION, GERMAN AEROSPACE CENTER (DLR), OBERPFAFFENHOFEN, GERMANY

1 Introduction

Different solutions exist to solve the pedestrian indoor localization problem, but many of them rely on additional infrastructure that either needs to be installed or is assumed to be available. An overview of different techniques that address the challenges in areas not covered by Global Navigation Satellite Systems (GNSS) is given in Liu et al. (2007), Gu et al. (2009), and Harle (2013).

Radio and satellite navigation solve the positioning problem using the piloting method. Piloting is the process of determining one's position based on external objects, such as antennas or satellites. Inertial navigation performs positioning using the dead-reckoning method. Dead-reckoning is the process of determining one's current position projecting course and speed or elapsed distance from a known previous position. Dead-reckoning is, therefore an infrastructure-free positioning system.

Pedestrian dead-reckoning is performed using inertial sensors that are usually combined with magnetometer and barometer. Inertial sensors are of high interest because of the possibility of providing positioning without touching privacy, unlike camera-based systems. Since medium- or low-cost inertial sensors are already available in every smartphone or other carry on electronic devices, the use of inertial sensors is especially interesting for mass market applications.

Example application areas suitable for inertial positioning are fire fighter rescuing, police men supervision, industrial inspections, or supervision of elderly people. Those professional applications require a system that is small-sized, light-weighted, has low power consumption, can be easily mounted on the body and is not dependent on infrastructure. For instance, in the case of fire, infrastructure might be disturbed. In addition, in some applications cameras might not be allowed as it is the case in industrial inspections. Integrating inertial sensors in clothes or footwear is an option to handle the fixation of sensors.

Geographical and Fingerprinting Data for Positioning and Navigation Systems. https://doi.org/10.1016/B978-0-12-813189-3.00016-2

Suited sensor locations are the shoe or the pocket (Munoz Diaz et al., 2014; Bousdar et al., 2016). Alternatively, sensor locations mounted, e.g., at the wrist (Diez et al., 2016), at the head (Windau and Itti, 2016), at the torso (Do et al., 2016), or in the backpack are also investigated. Nevertheless, it is still challenging to use the sensors of the smartphones for positioning without any other help due the fact that the sensor is usually not fixed at the body and is in different motion mode like texting, phoning, swinging, or even in irregular motion mode (Renaudin et al., 2012).

Besides using only inertial sensors, fusion with other sensors or maps of the environment (Beauregard et al., 2008; Krach and Robertson, 2008; Woodman and Harle, 2008; Kaiser et al., 2015) is in any case possible and recommendable. Additionally, the combination of sensors mounted on different locations of the human body is a promising solution for professional use cases (Bousdar et al., 2016).

In Section 2, inertial sensors as well as magnetometer sensors are explained. Pedestrian positioning systems based on inertial sensors are usually classified depending on the location of the body where they are mounted on. This classification is based on the algorithm they use to compute the position of the pedestrian. There are two types of algorithms, namely strapdown and step&heading. These two algorithms correspond with shoe-mounted positioning systems and non-shoe-mounted positioning systems, respectively. The estimation of the orientation is of key importance to perform inertial positioning and it is computed disregarding the location where the sensor is mounted on. Therefore, the orientation estimation is deeply described in Section 3, followed by the shoe-mounted inertial positioning in Section 4 and the non-shoe-mounted inertial positioning in Section 5. The positioning estimation based on medium- and low-cost inertial sensors suffers from propagation of errors and accumulated drift over time. Therefore, additional algorithms are necessary to compensate the drift. An extensive state of the art review is handled in Section 6. Finally, Section 7 is devoted to discussion of the present and future status of inertial positioning for pedestrians.

2 Inertial Sensors and Magnetometers

Inertial sensors are composed of accelerometers and gyroscopes, which measure specific force and turn rate, respectively. The so-called inertial measurement unit contains three mutually orthogonal accelerometers and three mutually orthogonal gyroscopes. Therefore, the acceleration and turn rate measurements are triads.

Inertial sensors based on micro-electromechanical (MEMS) technology have improved its performance over the last decades. However, using MEMS-based inertial sensors the resulting positioning is less accurate than using other technologies like solid state accelerometers or optical gyroscopes (Woodman, 2007). The most common error sources that disturb inertial measurements are biases, bias stability, and thermo-mechanical noise (Woodman, 2007).

The biases are the measured averaged value of the acceleration and turn rate when neither acceleration nor rotation, respectively, is undergoing. The biases introduce a systematic error in the integrated measurements, i.e., velocity, position, and orientation. The systematic error can be compensated by averaging acceleration and turn rate while the sensor is static and subtracting the averaged value from the measured acceleration and turn rate, respectively.

The bias stability describes how the biases change over time under stable conditions, usually at constant temperature. Temperature fluctuations due to changes in the environment and sensor self-heating modify the biases value. The change in the biases is caused by flicker noise, which is visible at low frequencies. Other slow changing errors also affect the bias values. The bias stability introduces a non systematic error in the integrated signals due to the fact that the biases wander over time.

The thermo-mechanical noise introduces a white noise sequence, namely a sequence of zero-mean uncorrelated random variables. In turn, such a sequence disturbs the integrated measurements, i.e., velocity, position, and orientation, by a random walk. A random walk is a process consisting of a series of steps, the direction and size of which are randomly determined (Woodman, 2007).

Magnetometers, which are commonly embedded together with the inertial sensors, measure magnetic fields, e.g., the Earth's magnetic field. Usually, a magnetometer unit is formed by three mutually orthogonal magnetometers. MEMS-based magnetometers are frequently found in smartphones and similar electronic devices. Nevertheless, other technologies exist to implement magnetic field sensors (Lenz and Edelstein, 2006).

The main phenomenon that affects the performance of magnetic field sensors is the temperature effect. The temperature error causes an increasing noise present in the measured magnetic field. Therefore, temperature compensation algorithms or specific electronic design (Beroulle et al., 2003) are required to limit the temperature effect.

For navigation purposes the Earth's magnetic field is widespread used. However, magnetometers are also affected by the presence of ferromagnetic materials, which are common in urban and indoor environments. The modifications introduced by these materials lead to erroneous orientation estimation. This is the reason why magnetic field sensors are historically ruled out from indoor navigation systems. Nevertheless, magnetometers can still be used to improve the performance of inertial sensors (Azfal et al., 2011; Zampella et al., 2012; Munoz Diaz et al., 2017).

3 Orientation Estimation

The orientation estimation aims at combining the measurements of gyroscopes, accelerometers, and magnetometers in an optimal way to obtain the orientation of the sensor. Along with the orientation angles, it is convenient to estimate also the biases of the gyroscopes. The biases, which were explained in Section 2, are therefore estimated in order to be subtracted from the turn rate measurements.

A common tool to compute the orientation in the state of the art is the Kalman filter. A Kalman filter is one implementation of Bayesian filters used to estimate the states of a system. In the particular case of the orientation estimation the states are the orientation angles, i.e., ϕ roll, θ pitch, and ψ yaw or heading, and the biases of the gyroscope.

The Kalman filter combines a prediction stage and an update stage. The prediction stage implements the system model, which represents how the system states evolve over time. The update stage incorporates measurements that relate to the system states. Usually the prediction stage is based on the integration over time of the turn rate measurements. The update stage typically incorporates the acceleration and magnetic measurements in order to reduce the error due to the integration over time of the turn rate measurements that contain biases and noise.

Both, the full state and the error state Kalman filter have similar performance, as shown in Wagner and Wieneke (2003). In this chapter, the full state vector is chosen, being $\boldsymbol{x}_{\mathrm{o}}^{k}$ composed of the orientation angles and the biases $\boldsymbol{b}^{k}$ of the gyroscopes:

$$\boldsymbol{x}_{\mathrm{o}}^{k} = [\phi^{k}, \theta^{k}, \psi^{k}, b_{\mathrm{x}}^{k}, b_{\mathrm{y}}^{k}, b_{\mathrm{z}}^{k}]^{T}. \quad (1)$$

In the following, α, ω, μ will be used to represent acceleration, turn rate, and magnetic measurements, respectively.

3.1 Prediction Stage

Gyroscopes measure in body frame the turn rate of the sensor with respect to the inertial frame. In order to have the turn rate measurements in body frame with respect to the navigation frame, the transport rate and the Earth rotation have to be subtracted. However, for pedestrian positioning, the transport rate is negligible and the Earth rotation, which is approximately $15^{\circ}\ \mathrm{h}^{-1}$, is usually not compensated. Therefore, it is assumed that the turn rate measured by the gyroscopes is approximately the turn rate of the sensor in body frame with respect to the navigation frame (see Fig. 1).

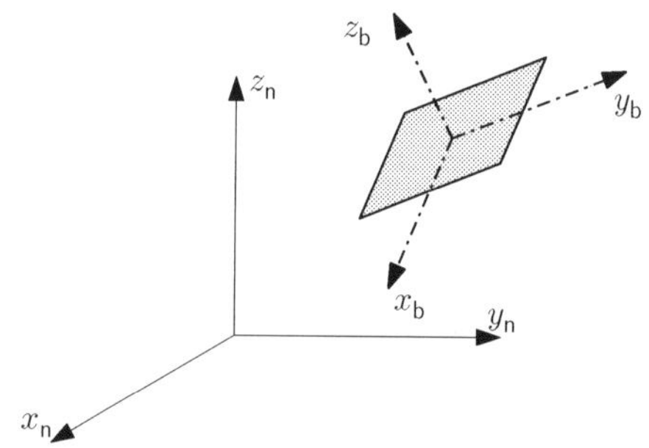

FIG. 1 Navigation frame, $\{x, y, z\}_{\mathrm{n}}$, and body frame of the sensor, $\{x, y, z\}_{\mathrm{b}}$. The navigation frame is fixed over time. The body frame changes with the sensor orientation.

The turn rate at the time k is defined as $\boldsymbol{\omega}^k = [\omega_x^k, \omega_y^k, \omega_z^k]^T$. To compute the orientation, the biases have to be subtracted from the turn rate. Then, the corrected turn rate measurements are integrated once over time.

The direction cosine matrix $\boldsymbol{C}^k$ is a 3×3 matrix in which each column is a unit vector along the body axes specified in terms of the navigation axes. That means, the matrix $\boldsymbol{C}^k$ represents the rotation of the body frame with respect to the navigation frame at the time k (see Fig. 1). From this rotation the orientation of the sensor is deduced.

The orientation at the time k is the orientation at the time $k-1$ modified by the rotation that took place within the last δt seconds, represented in a matrix form as $\boldsymbol{A}^k$:

$$\boldsymbol{C}^k = \boldsymbol{C}^{k-1} \cdot \boldsymbol{A}^k, \tag{2}$$

being $\boldsymbol{A}^k$

$$\boldsymbol{A}^k = \boldsymbol{I} + \frac{\sin(\sigma)}{\sigma} \cdot \boldsymbol{B}^k + \frac{1-\cos(\sigma)}{\sigma^2} \cdot \boldsymbol{B}^{k^2}, \tag{3}$$

where $\sigma = |\boldsymbol{\omega}^k \delta t|$ and

$$\boldsymbol{B}^k = \begin{pmatrix} 0 & -\omega_z^k \delta t & \omega_y^k \delta t \\ \omega_z^k \delta t & 0 & -\omega_x^k \delta t \\ -\omega_y^k \delta t & \omega_x^k \delta t & 0 \end{pmatrix}. \tag{4}$$

In order to tackle the estimation of the biases of the gyroscope, a noise model is presented in the following. The biases are predicted using the presented model. The turn rate measurements $\boldsymbol{\omega}^k$ can be represented as

$$\boldsymbol{\omega}^k = \tilde{\boldsymbol{\omega}}^k + \boldsymbol{e}^k, \tag{5}$$

being $\tilde{\boldsymbol{\omega}}^k$ the error free turn rate and $\boldsymbol{e}^k$ the measurement error. The turn rate error can be decomposed into two errors

$$\boldsymbol{e}^k = \boldsymbol{b}^k + \boldsymbol{v}, \tag{6}$$

where $\boldsymbol{v}$ is the sensor noise that can be modeled as Gaussian white noise. To determine the biases error an auto-regressive model of order one (AR1) (Munoz Diaz et al., 2013) is chosen. The AR1 model is defined as

$$\hat{\boldsymbol{b}}^k = \boldsymbol{c} \cdot \boldsymbol{b}^{k-1} + \boldsymbol{n}. \tag{7}$$

The biases follow an exponentially correlated noise term defined in the AR1 model as the constant $\boldsymbol{c}$, which is equal to the term $e^{-\frac{1}{\tau}}$, where $\boldsymbol{\tau}$ is the correlation coefficient for each axis and $\boldsymbol{n}$ can be modeled as Gaussian white noise with standard deviation $\boldsymbol{\sigma_n}$ for each axis.

3.2 Update Stage

There are several updates that can be implemented to improve the orientation estimation. Usually the updates are signals directly measured by the sensors, but also pseudo-measurements can act as updates. The term pseudo-measurement refers to non-directly measured but computed signals. In the following, the most important updates will be summarized.

3.2.1 Absolute Gravity Update

During the walk, there are periods in which the acceleration due to the movement of the sensor is zero or quasi-zero. During these periods only the gravity acceleration is measured. In such case, the orientation angles roll and pitch can be extracted at the time k as follows:

$$\bar{\phi}^k = \arctan\left(\frac{\alpha_{\mathrm{y}}^k}{\alpha_{\mathrm{z}}^k}\right) \tag{8}$$

and

$$\bar{\theta}^k = \arctan\left(\frac{-\alpha_{\mathrm{x}}^k}{\sqrt{\alpha_{\mathrm{y}}^{k^2} + \alpha_{\mathrm{z}}^{k^2}}}\right). \tag{9}$$

The measurement vector $\boldsymbol{z}_{\mathrm{o}}^k$ of the Kalman filter at the time k can be written as:

$$\boldsymbol{z}_{\mathrm{o}}^k = [\bar{\phi}^k, \bar{\theta}^k]^T. \tag{10}$$

3.2.2 Differential Gravity Update

Likewise, within the periods of zero or quasi-zero acceleration, the acceleration at the current time can be computed applying the rotation of the last δt seconds, $\boldsymbol{A}^k$, to the acceleration measured at the previous time $\boldsymbol{\alpha}^{k-1}$ as follows:

$$\bar{\boldsymbol{\alpha}}^k = \boldsymbol{A}^k \cdot \boldsymbol{\alpha}^{k-1}. \tag{11}$$

The pseudo-measurement $\bar{\boldsymbol{\alpha}}^k$ is used as update. This update has been proposed in Renaudin and Combettes (2014). The measurement vector $\boldsymbol{z}_{\mathrm{o}}^k$ of the Kalman filter at the time k can be written as:

$$\boldsymbol{z}_{\mathrm{o}}^k = [\bar{\alpha}_{\mathrm{x}}^k, \bar{\alpha}_{\mathrm{y}}^k, \bar{\alpha}_{\mathrm{z}}^k]^T. \tag{12}$$

3.2.3 Absolute Magnetic Field Update

During the walk there are periods in which the measured magnetic field is constant or quasi-constant. At the beginning of the quasi-constant magnetic field period, the measured magnetic field is projected onto the navigation frame and chosen as reference $\bar{\boldsymbol{\mu}}_{\mathrm{r}}$. It is assumed that, during quasi-constant magnetic field periods, the measured magnetic field does not change. Therefore, the reference magnetic field is used as pseudo-measurement for the update. This update has been proposed in Azfal et al. (2011) and further analyzed in Bancroft and Lachapelle (2012). The measurement vector $\boldsymbol{z}_{\mathrm{o}}^{k}$ of the Kalman filter at the time k can be written as:

$$\boldsymbol{z}_{\mathrm{o}}^{k} = [\bar{\mu}_{\mathrm{r_x}}, \bar{\mu}_{\mathrm{r_y}}, \bar{\mu}_{\mathrm{r_z}}]^{T}. \tag{13}$$

3.2.4 Differential Magnetic Field Update

Likewise, within these periods of constant or quasi-constant magnetic field, the magnetic field at the current time can be computed applying the rotation of the last δt seconds, $\boldsymbol{A}^{k}$, to the magnetic field measured at the previous time $\boldsymbol{\mu}^{k-1}$ as follows:

$$\bar{\boldsymbol{\mu}}^{k} = \boldsymbol{A}^{k} \cdot \boldsymbol{\mu}^{k-1}. \tag{14}$$

The pseudo-measurement $\bar{\boldsymbol{\mu}}^{k}$ is used as update. This update has been proposed in Zampella et al. (2012) and Azfal et al. (2011). The measurement vector $\boldsymbol{z}_{\mathrm{o}}^{k}$ of the Kalman filter at the time k can be written as:

$$\boldsymbol{z}_{\mathrm{o}}^{k} = [\bar{\mu}_{\mathrm{x}}^{k}, \bar{\mu}_{\mathrm{y}}^{k}, \bar{\mu}_{\mathrm{z}}^{k}]^{T}. \tag{15}$$

3.2.5 Absolute Compass Update

If the measured magnetic field is homogeneous, the yaw angle can be computed at each time k as follows:

$$\bar{\psi}^{k} = \arctan\left(\frac{-\mu_{\mathrm{h_x}}^{k}}{\mu_{\mathrm{h_y}}^{k}}\right) + \mathrm{D}, \tag{16}$$

being $\boldsymbol{\mu}_{\mathbf{h}_i}^{k}$ where $i = \{\mathrm{x,y}\}$ is the magnetic field intensity at the time k for the i-axis projected onto the horizontal plane. The declination angle, which is known for every location on the Earth, is represented by D. The measurement z_{o}^{k} of the Kalman filter at the time k can be written as:

$$z_{\mathrm{o}}^{k} = \bar{\psi}^{k}. \tag{17}$$

3.2.6 Zero Angular Rate Update

Within the periods where the sensor is not rotating, the turn rate measurements can be assumed to be zero. This assumption implies that any turn rate measured during these periods is due to errors, e.g., biases. This update has been proposed in Groves (2013). The measurement vector $\boldsymbol{z}_{\mathrm{o}}^{k}$ of the Kalman filter at the time k can be written as:

$$\boldsymbol{z}_{\mathrm{o}}^{k} = [0, 0, 0]^{T}. \tag{18}$$

4 Shoe-Mounted Inertial Positioning

Shoe-mounted positioning systems represent the first massively implemented positioning system for pedestrians. Shoe-mounted positioning is usually derived with the strapdown algorithm. The biomechanics of the foot allow performing re-calibrations at every step, thus limiting the rapidly growing positioning error. Back in 2005, Foxlin (2005) proposed to re-calibrate the strapdown algorithm for shoe-mounted systems performing Zero-velocity UPdaTes (ZUPT).

The strapdown algorithm is composed of two phases, namely orientation estimation and position estimation. Fig. 2 represents the block diagram of the strapdown algorithm.

The orientation estimation is performed as explained in Section 3, resulting in a direction cosine matrix $\boldsymbol{C}^k$ that represents the rotation of the body frame with respect to the navigation frame for each time k. The position estimation phase starts after the orientation is computed. The orientation is used to project the acceleration measurements onto the navigation frame.

$$\boldsymbol{\alpha}_{\mathrm{n}}^k = \boldsymbol{C}^k \cdot \boldsymbol{\alpha}^k. \tag{19}$$

Secondly, the gravity acceleration, $\boldsymbol{g}_{\mathrm{n}} = [0, 0, g]^T$, is subtracted from the projected acceleration. By doing so, the remaining acceleration corresponds only to the acceleration due to the movement of the body. Lastly, this remaining acceleration is integrated twice over time to compute the position. The algorithm strapdown requires an initial position p_0, an initial velocity v_0 and also an initial orientation. This is represented in the following equations:

$$\boldsymbol{v}_{\mathrm{n}}^k = \boldsymbol{v}_{\mathrm{n}}^{k-1} + \delta t \cdot (\boldsymbol{\alpha}_{\mathrm{n}}^k - \boldsymbol{g}_{\mathrm{n}}) \tag{20}$$

and

$$\boldsymbol{p}_{\mathrm{n}}^k = \boldsymbol{p}_{\mathrm{n}}^{k-1} + \delta t \cdot \boldsymbol{v}_{\mathrm{n}}^k, \tag{21}$$

being $\boldsymbol{p}^k$ the position at the time k, $\boldsymbol{v}$ the velocity at the time k, and δt the sampling time.

The shoe-mounted positioning is usually implemented using a Kalman filter, whose state vector $\boldsymbol{x}_{\mathrm{p}}$ is defined as follows:

$$\boldsymbol{x}_{\mathrm{p}}^k = [\boldsymbol{p}^k, \boldsymbol{v}^k, \boldsymbol{\Psi}^k, \boldsymbol{b}^k]^T, \tag{22}$$

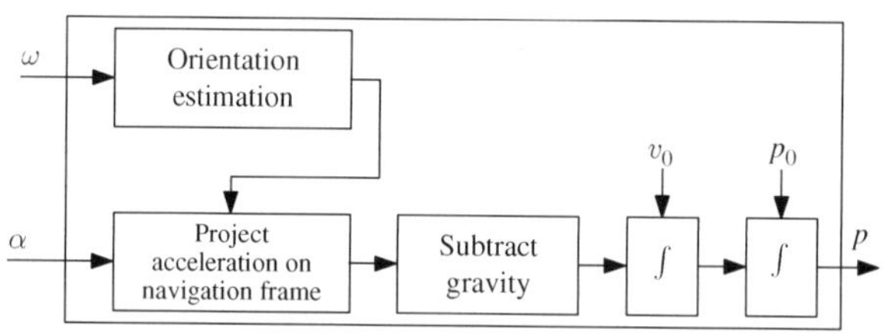

FIG. 2 Block diagram of the strapdown algorithm.

being $\boldsymbol{\Psi}^k$ the orientation at the time k and $\boldsymbol{b}^k$ the biases of the gyroscope at the time k. Also an error state Kalman filter is possible. The Kalman filter for shoe-mounted positioning systems is divided into two stages namely prediction and update. The equations described above represented by the block diagram of Fig. 2 represents the prediction stage. For error state Kalman filters, the equations described above do not form part of the filter, since the states consist of errors.

The aforementioned re-calibrations are performed during the update stage. The human gait cycle comprises eight phases (Streifeneder, 2016). Four of them can be observed using a foot-mounted sensor, i.e., the loading response, mid-stance, terminal stance, and swing as shown in Fig. 3.

In the case of shoe-mounted inertial positioning, the mid-stance is of high interest. The mid-stance phase is the period of the gait cycle when the foot is in contact with the ground. Fig. 4 represents the vertical acceleration measured by a shoe-mounted accelerometer. During the stance phases, indicated by the shadowed areas, the only acceleration measured by the shoe-mounted accelerometers is the gravity. Constant acceleration implies zero velocity, thus ZUPT corrections can be applied during these periods.

The stance phase detection is usually performed based on thresholds for the acceleration and turn rate (Foxlin, 2005). If both the acceleration and turn rate are within predefined thresholds during a minimum time, the stance phase is detected. The authors in Ruppelt et al. (2016) developed a stance phase detection based on a finite-state machine. In the latter, each phase of the human gait is a state of the finite-state machine. Such an approach can even detect the foot stance phase in challenging situations like walking the stairs.

Stance phase detection algorithms however, require these thresholds to be adapted to the particular inertial sensors, to the pedestrian or both. Furthermore, threshold-based algorithms perform optimally during walking on flat surfaces. In order to detect stance phases during other activities, e.g., walking stairs, a more sophisticated method is suggested by the authors in Ruppelt et al. (2016). The disadvantage of the latter is its complexity regarding both, design and implementation.

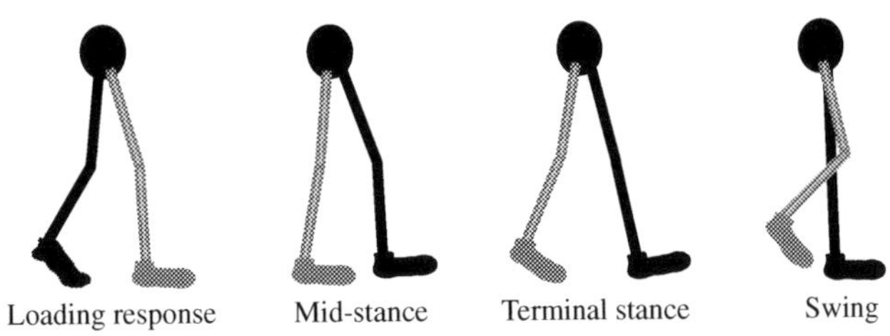

FIG. 3 Diagram of the gait phases visible using inertial sensors mounted on the right shoe *(grey leg)*.

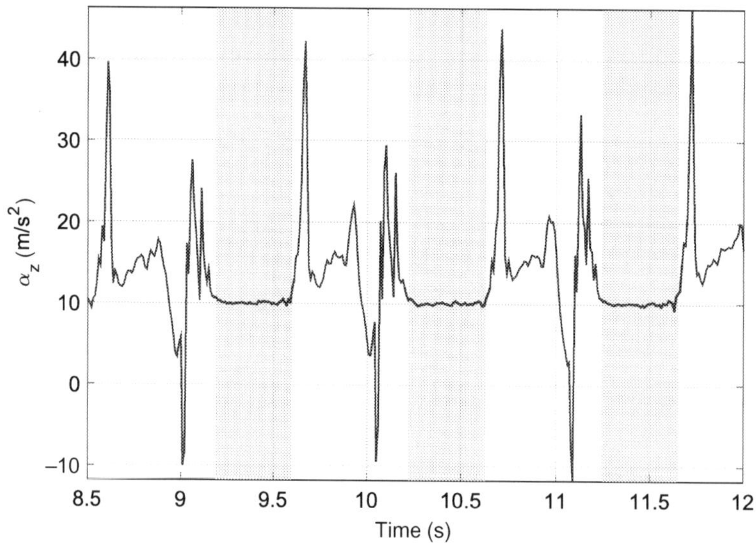

FIG. 4 Vertical acceleration measured by shoe-mounted accelerometers during three steps. The *shadowed areas* indicate the time when the foot is in mid-stance phase.

Upon detection of the stance phase, the states of the filter are updated with the ZUPT pseudo-measurement vector z_{p}^{k} at the time k that can be written as:

$$z_{\mathrm{p}}^{k} = [0, 0, 0]^{T}. \tag{23}$$

The use of ZUPT pseudo-measurements makes the error growth linear in time instead of cubic. The error accumulation, although linear, remains a challenge in pedestrian positioning based on shoe-mounted inertial sensors.

5 Non-shoe-Mounted Inertial Positioning

Non-shoe-mounted positioning systems are of high interest, because they can make use of the inertial sensors embedded in any wearable such as smart watch, smart glasses, and smart clothing among others. The error accumulated using these sensors cannot be mitigated with zero-velocity corrections, since the targeted body locations, i.e., head, wrist, etc. continuously move while walking. Therefore, in these cases the step&heading algorithm, represented in the block diagram of Fig. 5, is appropriate.

The step&heading algorithm is based on the following equations:

$$\begin{aligned} p_{\mathrm{x}}^{k} &= p_{\mathrm{x}}^{k-1} + s^{k} \cdot \cos(\psi^{k}), \\ p_{\mathrm{y}}^{k} &= p_{\mathrm{y}}^{k-1} + s^{k} \cdot \sin(\psi^{k}), \end{aligned} \tag{24}$$

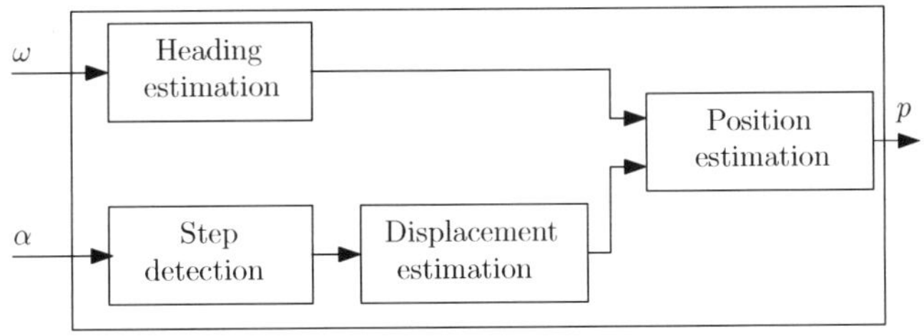

FIG. 5 Block diagram of the step&heading algorithm.

where p_x^k and p_y^k represent the position in the x- and y-axis at the time k, s^k stands for the step length at the time k and ψ^k is the heading of the pedestrian at the time k. Therefore, in order to compute the position of the pedestrian, two steps are necessary: the orientation estimation, to have the heading angle, and the displacement estimation.

The orientation estimation is carried out as described in Section 3. The different updates are not usually applied continuously, but only within particular periods, as explained in Section 3. From the complete orientation, usually only the heading angle ψ is used to compute the position of the pedestrian, as indicated in Eq. (24).

The step&heading algorithm is usually defined in 2D, as indicated in Eq. (24). However, for particular sensor locations it is possible to solve 3D positioning. In that case the position in the z-axis is as follows:

$$p_z^k = p_z^{k-1} + d_z^k, \tag{25}$$

where d_z^k represents the vertical displacement from the time $k-1$ to the time k. The authors in Munoz Diaz and Mendiguchia Gonzalez (2014) demonstrate that, if the inertial sensors are attached to the lower limb of the pedestrian, it is possible to differentiate between walking horizontally and climbing stairs by means of the orientation of the leg of the pedestrian. The information on the walking surface allows deriving the vertical displacement d_z.

The displacement estimation, being the step length or also the vertical displacement, is triggered every time a new step is detected, as shown in Fig. 5. The following subsections detail the step detection on horizontal surfaces, the step detection on stairs, the step length estimation and last but not least the vertical displacement estimation.

5.1 Step Detection on Horizontal Surfaces

The well known algorithm to detect steps is based on acceleration measurements. This algorithm is valid for all sensor locations. Fig. 6 shows the acceleration measured with the sensor introduced in the front pocket of the trousers. The dashed curve represents the norm of the acceleration $\|\alpha\|$ where the gravity has already been subtracted. A common procedure is to apply a low-pass filter (LPF) to this signal, in order to improve the performance regarding undetected steps and false detected steps. The LPF curve is shown

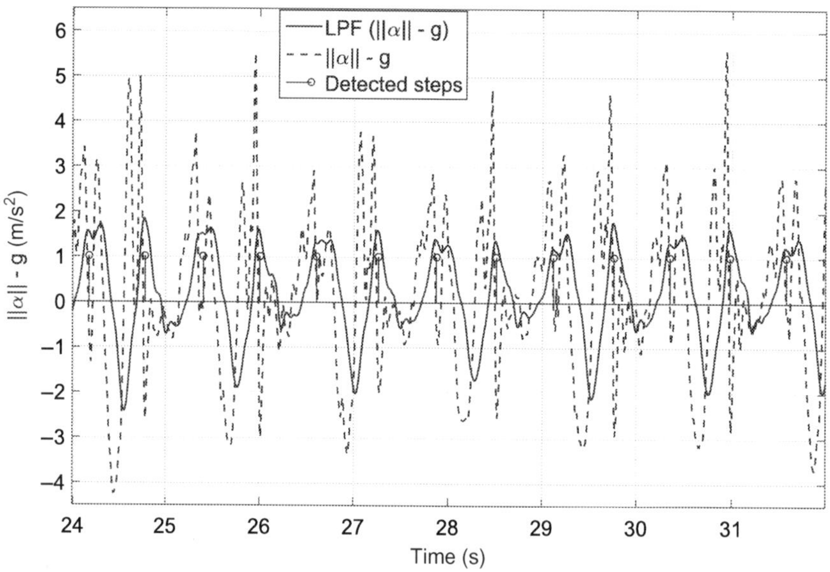

FIG. 6 The *dashed curve* represents the norm of the acceleration α with the gravity compensated, while the *solid curve* represents the low-pass filtered (LPF) acceleration. The detected steps are highlighted with *sticks*.

in solid and the detected steps are highlighted with sticks. In Munoz Diaz (2015), the performance of step detectors based on the norm of the acceleration and its low-pass filtered version has been analyzed. For horizontal surfaces, i.e., 2D walks, the false step detection rate decreases by using the filtered acceleration.

For some sensor locations, e.g., the lower limb, the step detection can also be performed using the pitch angle estimation, as suggested in Munoz Diaz and Mendiguchia Gonzalez (2014) and Xiao et al. (2014). Fig. 7 shows the pitch angle estimation for seven steps where the sensor has been introduced in the front pocket of the trousers. The detected steps are highlighted with sticks. In Munoz Diaz (2015), the performance of the step detector based on the pitch has been compared with the step detector based on the acceleration. The authors demonstrate that the step detection based on the pitch angle is more robust also for different walking speeds.

5.2 Step Detection on Stairs

Since the step&heading only based on inertial measurements is usually limited to horizontal displacements, i.e., 2D scenarios, the step detectors based on the acceleration do not offer reliable results when walking on stairs. The analysis carried out in Munoz Diaz (2015) for 3D scenarios shows that the undetected steps rate is high. However, the false detection rate is dramatically reduced by using the low-pass filtered acceleration. The authors in Munoz Diaz (2015) concluded that it is possible to successfully detect all steps

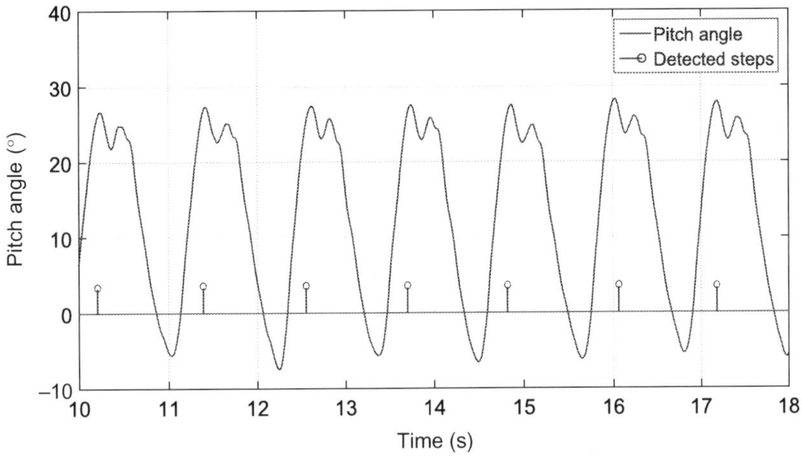

FIG. 7 The *curve* represents the pitch angle estimation over seven steps and the steps detected are highlighted with *sticks*.

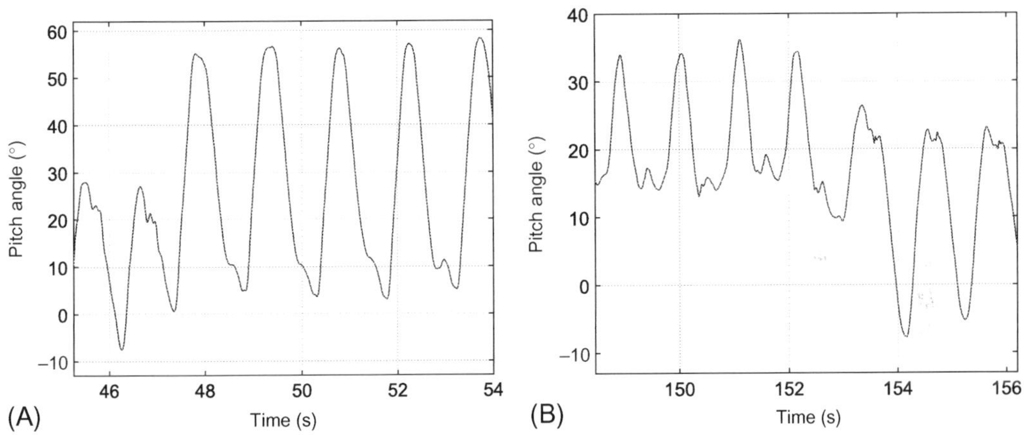

FIG. 8 (A) The pitch angle estimation when walking first two steps horizontally and then upstairs. (B) The pitch angle estimation when walking downstairs and the last three steps were taken walking horizontally.

while walking on stairs with the pitch-based step detector. Fig. 8 shows on the left side seven steps, where the first two steps were taken on an horizontal surface and the rest walking upstairs. The right side shows the pitch estimation during seven steps where the first four where taken walking downstairs and on an horizontal surface and the following three walking horizontally. The authors in Munoz Diaz (2015) show that detecting all steps is possible in 3D scenarios when using the pitch-based step detector. Additionally, it is possible to distinguish whether the pedestrian is walking horizontally, upstairs, or downstairs, which is the key to estimate the vertical displacement.

5.3 Step Length Estimation

The main current algorithms to compute the step length s^k at each time k can be classified depending on the sensor location, as specified in Alvarez et al. (2006) and Jahn et al. (2010).

If the sensor is attached to the body near the center of mass, two options exist:

- Based on a biomechanical model, where the kneeless biped is modeled as an inverted pendulum. The final estimation is scaled by a constant m that is calibrated for each user (Shih et al., 2012).

$$s^k = m \cdot \sqrt{2 \cdot L \cdot d_{Z_p}^k - {d_{Z_p}^k}^2}, \tag{26}$$

  where $d_{Z_p}^k$ represents the vertical displacement of the pelvis at the time k and L, the leg's length.
- Using an empirical relationship of the vertical acceleration and the step length (Jin et al., 2011; Goyal et al., 2011). The final estimation is scaled by a constant m that is calibrated for each user.

$$s^k = m \cdot \sqrt[4]{\alpha_{Z\max} - \alpha_{Z\min}}, \tag{27}$$

  where $\alpha_{Z\max}$ and $\alpha_{Z\min}$ are the maximum and minimum values of the vertical acceleration during each step.

If there are no restrictions on the sensor location on the human body (Shin et al., 2007, 2010; Gusenbauer et al., 2010; Renaudin et al., 2012, 2013), taking advantage of the relationship between step length, height of the user h, step frequency f_s^k at the time k and the calibration parameters (j, l, q) different for each user, the step length can be estimated as:

$$s^k = h \cdot (j \cdot f_s^k + l) + q. \tag{28}$$

In Munoz Diaz and Mendiguchia Gonzalez (2014), the step length estimator based on the pitch angle was presented. The authors assessed the relationship between the pitch amplitude and the step length with measurements recorded by 18 volunteers of different age, gender, height, and weight at different walking speeds. The authors propose a linear step length model based as follows:

$$s^k = j \cdot \Delta\theta_{\mathrm{H}}^k + l, \tag{29}$$

where $\Delta\theta_{\mathrm{H}}^k$ represents the pitch amplitude in horizontal surfaces at the time k. The parameters (j, l) can be universal or personalized for each pedestrian.

The authors in Munoz Diaz (2015) show an analysis comparing the step length estimator based on the step frequency and based on the pitch angle for a sensor introduced in the front pocket of the trousers. The results reveal that, for normal and constant walking speed, both estimators offer similar performance. However, for very low walking speed including stops and for high walking speeds, the detector based on the pitch angle offers more accurate results than the detector based on the step frequency.

5.4 Vertical Displacement Estimation

The authors in Munoz Diaz (2015) estimate, for the first time, the vertical displacement in a step&heading algorithm using only inertial sensors. This is possible thanks to the pitch angle estimation that allows identifying whether the pedestrian is walking horizontally or climbing stairs. The pitch angle estimation allows also distinguishing between walking downstairs and upstairs. In order to estimate the vertical displacement d_V, the authors carried out a set of experiments with the objective of finding a relationship between the pitch angle and the height of the steps of the staircase. The results show that there is a relationship between the amplitude of the pitch angle and the height of the steps. The authors propose a linear model that relates these two variables for walking downstairs and upstairs as follows:

$$\begin{aligned} d^k_{\text{VU}} &= j \cdot \Delta\theta^k_{\text{U}} + l, \\ d^k_{\text{VD}} &= q \cdot \Delta\theta^k_{\text{D}} + w, \end{aligned} \tag{30}$$

where d^k_{VU} and d^k_{VD} are the estimated vertical displacement for up and downstairs, respectively, at the time stamp k. $\Delta\theta^k_{\text{U}}$ and $\Delta\theta^k_{\text{D}}$ represent the pitch amplitude for steps up and down, respectively, at the time stamp k. The parameters (j, l, q, w) can be universal or personalized for each pedestrian.

The presented model is valid to estimate the height of the steps up and down of the vast majority of the staircases that can be found in every building, because it assumes a standard depth and focuses on the height of the step. Thus, the horizontal displacement is assumed.

Fig. 9 shows a 3D trajectory corresponding to a walk recorded in the German museum in Munich with the sensor introduced in the front pocket of the trousers and using only inertial sensors. The walk starts at (0, 0, 0) and at the point (90, −20, 0) the pedestrian takes the stairs until the first floor. Then the pedestrian walks a round on the first floor and takes

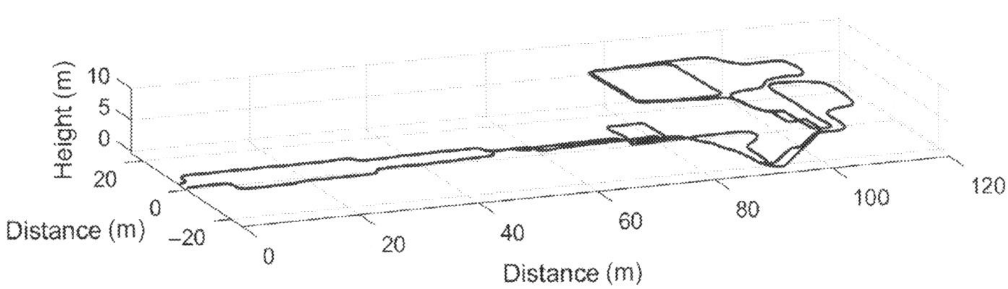

FIG. 9 The *curve* shows the 3D trajectory estimated using only inertial sensors introduced in the front pocket of the trousers.

the stairs to the second floor. After some rounds on the second floor the pedestrian walks downstairs two floors and comes back to the initial position.

The algorithm proposed by the authors assumes that the height of all steps of the staircase is the same, i.e., υ_U and υ_D are equal. Thus, the algorithm gathers data when walking up and downstairs and the height of the steps of the staircase is estimated with the average of the pitch amplitude of all steps up and down using Eq. (30).

6 Drift Reduction Methods

Regarding pedestrian inertial navigation systems using medium- and low-cost MEMS sensors, the accumulated error in the yaw angle estimation is still an unsolved issue. This error, commonly called drift, should be computed and used to prevent positioning errors.

The authors in Munoz Diaz et al. (2017) concluded that the drift error is mainly composed of biases, particularly the bias of the z-axis gyroscope. The biases of the x- and y-axes gyroscopes can be estimated through the gravitational field, as assessed in Munoz Diaz et al. (2017). Therefore, the error in roll and pitch angles can be corrected with the estimation of the x- and y-axes biases. On the contrary, the yaw angle suffers from ever growing errors that mainly arise from a poor estimation of the bias of the z-axis gyroscope (Munoz Diaz et al., 2017).

Additionally, there is an accumulating error in the vertical axis, i.e., height error. This severe error is in most of the cases mitigated with the use of barometers. However, there are also solutions for only inertial-based systems that are described in this section.

6.1 Heuristic Drift Elimination Algorithms

Heuristic drift elimination algorithms assume that pedestrians walk on a straight line in the building in directions which are parallel to the outer walls of the building. If the pedestrian does not move on a straight line, these corrections are suspended (Borenstein and Ojeda, 2010). After the first heuristic drift elimination algorithm was published, many authors in the literature have proposed similar ideas or improvements, such as coping with complex buildings including curved corridors or wide areas not restricted by corridors (Abdulrahim et al., 2010, 2012; Jimenez et al., 2011).

Additionally, the heuristic drift elimination has been suggested in combination with other heading corrections such as zero angular rate updates and magnetic measurements (Jimenez et al., 2010, 2012). The combination with available maps has also been proposed to restrict the possible heading angles by taking into account the walls of the buildings (Aggarwal et al., 2011; Pinchin et al., 2012). The high nonlinearities of the maps force the use of particle filters that weight the particles according to the similarity of their heading with the direction of the walls. The main drawback of these approaches is that previous knowledge is necessary, e.g., the map or the shape of the corridors.

6.2 SLAM-Based Algorithms

A suitable solution to drift reduction is the use of the simultaneous localization and mapping (SLAM) algorithm, which has been used for decades in robotics. The SLAM algorithm simultaneously generates a map of the desired landmarks and locates the user/robot within this map. These landmarks can be detected with any sensor, such as a laser scanner or a camera. The automatic vacuum cleaner, for example, generates a map of the room and locates itself within this map where the interesting landmarks, i.e., sofa, table, doors are included.

The SLAM algorithm has also been adapted to pedestrian navigation aiming at reducing the drift error. In order for the SLAM algorithm to successfully reduce the drift, a re-visit is necessary. That means, the pedestrian detects landmarks during the trajectory and, when part of the trajectory is re-visited, the landmark is again detected. The same landmark detected twice is an indicative of being again at the same position, therefore, corrections can be applied.

Commonly a particle filter is used that generates particles that move with different errors. When landmarks are re-visited, all particles are weighted depending on the landmarks position. Thus, particles that followed a trajectory with the current drift are high weighted, because they most likely correspond to the detected position. In Robertson et al. (2009), the 2D space is divided into a grid of uniform and adjacent hexagons, which can be considered as landmarks. When the same hexagons are re-visited the aforementioned corrections are carried out. The same procedure is applied for 3D trajectories but dividing the volume into hexagonal prisms with eight faces (Garcia Puyol et al., 2014). This procedure can also be applied if the hexagons are identified by the magnetic field intensity (Robertson et al., 2013). The main drawback of these algorithms is the complexity and processing time to manage the numerous hexagons or hexagonal prisms.

In Hardegger et al. (2012), the proposed landmarks are some location-related activities carried out by the pedestrian, such as sitting, lying, or opening doors. Based on the assumption that these activities are always performed at the same place, their repeated detection leads to the aforementioned corrections. The main drawback of these methods is that the heading estimation is not explicitly corrected, just corrections on the position are applied.

6.3 Multi-inertial Sensor Fusion

Multi-inertial sensor fusion combines two or more inertial sensors to reduce the drift in inertial positioning systems. Multi-inertial sensor fusion algorithms can be classified into two types: loose coupling and tight coupling.

Loose coupling algorithms combine the output of different inertial positioning systems. The aim is to generate a combined position estimation with less drift than the individual position estimations. The authors in Bousdar Ahmed and Munoz Diaz (2017) propose a loose coupling algorithm to combine the outputs of a shoe-mounted and a pocket-mounted inertial sensors. The so-called smart update approach is followed, i.e., the individual position estimations are combined favoring automatically the one containing less drift. A novel metric, the quality factor, is proposed to seamlessly identify which position estimation contains less drift.

Tight coupling algorithms target drift reduction by combining the raw data from two or more inertial sensors. On the one hand, inertial sensor arrays can be used to process all individual acceleration and turn rate measurements. In Skog et al. (2016), a maximum likelihood estimator combines the measurements from an array and the authors state that the information gained is proportional to the square of the array dimension. On the other hand, human biomechanics can also be used to reduce the drift. The body mounted sensors in Ahmadi et al. (2015) are combined to reduce the drift in a gait monitoring system. The authors use a kinematic leg model to ensure that the motion is coherent with the biomechanical behavior of the leg.

6.4 Landmark-Based Algorithms

In Millonig and Schechtner (2007), a study has been carried out concluding that landmarks play an important role for pedestrian navigation, therefore, it is recommendable to develop methods to include landmarks information in pedestrian navigation systems.

One of the most intuitive ways of detecting landmarks during the trajectory is using visual information. The chosen landmarks are tracked over time in order to use this motion to constrain the drift. In Griesbach et al. (2014), a stereo vision camera is used to extract the optical information of the landmarks. The heuristic drift elimination algorithms can also be seen as landmark-based algorithms, since the manmade straight corridors can be interpreted as landmarks. The main difference is that the landmarks of the heuristic drift elimination algorithms do not need to be tracked over time.

In Jimenez R. et al. (2011), an algorithm that makes use of detected ramps in buildings for correcting the drift is presented. In the article, foot-mounted inertial sensors are used and the position of the ramps of the target building is previously known. Ramps are detected through the slope of the terrain and corrections of the position of the pedestrian are applied. However, this approach does not compute the drift value. Therefore, although the position is corrected, the proposed approach does not bind the error of the yaw angle.

In Munoz Diaz et al. (2017) and Munoz Diaz and Caamano (2017), the authors propose the use of landmarks to compensate the drift error. The proposed landmarks are corners and stairs, thus, plentiful in indoor environments. These landmarks are seamlessly detected using only inertial sensors. The algorithm is based on re-visiting these landmarks in order to perform corrections. The trajectory of the landmark during the re-visit can be fully or partially overlapped or even with no overlap. The novelty of this contribution relies on computing the drift value. This accumulated drift is fed back to the orientation filter. This approach allows the yaw angle to be corrected and also prevent future positioning errors due to a drifted yaw angle estimation. Additionally, position corrections are also carried out. The authors recommend, depending on the positioning system requirements, to perform these corrections online, while the landmarks are re-visited, or offline, computing the overall drift value and using it to postprocess the recorded data.

6.5 Height Error Correction

The error in the height computation is the source of the confusion between different floors. This error is usually mitigated using barometers. The barometer sensor relates the change of height with the atmospheric pressure changes. The height error is mainly affecting the shoe-mounted inertial systems, because the step&heading approach is usually 2D defined.

There are several approaches for only inertial-based systems using medium- and low-cost MEMS sensors to correct the height error. The authors in Munoz Diaz et al. (2018) use the pitch angle of the foot to identify if the pedestrian is walking on horizontal surfaces or climbing stairs. The authors in Abdulrahim et al. (2012) apply an empirical threshold to assume that the user is walking horizontally and apply height corrections. These corrections act keeping the height at the same value, when the pedestrian is walking horizontally and only during the mid-stance phase (see Fig. 4). The authors in Ruppelt et al. (2016) apply also height constraints based on a finite state machine step detector.

7 Conclusions

In this chapter a review of the methods applied for pedestrian positioning using inertial sensors has been presented. Pedestrian inertial positioning is usually derived in two different ways depending on the location of the sensor on the human body: (i) for shoe-mounted sensors the strapdown algorithm is used, due to the possibility of perform re-calibrations at every step; (ii) for the rest of body locations the step&heading algorithm is preferred.

The big advantage of medium- and low-cost MEMS inertial sensors relies on their low price, small size, and widespread. Additionally, inertial positioning constitutes an infrastructure-less positioning system. Their clear disadvantage, however, is the remaining drift error on the estimated positioning.

There are many publications tackling the compensation of the drift error resulting when using inertial sensors. Drift affects inertial positioning disregarding the body location where the sensor is mounted on. Nowadays the trend is clearly pointing at sensor fusion. That means, combining the information of all sensors available. Especially recommended is the fusion with satellite and radio positioning systems. The piloting method does not suffer from drift, unlike the dead-reckoning method that propagates and accumulates error over time. This chapter, however, is focused on inertial sensors, thus, a review of the latest drift reduction methods using only inertial sensors is provided.

However, the current research on pedestrian inertial positioning is slowly approaching a static stage. While the latest proposals on sensor fusion and drift reduction algorithms greatly contribute to a more accurate positioning, the fact is that the drift error is low-bounded by the sensor technology. The breakthrough will be driven by the next generation of compact inertial sensors. In future, the inertial pedestrian dead-reckoning performed with the strapdown algorithm will be possible disregarding the body location where the sensor is mounted on. The next generation of high-quality compact inertial sensors will eliminate the current strong need of performing constant re-calibrations by offering a more steady bias stability.

References

Abdulrahim, K., Hide, C., Moore, T., Hill, C., 2010. Aiding MEMS IMU with building heading for indoor pedestrian navigation. In: Ubiquitous Positioning Indoor Navigation and Location Based Service (UPINLBS). IEEE.

Abdulrahim, K., Hide, C., Moore, T., Hill, C., 2012. Using constraints for shoe mounted indoor pedestrian navigation. R. Inst. Navig. 65, 15–28.

Aggarwal, P., Thomas, D., Borenstein, J., Ojeda, L., 2011. Map matching and heuristic elimination of gyro drift for personal navigation systems in GPS-denied conditions. J. Meas. Sci. Technol. 22.

Ahmadi, A., Destelle, F., Monaghan, D., Moran, K., O-Connor, N.E., Unzueta, L., Linaza, M.T., 2015. Human gait monitoring using body-worn inertial sensors and kinematic modelling. In: 2015 IEEE SENSORS. IEEE. https://doi.org/10.1109/icsens.2015.7370310.

Alvarez, D., González, R.C., López, A., Alvarez, J.C., 2006. Comparison of step length estimators from wearable accelerometer devices. In: 28th Annual International Conference of the IEEE Engineering in Medicine and Biology Society, 2006. EMBS'06. IEEE, pp. 5964–5967.

Azfal, M.H., Renaudin, V., Lachapelle, G., 2011. Use of Earth's magnetic field for mitigating gyroscope errors regardless of magnetic perturbation. Sensors 11, 11390–11414.

Bancroft, J.B., Lachapelle, G., 2012. Use of magnetic quasi static field (QSF) updates for pedestrian navigation. In: IEEE/ION Position Location and Navigation Symposium (PLANS).

Beauregard, S., Widyawan, Klepal, M., 2008. Indoor PDR performance enhancement using minimal map information and particle filters. In: Proceedings of the IEEE/ION Position Location and Navigation Symposium (PLANS) 2008, Monterey, USA.

Beroulle, V., Bertrand, Y., Latorre, L., Nouet, P., 2003. Monolithic piezoresistive CMOS magnetic field sensors. Sens. Actuators A: Phys. 103 (1), 23–32. https://doi.org/10.1016/S0924-4247(02)00317-5. http://www.sciencedirect.com/science/article/pii/S0924424702003175.

Borenstein, J., Ojeda, L., 2010. Heuristic drift elimination for personnel tracking systems. R. Inst. Navig. 63, 591–606.

Bousdar Ahmed, D., Munoz Diaz, E., 2017. Loose coupling of wearable-based INSs with automatic heading evaluation. Sensors 17, 2534. https://doi.org/10.3390/s17112534. http://www.mdpi.com/1424-8220/17/11/2534.

Bousdar, D., Munoz Diaz, E., Kaiser, S., 2016. Performance comparison of foot- and pocket-mounted inertial navigation systems. In: 7th International Conference on Indoor Positioning and Indoor Navigation, 4–7 October, 2016, Madrid, Spain.

Diez, L.E., Alfonso Bahillo, A., Bataineh, S., Masegosa, A.D., Perallos, A., 2016. Enhancing improved heuristic drift elimination for wrist-worn PDR systems in buildings. In: 2016 IEEE 84th Vehicular Technology Conference (VTC-Fall), pp. 1–5.

Do, T.N., Liu, R., Yuen, C., Zhang, M., Tan, U.X., 2016. Personal dead reckoning using IMU mounted on upper torso and inverted pendulum model. IEEE Sens. J. 16 (21), 7600–7608.

Foxlin, E., 2005. Pedestrian tracking with shoe-mounted inertial sensors. IEEE Comput. Graph. Appl. 25 (6), 38–46.

Garcia Puyol, M., Bobkov, D., Robertson, P., Jost, T., 2014. Pedestrian simultaneous localization and mapping in multistory buildings using inertial sensors. IEEE Trans. Intell. Trans. Syst. 15, 1714–1727.

Goyal, P., Ribeiro, V.J., Saran, H., Kumar, A., 2011. Strap-down pedestrian dead-reckoning system. In: 2011 International Conference on Indoor Positioning and Indoor Navigation (IPIN). IEEE, pp. 1–7.

Griesbach, D., Baumbach, D., Zuev, S., 2014. Stereo-vision-aided inertial navigation for unknown indoor and outdoor environments. In: IEEE International Conference on Indoor Positioning and Indoor Navigation (IPIN). IEEE, pp. 709–716.

Groves, P.D., 2013. Principles of GNSS, inertial, and multisensor integrated navigation systems, second ed. Artech House. ISBN 9781608070053.

Gu, Y., Lo, A., Niemegeers, I., 2009. A survey of indoor positioning systems for wireless personal networks. IEEE Commun. Surv. Tutorials 11 (13–32), 1281–1293.

Gusenbauer, D., Isert, C., Krosche, J., 2010. Self-contained indoor positioning on off-the-shelf mobile devices. In: 2010 International Conference on Indoor Positioning and Indoor Navigation (IPIN). IEEE, pp. 1–9.

Hardegger, M., Roggen, D., Mazilu, S., Troster, G., 2012. ActionSLAM: using location-related actions as landmarks in pedestrian SLAM. In: IEEE International Conference on Indoor Positioning and Indoor Navigation (IPIN). IEEE.

Harle, R., 2013. A survey of indoor inertial positioning systems for pedestrians. IEEE Commun. Surv. Tutorials 15 (3), 1281–1293.

Jahn, J., Batzer, U., Seitz, J., Patino-Studencka, L., Gutiérrez Boronat, J., 2010. Comparison and evaluation of acceleration based step length estimators for handheld devices. In: 2010 International Conference on Indoor Positioning and Indoor Navigation (IPIN). IEEE, pp. 1–6.

Jimenez, A.R., Seco, F., Prieto, J.C., Guevara, J., 2010. Indoor pedestrian navigation using an INS/EKF framework for yaw drift reduction and a foot-mounted IMU. In: 7th Workshop on Positioning Navigation and Communication (WPNC). IEEE, pp. 135–143.

Jimenez, A.R., Seco, F., Zampella, F., Prieto, J.C., 2011. Improved heuristic drift elimination (iHDE) for pedestrian navigation in complex buildings. In: IEEE International Conference on Indoor Positioning and Indoor Navigation (IPIN). IEEE, pp. 1–8.

Jimenez, A.R., Seco, F., Zampella, F., Prieto, J.C., Guevara, J., 2012. Improved heuristic drift elimination with magnetically-aided dominant directions (MiHDE) for pedestrian navigation in complex buildings. J. Locat. Based Serv. 6, 186–210.

Jimenez A.R., Seco, F., Zampella, F., Prieto, J.C., Guevara, J., 2011. PDR with a foot-mounted IMU and ramp detection. Sensors 11, 9393–9410.

Jin, Y., Toh, H.S., Soh, W.S., Wong, W.C., 2011. A robust dead-reckoning pedestrian tracking system with low cost sensors. In: 2011 IEEE International Conference on Pervasive Computing and Communications (PerCom). IEEE, pp. 222–230.

Kaiser, S., Khider, M., Garcia Puyol, M., Bruno, L., Robertson, P., 2015. Map aided indoor navigation. In: Karimi, H. (Ed.), Indoor Wayfinding and Navigation. Taylor and Francis, pp. 107–140.

Krach, B., Robertson, P., 2008. Cascaded estimation architecture for integration of foot-mounted inertial sensors. In: Proceedings of the IEEE/ION Position Location and Navigation Symposium (PLANS) 2008, Monterey, USA.

Lenz, J., Edelstein, S., 2006. Magnetic sensors and their applications. IEEE Sens. J. 6 (3), 631–649. https://doi.org/10.1109/JSEN.2006.874493.

Liu, H., Darabi, H., Banerjee, P., Liu, J., 2007. Survey of wireless indoor positioning techniques and systems. IEEE Trans. Syst. Man Cybern. C: Appl. Rev. 37 (6), 1067–1080.

Millonig, A., Schechtner, K., 2007. Developing landmark-based pedestrian-navigation systems. IEEE Trans. Intell. Transp. Syst. 8 (1), 43–49.

Munoz Diaz, E., 2015. Inertial pocket navigation system: unaided 3D positioning. Sensors 15, 9156–9178. https://doi.org/10.3390/s150409156. http://www.mdpi.com/1424-8220/15/4/9156.

Munoz Diaz, E., Caamano, M., 2017. Landmark-based online drift compensation algorithm for inertial pedestrian navigation. In: IEEE International Conference on Indoor Positioning and Indoor Navigation (IPIN). IEEE.

Munoz Diaz, E., Mendiguchia Gonzalez, A.L., 2014. Step detector and step length estimator for an inertial pocket navigation system. In: IEEE International Conference on Indoor Positioning and Indoor Navigation (IPIN). IEEE.

Munoz Diaz, E., Heirich, O., Khider, M., Robertson, P., 2013. Optimal sampling frequency and bias error modeling for foot-mounted IMUs. In: IEEE International Conference on Indoor Positioning and Indoor Navigation (IPIN). IEEE.

Munoz Diaz, E., Mendiguchia Gonzalez, A.L., de Ponte Müller, F., 2014. Standalone inertial pocket navigation system. In: IEEE/ION Position Location and Navigation Symposium (PLANS). IEEE.

Munoz Diaz, E., Caamano, M., Fuentes Sanchez, F.J., 2017. Landmark-based drift compensation algorithm for inertial pedestrian navigation. Sensors 17, 1555. https://doi.org/10.3390/s17071555. http://www.mdpi.com/1424-8220/17/7/1555.

Munoz Diaz, E., de Ponte Müller, F., García Domínguez, J.J., 2017. Use of the magnetic field for improving gyroscopes' biases estimation. Sensors 17, 832. https://doi.org/10.3390/s17040832. http://www.mdpi.com/1424-8220/17/4/832.

Munoz Diaz, E., Kaiser, S., Bousdar Ahmed, D., 2018. Height error correction for shoe-mounted inertial sensors exploiting foot dynamics. Sensors 18, 888. https://doi.org/10.3390/s18030888. http://www.mdpi.com/1424-8220/18/3/888.

Pinchin, J., Hide, C., Moore, T., 2012. A particle filter approach to indoor navigation using a foot-mounted inertial navigation system and heuristic heading information. In: IEEE International Conference on Indoor Positioning and Indoor Navigation (IPIN). IEEE, pp. 1–10.

Renaudin, V., Combettes, C., 2014. Magnetic, acceleration fields and gyroscope quaternion (MAGYQ)-based attitude estimation with smartphone sensors for indoor pedestrian navigation. Sensors 14, 22864–22890.

Renaudin, V., Susi, M., Lachapelle, G., 2012. Step length estimation using handheld inertial sensors. Sensors 12 (7), 8507–8525.

Renaudin, V., Demeule, V., Ortiz, M., 2013. Adaptive pedestrian displacement estimation with a smartphone. In: International Conference on Indoor Positioning and Indoor Navigation, pp. 916–924.

Robertson, P., Angermann, M., Krach, B., 2009. Simultaneous localization and mapping for pedestrians using only foot-mounted inertial sensors. In: International Conference on Ubiquitous Computing, pp. 93–96.

Robertson, P., Frassl, M., Angermann, M., Doniec, M., Julian, B.J., Garcia Puyol, M., Khider, M., Lichtenstern, M., Bruno, L., 2013. Simultaneous localization and mapping for pedestrians using distortions of the local magnetic field intensity in large indoor environments. In: IEEE International Conference on Indoor Positioning and Indoor Navigation (IPIN). IEEE, pp. 1–10.

Ruppelt, J., Kronenwett, N., Scholz, G., Trommer, G.F., 2016. High-precision and robust indoor localization based on foot-mounted inertial sensors. In: 2016 IEEE/ION Position, Location and Navigation Symposium (PLANS), pp. 67–75. https://doi.org/10.1109/PLANS.2016.7479684.

Shin, S.H., Park, C.G., Kim, J.W., Hong, H.S., Lee, J.M., 2007. Adaptive step length estimation algorithm using low-cost MEMS inertial sensors. In: Sensors Applications Symposium, 2007. SAS'07. IEEE, pp. 1–5.

Shin, S.H., Lee, M.S., Park, C.G., Hong, H.S., 2010. Pedestrian dead reckoning system with phone location awareness algorithm. In: 2010 IEEE/ION Position Location and Navigation Symposium (PLANS). IEEE, pp. 97–101.

Shih, W.Y., Chen, L.Y., Lan, K.C., 2012. Estimating walking distance with a smart phone. In: 2012 Fifth International Symposium on Parallel Architectures, Algorithms and Programming (PAAP). IEEE, pp. 166–171.

Skog, I., Nilsson, J.O., Handel, P., Nehorai, A., 2016. Inertial sensor arrays, maximum likelihood, and Cramér-Rao bound. IEEE Trans. Signal Process. 64 (16), 4218–4227. https://doi.org/10.1109/tsp.2016.2560136.

Streifeneder, 2016. The eight phases of human gait cycle. https://www.streifeneder.com/downloads/o.p./400w43_e_poster_gangphasen_druck.pdf.

Wagner, J.F., Wieneke, T., 2003. Integrating satellite and inertial navigation—conventional and new fusion approaches. Control Eng. Pract. 11 (5), 543–550.

Windau, J., Itti, L., 2016. Walking compass with head-mounted IMU sensor. In: 2016 IEEE International Conference on Robotics and Automation (ICRA), pp. 5542–5547.

Woodman, O.J., 2007. An introduction to inertial navigation. Tech. Rep., University of Cambridge. Computer Laboratory, uCAM-CL-TR-696. ISSN 1476-2986.

Woodman, O., Harle, R., 2008. Pedestrian localisation for indoor environments. In: Proceedings of UbiComp 2008, Seoul, South Korea.

Xiao, Z., Wen, H., Markham, A., Trigoni, N., 2014. Robust pedestrian dead reckoning (R-PDR) for arbitrary mobile device placement. In: IEEE International Conference on Indoor Positioning and Indoor Navigation (IPIN). IEEE.

Zampella, F., Khider, M., Robertson, P., Jimenez, A., 2012. Unscented Kalman filter and magnetic angular rate update (MARU) for an improved pedestrian dead-reckoning. In: IEEE/ION Position Location and Navigation Symposium (PLANS). IEEE.

17

Fundamentals of Airborne Acoustic Positioning Systems

Fernando J. Álvarez Franco

SENSORY SYSTEMS RESEARCH GROUP, UNIVERSITY OF EXTREMADURA, BADAJOZ, SPAIN

1 Introduction

Active acoustics has been extensively used in the design of airborne local positioning systems for the last two decades, and it is today considered a classical and reliable solution to this technological challenge (Mautz, 2012). Using typical working frequencies within the high sonic and very low ultrasonic bands (15,000–50,000 Hz), this solution is characterized by a centimeter precision with coverage ranges of some tens of meters.

Although acoustics had been already used to locate and track submerged objects since the early 1960s (Milne, 1983), it is not after more than thirty years that the first airborne acoustic positioning systems (AAPS) were developed. These pioneering systems were based on the emission of short and constant-frequency ultrasonic pulses whose arrival was detected by following a simple amplitude or energy thresholding procedure. This was, for example, the case of the Active Bat System (Ward et al., 1997) where wireless badges (bats) carried by personnel or attached to certain equipment emitted 40-kHz ultrasonic pulses of 50 μs-duration after being triggered over a wireless link. These pulses were received by a set of ceiling-mounted sensors that measured the pulses Time-of-Arrival (TOA) and computed the badge three-dimensional position by spherical lateration. An alternative approach was proposed in the Constellation System (Foxlin and Harrington, 1998), where a set of ultrasonic emitters was deployed at known locations in the environment. These beacons emitted a 40-kHz ultrasonic pulse after receiving an infrared trigger code from the unit to be located, which communicated with the beacons one-at-a-time. The receiving unit needed to compute at least three TOAs from the emissions of different beacons to obtain its position by spherical lateration. Note that, in this case, the device to be located was in charge of computing its own position using the signals emitted from different beacons. This architecture has been defined by some authors as privacy-oriented, in contrast to the centralized architecture of the Active Bat system. An additional example of a narrowband AAPS that cannot be considered centralized or privacy-oriented is the Cricket system (Priyantha et al., 2000). This system was based on a set of independent beacons

Geographical and Fingerprinting Data for Positioning and Navigation Systems. https://doi.org/10.1016/B978-0-12-813189-3.00017-4

that incorporated an RF transceiver, an ultrasonic emitter, and an ultrasonic receiver. Each beacon could compute its own position by measuring the TOAs of the 40-kHz ultrasonic pulses of 125 μs-duration emitted by nearby beacons, which also broadcast their own position through a 433-MHz RF signal. Again, spherical lateration was used to compute the desired location. All these systems featured very simple emitter and receiver acoustic modules, but at the expense of providing a limited positioning accuracy of some decimeters with high sensitivity to in-band noise. Moreover, special attention had to be paid to avoid interference between different emitters, either by making use of time-multiplexing strategies (Foxlin and Harrington, 1998) by developing specific algorithms (Priyantha et al., 2000).

A solution to these limitations was soon provided by the pulse compression technique extensively used in radar systems (Skolnik, 1980). A new generation of broadband AAPS started being developed in the early 2000s, based on the emission of Binary Phase Coded signals that, as will be detailed later, were detected by matched filtering. This spread spectrum technique had been successfully incorporated in the development of high precision airborne sonars some years before (Peremans et al., 1993; Jörg and Berg, 1998; Ureña et al., 1999), so its application in the field of AAPS was rather straightforward. One of the first broadband AAPS is presented in Hazas and Ward (2002, 2003), where the authors propose the use of 511-bit Gold codes to modulate a 50-kHz ultrasonic carrier with a bit period of 50 μs, thus giving a total emission duration of 25.55 ms. In Hazas and Ward (2002), eight receivers are installed in the ceiling of an office room to configure a centralized architecture that measures the TOAs of the signals emitted by a set of synchronized transmitters. The authors reported positioning accuracies slightly above 2 cm when using spherical lateration in a noisy environment. In Hazas and Ward (2003), a privacy-oriented architecture is presented with positioning accuracies around 5 cm. A similar privacy-oriented architecture is presented in Villadangos et al. (2005), based on the modulation of a 50-kHz carrier with 127-bit Gold codes and a bit period of 20 μs, for a significantly shorter emission duration of 2.54 ms. The main contribution of this broadband AAPS is the proposal of a hyperbolic lateration algorithm, based on the measurement of the Time-Difference-of-Arrival (TDOA) between the first detected signal, emitted by the nearest beacon, and the other signals detected subsequently. This positioning strategy avoids the need for a synchronized triggering signal between the beacons and the receiver. An improved version of this system is presented in Ureña et al. (2007), where the 50 kHz carrier is modulated by 255-bit Kasami codes to obtain accuracies below 1 cm in the horizontal positioning of the receiver. Since the appearance of these initial works, other pseudo-random sequences have been proposed to encode the emissions of more computationally efficient broadband AAPS, such as LS (Pérez et al., 2007) or CSS (Pérez et al., 2012). In all these broadband systems, the TOA or TDOA of the received signal is measured when the auto-correlation peak of this signal exceeds a detection threshold, what improves the precision of the range measurement between one and two orders of magnitude with respect to that of the previous narrowband systems. Nevertheless, the longer duration of AAPS signals leads to the appearance of new problems that may hinder their detection.

Some of the most recent works in the field of broadband AAPS are focused on the proposal of solutions to compensate for some of these problems, such as Doppler shift (Álvarez et al., 2013), multipath propagation (Álvarez et al., 2017a), or Multiple Access Interference (MAI) (Aguilera et al., 2018b).

This chapter has been written to give the reader a general overview of AAPS design fundamentals. With this idea in mind, next section includes some basic concepts and results related with acoustic wave propagation in air, which eventually determine the limits of this technology performance. Next, Section 3 deals with the signal detection process and describes the different positioning observables. Section 4 reviews those positioning strategies directly related with the most used observables described in the previous section. Finally, the above mentioned signal detection hindering phenomena and compensation techniques are described in some detail in Section 5.

2 Acoustic Wave Propagation in Air

As stated before, the performance of acoustic technology is directly related to the propagation characteristics of these waves in air. Some of these characteristics are briefly reviewed next.

2.1 Absorption

Acoustic waves are longitudinal mechanical waves that propagate through a material medium by pressure variations. When they propagate, part of the energy is dissipated as heat inside the medium, thus causing an exponential decay or the pressure amplitude P with the traveled distance r as,

$$P(r) \propto e^{-\alpha \cdot r}, \tag{1}$$

where α is the absorption coefficient. In the air, there are basically two different mechanisms that cause absorption of acoustic waves: the viscothermal losses (or classical absorption) and the oxygen and nitrogen molecular relaxation processes. The theoretical analysis of both processes led to a set of equations that have been later experimentally adjusted to increase agreement with real data. Today, these equations are grouped into the ISO-9613 standard (ISO/TC 43 Technical Committee, Acoustics, Sub-Committee SC1, Noise, 1993). This norm establishes that the absorption coefficient can be calculated as,

$$\begin{aligned}\alpha(\mathrm{dB/m}) = 8.69 f^2 \Bigg\{ & 1.84 \cdot 10^{-11} \left(\frac{P}{P_{ref}}\right)^{-1} \cdot \left(\frac{T}{T_{ref}}\right)^{\frac{1}{2}} \\ & + \left(\frac{T}{T_{ref}}\right)^{\frac{-5}{2}} \cdot \left[0.01275 \frac{e^{\frac{-2239.1}{T}}}{f_{rO} + \frac{f^2}{f_{rO}}} + 0.1068 \frac{e^{\frac{-3352}{T}}}{f_{rN} + \frac{f^2}{f_{rN}}} \right] \Bigg\},\end{aligned} \tag{2}$$

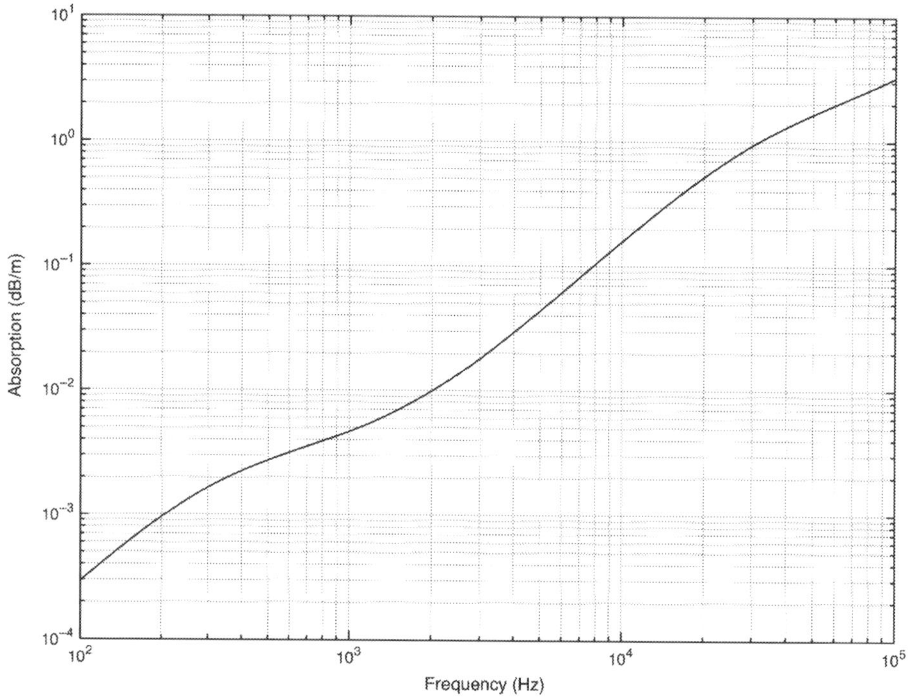

FIG. 1 Frequency dependence of the absorption coefficient in air ($T = 20°\text{C}$, $H = 70\%$ and $P = 1$ atm).

where f is the wave frequency in Hz, P is the atmospheric pressure in Pa ($P_{ref} = 101{,}325$ Pa), T is the absolute temperature in K ($T_{ref} = 293.15$ K), and f_{rO}, f_{rN} represent the relaxation frequencies of oxygen and nitrogen respectively, which also depend on temperature, pressure, and relative humidity. Fig. 1 represents the frequency dependence of this absorption coefficient, which has been generated from Eq. (2) assuming typical indoor values for the temperature ($T = T_{ref}$), pressure ($P = P_{ref}$), and humidity ($H = 70\%$). As can be seen, absorption increases very fast with frequency and it is above 0.5 dB/m in the ultrasonic region ($f > 20$ kHz), thus limiting the maximum range of these inaudible signals to some tens of meters in practice.

2.2 Propagation Speed

The speed of acoustic waves in air is an increasing function of temperature that can be approximated as,

$$c\,(\text{m/s}) = 331.6 \cdot \sqrt{1 + \frac{T}{273.15}}, \tag{3}$$

where T represents temperature in Celsius. As can be easily inferred from Eq. (3), this speed is about six orders of magnitude lower than that of electromagnetic waves, what determines the high spatial resolution capability of this technology. On the negative side,

Doppler shift is a much more pronounced effect when using acoustic signals. Assuming a static receiver, the relative change of frequency caused by a moving emitter is given by,

$$\frac{\Delta f}{f_0} = \frac{v_e}{c}, \tag{4}$$

where $v_e \ll c$ is the emitter speed. From Eq. (4) it can be shown that the same shift caused in a radar signal by the fastest unmanned aerial vehicle ($v_e \approx 8000\,\text{m/s}$) is caused in an AAPS by a garden snail! ($v_e \approx 0.01\,\text{m/s}$).

2.3 Impedance

A third parameter that deserves some attention from AAPS designers is the specific acoustic impedance, which is a measure of the opposition that the medium presents to the acoustic flow, defined as,

$$Z_0\,(\text{rayls}) = \frac{p}{u}, \tag{5}$$

where p is the pressure at a certain point and u is the particle velocity at that point. Table 1 shows the acoustic impedance of air and typical building materials. Since the acoustic impedance of air is much lower than that of the building materials, acoustic waves propagating in air are almost perfectly reflected without phase change by those materials (reflection coefficient $R \approx 1$). Consequently, as opposed to radiofrequency-based positioning systems, signals generated by AAPS are confined within the room where the emitters have been deployed. This particularity also implies a strong multipath effect, as it is well known by acoustical engineers.

2.4 Outdoor Propagation

Most AAPS have been designed to operate indoors, with well controlled ambient conditions. However, these systems can be similarly deployed in outdoor environments if additional meteorological phenomena are considered.

To begin with, the atmospheric absorption represented in Eq. (2) can take a wider range of values because of the larger variability of temperature and humidity outdoors. Hence, this absorption could be more than five times greater during a hot and dry

Table 1 Specific Acoustic Impedance of Air and Typical Building Materials

Material	Z_0 (Rayls)
Air	413
Wood	1.57×10^6
Brick	7.40×10^6
Marble	10.5×10^6
Glass	13.0×10^6

summer evening than in a cold and humid winter morning. Also, additional attenuation phenomena such as the presence of fog or rain must be considered outdoors. A thick fog can be the main cause of attenuation for frequencies below 10 kHz (Cole and Dobbins, 1970; Davidson, 1975), and an intense rain can cause a nonnegligible attenuation for frequencies above 50 kHz (Shamanaeva, 1988).

Secondly, and due to the mechanical character of acoustic waves, wind modifies the apparent sound propagation speed as:

$$c\,(\mathrm{m/s}) = 331.6 \cdot \sqrt{1 + \frac{T}{273.15}} + v_l, \tag{6}$$

where v_l is the component of the wind parallel to the direction of propagation. In the atmosphere, both temperature and wind speed are functions of the altitude, and this dependence is inherited by the apparent propagation speed $c = c(z)$. This change of speed with altitude causes the refraction of acoustic waves, which in practice translates into a slight decrease in the received amplitude.

Atmospheric turbulence is the most problematic phenomenon when transmitting encoded acoustic signals outdoors, as it is a strongly random and nonlinear phenomenon. If an acoustic wave propagates through a turbulent medium, it finds a variety of strongly rotational fluxes (eddies) with different sizes, velocities, and temperatures. Each one of these eddies acts as a strong scatterer of acoustic energy and their combined effect alters the initial coherence of the wavefront, which will no longer be spherical and with identical amplitude after crossing the turbulent region. A receiver placed at a certain distance from the emitter will record random fluctuations in the amplitude and phase of the acquired signals. After a while, this effect can make the encoded signal completely unrecognizable by its matched receiver (Álvarez et al., 2006b).

3 Acoustic Signal Detection and Positioning Observables

As already mentioned in Section 1, one of the simplest methods to detect the arrival of an air-propagating acoustic pulse is the envelope detection technique represented in Fig. 2. This method rectifies and integrates the incoming signal, which is detected after exceeding a certain threshold. Envelope detection has been successfully used in many systems that do not require a very precise range measurement, as the popular Polaroid Ultrasonic Ranging System (Polaroid Corporation, 1991). Most of these systems incorporated an automatic control gain circuit in the receiver to compensate for the attenuation of ultrasound in air.

Once the received signal has been detected, there are different features of this signal that can be used for positioning purposes, commonly referred to as *positioning observables*,

- Power or Received Signal Strength (RSS), from where the traveled distance can be obtained if we know the emitted power and the attenuation model.

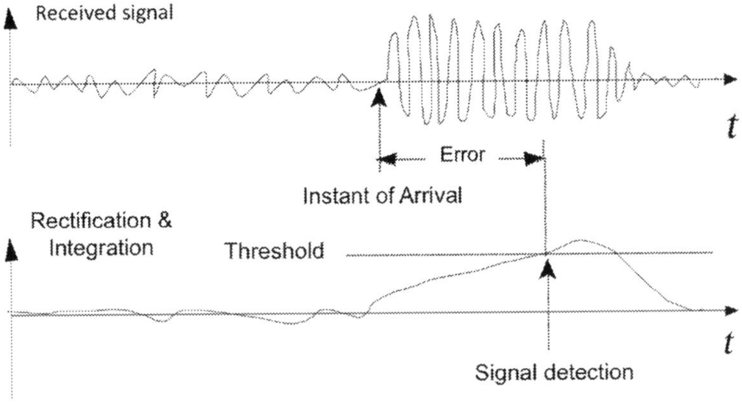

FIG. 2 Envelope detection technique.

- Time-of-Arrival (TOA), defined as the time at which the emitted signal reaches the receiver, from where the traveled distance can be easily obtained just by multiplying by the signal propagation speed.
- Time Difference of Arrival (TDOA), defined as the time difference between the moment the emitted signal is received and the moment a reference signal is received. As before, the difference in distances can be obtained by multiplying by the propagation speed.
- Angel of Arrival (AOA), from where the direction of the emitter can be inferred.

RSS is not commonly used in AAPS mainly due to nonuniform emission pattern of acoustic transducers and the strong dependence of air absorption with environmental parameters already described in Eq. (2). Besides, AOA is usually determined by measuring the different arrival times at an array of transducers. Hence, it can be stated that TOA and TDOA are the most common positioning observables when dealing with acoustic signals in air.

The precision of a time-delay measurement in an additive white Gaussian noise (AWGN) channel is given by (Kay, 1993)

$$\delta t \geq \frac{1}{\beta \cdot \sqrt{SNR}}, \tag{7}$$

where *SNR* is the signal-to-noise ratio at the output of the receiver and β is the effective or root mean square bandwidth. Eq. (7) states that to achieve a certain precision in a time-delay measurement a minimum value of the product bandwidth-SNR is required. Assuming a constant value for the SNR, this minimum bandwidth can be obtained by shortening the duration of the continuous frequency pulse, but then the amplitude of the signal should be increased to keep constant the signal energy. Actually, there is a physical limit for this amplitude imposed by both real transducers and electronics. An alternative is to modulate the original waveform to extend its bandwidth while keeping a constant energy, and use a matched filter in the receiver to detect this signal.

Two different alternatives have been proposed to modulate the emitted waveform with the aim of increasing its effective bandwidth. The first one is based on Linear Frequency Modulation (LFM), where the frequency of a pulsed waveform is linearly increased from f_1 to f_2 over the duration of the pulse. The second option is based on Binary Phase Coding, where a long pulse is divided into N subpulses whose phase is selected to be either 0 or π radians according to the bits of a certain code. If this code is a pseudo-random (PR) sequence, the waveform approximates a noise-modulated signal with a delta-like auto-correlation function. Fig. 3 shows a schematic representation of this broadband system. As can be seen, the TOA of the received signal is measured when the auto-correlation peak exceeds a certain threshold, thus improving the precision of the range measurement between one and two orders of magnitude with respect to that of narrowband systems. The main advantage of this approach is that different sequences from the same family can be generated with nearly null cross-correlation properties, thus allowing the simultaneous emission of different emitters with very low interference among them. To date, several broadband AAPS have been designed that propose the use of different families of PR sequences, such as Gold (Hazas and Ward, 2002), Kasami (Ureña et al., 2007), LS (Pérez et al., 2007), or CSS (Pérez et al., 2012). All these sequences exhibit very similar correlation properties, and the main difference between them is the possibility to design the matched filter shown in Fig. 3 as an efficient architecture that notably reduces the total number of arithmetic elements required to perform the correlations, thus allowing the actual implementation of a real-time operating system in a hardware platform. Examples of such efficient architectures in the receiver module of an airborne acoustic system can be found in Pérez et al. (2007, 2012) and Álvarez et al. (2006a)

4 Positioning Strategy

Two main positioning strategies have been developed in the design of AAPS, which are directly related to the main signal observables described in the previous section, namely, spherical and hyperbolic lateration. Although these techniques are not specific to acoustic

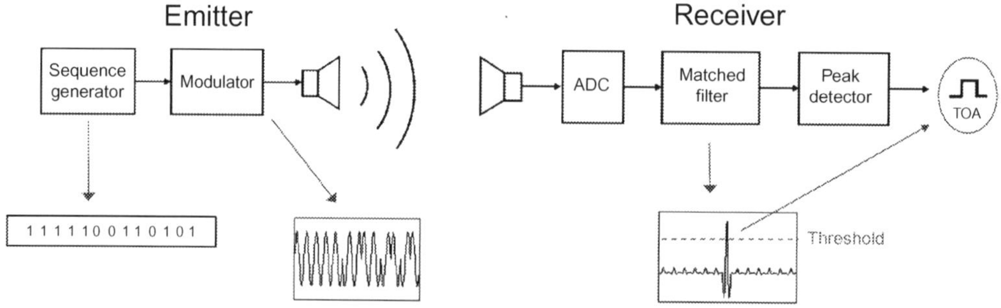

FIG. 3 Schematic representation of a broadband system based on Binary Phase Coding.

positioning systems, and detailed descriptions can be found elsewhere, they have been included in this chapter for completeness.

4.1 Spherical Lateration

Also known as trilateration, it can be defined as the method of determining the relative position of points by measuring absolute distances and using the geometry of spheres (see Fig. 4A). As mentioned before, absolute distance from the receiver to the ith beacon (d_i) can be easily obtained from the Time-of-Arrival of the signal emitted by this beacon (TOA_i) as,

$$d_i = c \cdot (TOA_i - t_{TX}), \tag{8}$$

where t_{TX} represents the instant of emission. Location is then obtained as the intersection of the spherical surfaces defined by these distances,

$$d_i = \sqrt{(x_i - x)^2 + (y_i - y)^2 + (z_i - z)^2}, \tag{9}$$

where (x_i, y_i, z_i) and (x, y, z) are the coordinates of the ith beacon and the receiver respectively. In three-dimensional geometry, a minimum of three TOAs (three beacons) are necessary to narrow the possible locations down to two, from which only one is usually

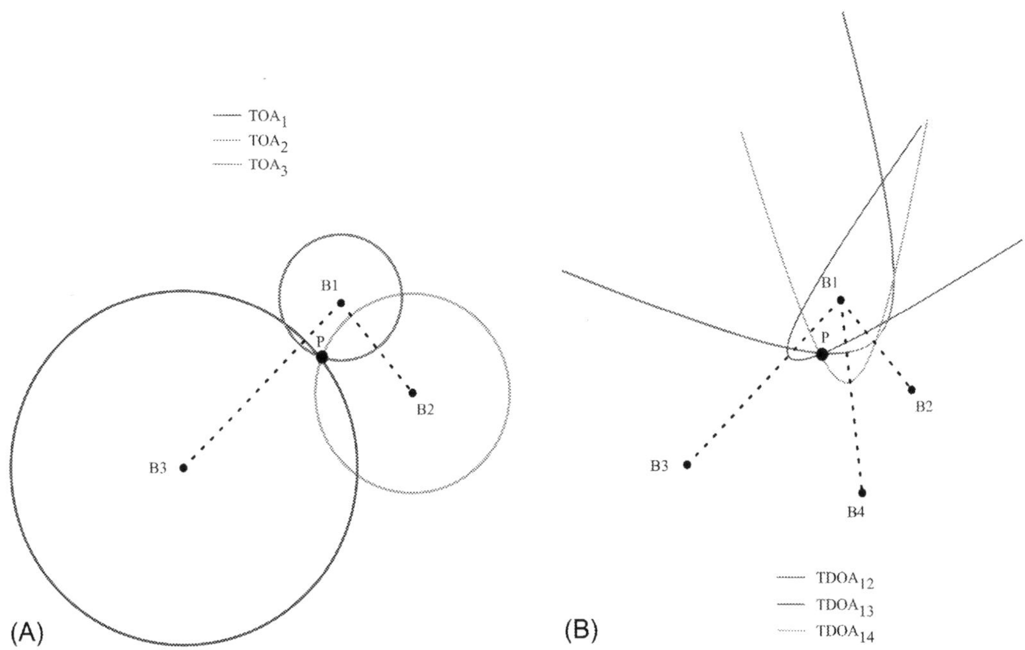

FIG. 4 Planar representation of the spherical (A) and hyperbolic (B) lateration techniques.

coherent with the geometry of the problem. Two families of methods are usually employed to solve the system of nonlinear equations represented in Eq. (9):

- Closed solution: the system is linearized by squaring the equations and introducing new variables. These methods are faster but less precise because the noise contribution is also squared. Besides, they require the participation of additional beacons to deal with the increased dimensionality of the problem. A representative example of this family is the Bancroft method (Bancroft, 1985).
- Iterative solution: equations are linearized by performing a Taylor expansion around an initial position estimate. The system is then iteratively solved to update the position until a final solution within a certain tolerance margin is obtained. These methods are slower than the previous ones and they also require an initial estimate, but in return they are more precise. A good example of this family is the Gauss-Newton algorithm (Bjorck, 1996).

The main inconvenient of spherical lateration is that the receiver must know the precise instant of emission (t_{TX}), and the emitters and receiver clocks must be synchronized, although this latter problem is not in acoustic systems as critical as it is in RF-based systems.

4.2 Hyperbolic Lateration

This strategy is also known as multilateration, and can be defined as the method of determining the relative position of points by measuring differential distances and using the geometry of hyperboloids (see Fig. 4B). This differential distance to the ith and jth beacons (d_{ij}) can be obtained from the time-difference-of-arrival of the signals emitted by them ($TDOA_{ij}$) as,

$$d_{ij} = c \cdot TDOA_{i,j}. \tag{10}$$

Location is in this case obtained as the intersection of the hyperbolic surfaces defined by these differential distances,

$$d_i - d_r = \sqrt{(x_i - x)^2 + (y_i - y)^2 + (z_i - z)^2} - \sqrt{(x_r - x)^2 + (y_r - y)^2 + (z_r - z)^2}, \tag{11}$$

where (x_i, y_i, z_i), (x_r, y_r, z_r), and (x, y, z) are the coordinates of the ith beacon, the reference beacon, and the receiver respectively. In three-dimensional geometry, a minimum of three TDOAs (four beacons) are necessary to narrow the possible locations down to 2 (again from which only one is usually coherent). The same methods employed to solve the set of spherical lateration equations can be used to solve the set of hyperbolic lateration equations. Note that in this positioning strategy only the emitters must be synchronized between them, but not with the receiver.

More beacons than the minimum required in each case can be used to introduce redundancy, thus obtaining an overdetermined system of equations. In this case, the system can be solved as a least squares problem.

5 Detection Hindering Phenomena and Compensation Strategies

The use of simultaneous and encoded emissions described at the end of Section 3, which notably improves the precision of the range measurements and robustness to in-band noise of broadband AAPS, comes hand-in-hand with new problems that may hinder the detection of these signals. Most recent research in the field of AAPS is partially devoted to the development of compensation algorithms for these phenomena. The most relevant are briefly described next.

5.1 Multiple Access Interference

The signal $r(t)$ received in a broadband AAPS, with N beacons simultaneously emitting binary phased coded signals $g_j(t)$ is given by,

$$r(t) = \sum_{j=1}^{N} A_j \cdot (h_j * g_j)(t - t_j) + n(t), \tag{12}$$

where t_j and A_j are respectively the TOA and amplitude of the signal to be estimated, and $n(t)$ represents the channel additive noise. The effect on the signal $g_j(t)$ of the acoustic channel is introduced by the convolution term $h_j(t)$, which represents the a priori unknown channel impulse response. As mentioned before, the output of the receiver is formed by correlation of $r(t)$ with all the signal patterns. For the kth beacon we have,

$$R_{rg_k}(t) = A_k \cdot (h_k * R_{g_k g_k})(t - t_k) + \sum_{j \neq k} A_j \cdot (h_j * R_{g_k g_j})(t - t_j) + \eta(t), \tag{13}$$

where $R_{g_k g_j}(t)$ represents the cross-correlation of codes $g_k(t)$ and $g_j(t)$ and $\eta(t)$ is the convolved noise. As Eq. (13) indicates, there are two effects which deteriorate the estimation of the TOAs: Intersymbol Interference (ISI), which is due to the fact that the limited bandwidth of the acoustic channel lowers the correlation peaks and degrades the signal detection and ranging; and Multiple-Access Interference (MAI) among all the emitted codes in which larger amplitude signals make difficult the detection of weaker signals emitted simultaneously. Combined, both effects can lead to large deviations of the TOAs estimates from their true values. This effect can be compensated by using recursive subtractive techniques, such as the Parallel Interference Cancelation algorithm (Aguilera et al., 2018b), but at the expense of increasing the time required to compute a new position estimate. It could also be attenuated by recovering the time-multiplexing strategy used in narrowband systems but with encoded emissions that can help identify the corresponding emitter, as proposed in Álvarez et al. (2017b).

5.2 Strong Multipath Propagation

As we have already stated in Section 2.3, multipath propagation is a common effect in indoor AAPS due to the specular reflections of the acoustic emissions at the room boundaries. This phenomenon gives rise to typical room impulse responses where the direct path is followed by a pattern of early reflections and then by a late-field reverberant tail, as the one shown in Fig. 5A. Since the pattern of early reflections is basically the representation of a sparse channel whose number of coefficients with nonnegligible magnitude is much lower than the total number of coefficients (see Fig. 5B), a matching pursuit (MP) algorithm can be used as a low complexity approximation to the maximum likelihood solution to estimate the TOA of the direct wave (Kim and Iltis, 2004).

If we consider N different beacons, the digitized samples of the received signal $\mathbf{r}$ can be represented by,

$$\mathbf{r} = \sum_{l=1}^{N} \mathbf{E}^l \mathbf{h}^l + \mathbf{n}, \tag{14}$$

where $\mathbf{h}^l$ is the lth channel coefficient vector, $\mathbf{E}^l$ is the characteristic signal matrix containing samples of the lth beacon emission, and $\mathbf{n}$ is a vector of zero-mean white Gaussian

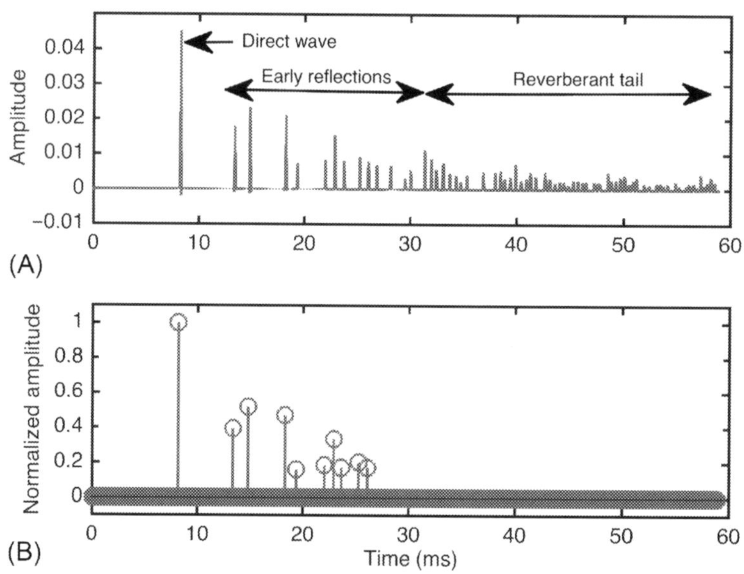

FIG. 5 (A) Acoustic impulse response of a room with highly reflective walls and (B) equivalent 10-coefficient sparse channel model.

noise samples. The MP algorithm estimates the channel coefficients $\hat{h}^l_{q_j}$ one at a time, using a greedy approach in which the detected path index q^l_j and corresponding coefficient $\hat{h}^l_{q_j}$ are computed from the following set of equations,

$$q^l_j = \underset{i \neq q^l_1, \ldots, q^l_{j-1}}{\mathrm{argmax}} \left\{ \frac{\left|(\mathbf{E}^l_i)^H \mathbf{r}^j\right|^2}{\left\|\mathbf{E}^l_i\right\|^2} \right\}, \tag{15}$$

and

$$\hat{h}^l_{q_j} = \frac{(\mathbf{E}^l_{q_j})^H \mathbf{r}^j}{\left\|\mathbf{E}^l_{q_j}\right\|^2}, \tag{16}$$

with

$$\mathbf{r}^{j+1} = \mathbf{r}^j - \frac{(\mathbf{E}^l_{q_j})^H \mathbf{r}^j \mathbf{E}^l_{q_j}}{\left\|\mathbf{E}^l_{q_j}\right\|^2}, \tag{17}$$

where $\mathbf{E}^l_i$ represents the ith column vector of matrix $\mathbf{E}^l$ and $\mathbf{r}^1 = \mathbf{r}$. Every new iteration of the algorithm $j = 1, 2, \ldots, N_f$, computes Eqs. (15) and (16) N times (one per channel), and only the largest coefficient $\hat{h}^l_{q_j}$ is stored. Next, the newly estimated signal $\hat{h}^l_{q_j}\mathbf{E}^l_{q_j}$ is subtracted from the current residue $\mathbf{r}^j$ to obtain the updated signal $\mathbf{r}^{j+1}$ as indicated by Eq. (17). This multipath cancelation technique has proven to notably decrease the mean positioning error measured under strong multipath conditions in an AAPS where a 16 kHz sonic carrier was modulated with 63-bit Kasami sequences (Álvarez et al., 2017a), and an AAPS based on a time-multiplexing strategy where a 41.67 kHz ultrasonic carrier was modulated with 255-bit Kasami sequences (Aguilera et al., 2018a).

5.3 Doppler Shift

In Section 2.2 we have seen that the low speed of sound in air responsible for the relatively high resolution of AAPS is also the cause of a strong Doppler shift. This phenomenon has no effect on envelope detection receivers, but can make broadband signals completely unrecognizable to the matched receiver. A straightforward solution to this problem would be to replace the single correlator at the receiver for a bank of them, each one matched to different frequency-shifted versions of the code to be detected. However, this simple implementation would require a very high operation frequency at the receiver if a fine Doppler resolution is required.

An alternative approach consists in using a multirate filter bank to compensate for the Doppler shift caused by the receiver's movement, following the scheme shown in Fig. 6. Let assume that the received signal $r(t)$ is sampled at f_s, thus giving the discrete-time signal $r[n]$ with spectrum $R(e^{j\omega})$. This signal is first expanded by a factor Q by including $Q - 1$ zeroes between consecutive samples of $r[n]$. A new signal $r_i[n] = r[n/Q]$ for

$n = 0, \pm Q, \pm 2Q, \ldots$ and 0 otherwise, is thus obtained whose spectrum $R_i(e^{j\omega})$ is a Q-fold compressed version of $R(e^{j\omega})$, i.e.,

$$R_i(e^{j\omega}) = R(e^{j\omega Q}). \tag{18}$$

As shown in Fig. 6, the expanded signal $r_i[n]$ is next filtered by a low-pass (LP) filter whose transfer function is given by,

$$H_{LP}(e^{j\omega}) = \begin{cases} Q & 0 \leq |\omega| \leq \pi/Q \\ 0 & \text{otherwise} \end{cases} \tag{19}$$

to obtain a new signal $r_f[n]$ with spectrum $R_f(e^{j\omega})$ given by,

$$R_f(e^{j\omega}) = \begin{cases} Q \cdot R(e^{j\omega Q}) & 0 \leq |\omega| \leq \pi/Q \\ 0 & \text{otherwise.} \end{cases} \tag{20}$$

This signal is then fed into a bank of K decimators with decimation factor of $Q - k + (K+1)/2$, for $k = [1, 2, \ldots, K]$, from where K signals $r_{d,k}[n] = r_f[n \cdot (Q - k + (K+1)/2)]$ are obtained. Assuming the absence of aliasing, the spectra $R_{d,k}(e^{j\omega})$ of these signals are given by,

$$R_{d,k}(e^{j\omega}) = \frac{Q}{Q - k + (K+1)/2} R(e^{j\omega \frac{Q}{Q+(K+1)/2-k}}), \quad 0 \leq |\omega| \leq \pi, \quad k = [1, 2, \ldots, K]. \tag{21}$$

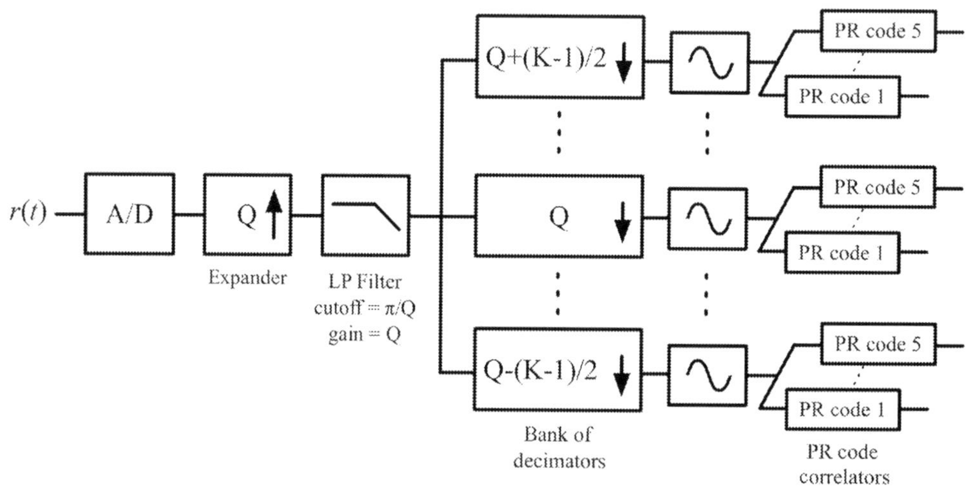

FIG. 6 Block diagram of the Doppler-tolerant receiver based on a multirate filter bank.

If the received signal $r(t)$ has been emitted as the pattern $p(t)$ by a certain beacon, and the receiver is moving toward this beacon with speed v_{r_k}, the spectra of the discrete-time versions of these signals $r[n]$ and $p[n]$ are related as,

$$R(e^{j\omega}) = P\left(e^{j\omega \frac{c}{c+v_{r_k}}}\right), \tag{22}$$

where c is again the speed of sound. Combining Eqs. (21) and (22) the spectra of the signals coming out from the decimators can be finally expressed as,

$$R_{d,k}(e^{j\omega}) \propto R\left(e^{j\omega \frac{Q}{Q+(K+1)/2-k}}\right) = P\left(e^{j\omega \frac{Q}{Q+(K+1)/2-k}\frac{c}{c+v_{r_k}}}\right). \tag{23}$$

From this latter expression, it is clear that the frequency shift caused by the receiver's movement is canceled if,

$$\frac{Q}{Q+(K+1)/2-k}\frac{c}{c+v_{r_k}} = 1. \tag{24}$$

Assuming $|v_{r_k}| \ll c$, the factor Q can be obtained from Eq. (24) as,

$$Q \approx \left(k - \frac{K+1}{2}\right)\frac{c}{v_{r_k}}. \tag{25}$$

This Doppler tolerant receiver was used in Álvarez et al. (2013) to improve the detection rate of broadband signals (40 kHz carrier modulated with 255-Kasami sequences), which were acquired by the receiver when moving in a horizontal plane with linear velocities of up to 6.82 m/s.

A different solution to compensate for this phenomenon consists in using intrinsically Doppler resilient polyphase codes to modulate the emissions of the AAPS. Although some works have already explored this interesting alternative (Paredes et al., 2011), their experimental results are limited by the bandwidth constraints of current acoustic transducer.

6 Conclusions

This chapter has presented the fundamentals of airborne acoustic positioning systems. First, the main phenomena affecting the propagation of acoustic waves in air have been reviewed to explain some specific features of this technology, such as high precision (centimeter), limited coverage (tens of meters), and room confined emissions. Also, additional phenomena that should be considered outdoors have been briefly described.

The most important positioning observables, namely, Time-of-Arrival (TOA) and Time-Difference-Of-Arrival (TDOA) have been then presented together with different methods to detect the arrival of acoustic signals. It has been seen that Binary Phase Coding can be used to extend the bandwidth of the emitted signals and improve the precision of the range measurements, at the expense of increasing the complexity of the receiver. In the last years, an important effort in this field of research has been devoted to the proposal of better encoding schemes and the design of the corresponding efficient correlation architectures. The positioning strategies directly related to these observables have been reviewed next, placing the emphasis on the different methods usually employed to solve the spherical and hyperbolic system of equations derived from the TOA and TDOA measurements respectively.

Finally, three different detection hindering phenomena and some of the strategies developed to compensate them, have been included in the last section of this chapter to provide the reader an insight into the most recent research in this active field of research.

References

Aguilera, T., Álvarez, F.J., Gualda, D., Villadangos, J.M., Hernéndez, A., Ureña, J.U., 2018a. Multipath compensation algorithm for TDMA-based ultrasonic local positioning systems. IEEE Trans. Instrum. Meas. 67 (5), 984–991.

Aguilera, T., Seco, F., Álvarez, F.J., Jiménez, A., 2018b. Broadband acoustic local positioning system for mobile devices with multiple access interference cancellation. Measurement 116, 483–494.

Álvarez, F.J., Hernández, A., Ureña, J.U., Mazo, M., García, J.J., Jiménez, J.A., Jiménez, A., 2006a. Real-time implementation of an efficient correlator for complementary sets of four sequences applied to ultrasonic pulse compression systems. Microprocess. Microsyst. 30, 43–51.

Álvarez, F.J., Urena, J., Mazo, M., Hernandez, A., Garcia, J.J., Marziani, C.D., 2006b. High reliability outdoor sonar prototype based on efficient signal coding. IEEE Trans. Ultrason. Ferroelectr. Freq. Control 53 (10), 1862–1871.

Álvarez, F.J., Hernández, A., Moreno, J.A., Pérez, M.C., Ureña, J.U., Marziani, C.D., 2013. Doppler-tolerant receiver for an ultrasonic LPS based on Kasami sequences. Sens. Actuators A: Phys. 189 (Supplement C), 238–253.

Álvarez, F.J., Aguilera, T., López-Valcarce, R., 2017a. CDMA-based acoustic local positioning system for portable devices with multipath cancellation. Digital Signal Process. 62 (Suppl. C), 38–51.

Álvarez, F.J., Esteban, J., Villadangos, J.M., 2017b. High accuracy APS for unmanned aerial vehicles. In: 13th International Conference on Advanced Technologies, Systems and Services in Telecommunications, Nis, Serbia, pp. 1–6.

Bancroft, S., 1985. An algebraic solution to the GPS equations. IEEE Trans. Aerosp. Electron. Syst. AES-21 (7), 56–59.

Bjorck, A., 1996. Numerical Methods for Least Squares Problems. SIAM.

Cole, J.E., Dobbins, R.A., 1970. Propagation of sound through atmospheric fog. J. Atmos. Sci. 27 (3), 426–434.

Davidson, G.A., 1975. Sound propagation in fogs. J. Atmos. Sci. 32 (11), 2201–2205.

Foxlin, E., Harrington, M., 1998. Constellation: A Wide-Range Wireless Motion-Tracking System for Augmented Reality and Virtual Set Applications. ACM Press.

Hazas, M., Ward, A., 2002. A Novel Broadband Ultrasonic Location System. In: UbiComp 2002: Ubiquitous Computing: 4th International Conference Göteborg, Sweden, September 29–October 1, 2002 Proceedings, pp. 264–280.

Hazas, M., Ward, A., 2003. A high performance privacy-oriented location system. In: Proceedings of the First IEEE International Conference on Pervasive Computing and Communications, 2003 (PerCom 2003), pp. 216–223.

ISO/TC 43 Technical Committee, Acoustics, Sub-Committee SC1, Noise, 1993. Attenuation of sound during propagation outdoors. Part 1: Calculation of the absorption of sound by the atmosphere. Tech. Rep. ISO 9613-1:1993(E). International Organization for Standardization, Geneve, Switzerland.

Jörg, K.-W., Berg, M., 1998. Sophisticated mobile robot sonar sensing with pseudo-random codes. Robot. Auton. Syst. 25 (3), 241–251.

Kay, S.M., 1993. Fundamentals of Statistical Signal Processing: Estimation Theory (Signal Processing Series). Englewood Cliffs, NJ, USA: Prentice-Hall.

Kim, S., Iltis, R.A., 2004. A matching-pursuit/GSIC-based algorithm for DS-CDMA sparse-channel estimation. IEEE Signal Process. Lett. 11 (1), 12–15.

Mautz, R., 2012. Indoor positioning technologies. Ph.D. thesis, Institute of Geodesy and Photogrammetry, Department of Civil, Environmental and Geomatic Engineering, ETH Zurich.

Milne, P.H., 1983. Underwater Acoustic Positioning Systems. Gulf Publishing Company, Houston, TX.

Paredes, J.A., Aguilera, T., Álvarez, F.J., Lozano, J., Morera, J., 2011. Analysis of Doppler effect on the pulse compression of different codes emitted by an ultrasonic LPS. Sensors 11, 10765–10784.

Peremans, H., Audenaert, K., Campenhout, J.M.V., 1993. A high-resolution sensor based on tri-aural perception. IEEE Trans. Robot. Autom. 9 (1), 36–48.

Pérez, M.C., Ureña, J.U., Hernández, A., de Marziani, C., Jimenez, A., Villadangos, J.M., Álvarez, F.J., 2007. Ultrasonic beacon-based local positioning system using loosely synchronous codes. In: 2007 IEEE International Symposium on Intelligent Signal Processing, pp. 1–6.

Pérez, M.C., Serrano, R.S., Ureña, J.U., Hernández, A., Marziani, C.D., Álvarez, F.J., 2012. Correlator implementation for orthogonal CSS used in an ultrasonic LPS. IEEE Sens. J. 12 (9), 2807–2816.

Polaroid Corporation, 1991. Ultrasonic Ranging Systems.

Priyantha, N.B., Chakraborty, A., Balakrishnan, H., 2000. The cricket location-support system. In: Proceedings of the 6th Annual International Conference on Mobile Computing and Networking, MobiCom '00. ACM, New York, pp. 32–43.

Shamanaeva, L.G., 1988. Acoustic sounding of rain intensity. J. Acoust. Soc. Am. 84 (2), 713–718.

Skolnik, M.I., 1980. Introduction to Radar Systems, second ed. McGraw Hill Book Co., New York.

Ureña, J.U., Mazo, M., García, J.J., Hernández, A., Bueno, E., 1999. Classification of reflectors with an ultrasonic sensor for mobile robot applications. Robot. Auton. Syst. 29 (4), 269–279.

Ureña, J.U., Hernández, A., Jiménez, A., Villadangos, J.M., Mazo, M., García, J.C., García, J.J., Álvarez, F.J., de Marziani, C., Pérez, M.C., Jiménez, J.A., Jiménez, A.R., Seco, F., 2007. Advanced sensorial system for an acoustic LPS. Microprocess. Microsyst. 31 (6), 393–401.

Villadangos, J.M., Urena, J., Mazo, M., Hernandez, A., Alvarez, F.J., Garcia, J.J., Marziani, C.D., Alonso, D., 2005. Improvement of ultrasonic beacon-based local position system using multi-access techniques. In: IEEE International Workshop on Intelligent Signal Processing, 2005, pp. 352–357.

Ward, A., Jones, A., Hopper, A., 1997. A new location technique for the active office. IEEE Pers. Commun. 4 (5), 42–47.

18

Indoor Positioning System Based on PSD Sensor

David Rodríguez-Navarro, José Luis Lázaro-Galilea, Alfredo Gardel-Vicente, Ignacio Bravo-Muñoz, Álvaro De-La-Llana-Calvo
DEPARTMENT OF ELECTRONICS, UNIVERSITY OF ALCALÁ, MADRID, SPAIN

1 Introduction

The chapter presents an indoor positioning system (IPS) based on the determination of the signal angle of arrival (AoA). It is focused in the use of infrared and luminous signals for positioning. This type of technology is possibly the least developed or used to date. The main reason is the lack of commercial sensors focused for positioning. Typically, the measurement methods for light sensor systems are based on point-to-point telemetry as is the case of LIDAR applications (light detection and ranging) (Kashani et al., 2015).

One of the most referenced works using this kind of technology for indoor applications is the so-called *the Active Badge System* (Want et al., 1992), which was implemented in a hospital where several staff people carried a device that emitted coded infrared signals. The positioning method was based on Cell-ID, a system which determines the room where each staff member was in.

AoA and time difference of arrival (TDoA) are two techniques widely used in obtaining agents position. The AoA technique is based on determining the AoA of the signal from the emitter to one or more receivers (if the geometry of the system is known, a single emitter would be enough).

An example of the use of AoA is presented in Lee et al. (2004), is based on three receivers composed of two photodiodes each, located in such a way that the mobile agent is enclosed within a small triangular coverage area. It is possible to measure the AoA differences between pairs of receivers and using an iterative method that minimizes a cost function, the position of the mobile agent is then obtained. As a disadvantage, the coverage of this positioning system (Lee et al., 2004) is limited to the overlapped area seen from the three receivers, which is only 40 cm wide.

On the other hand, TDoA technique is based on determining the difference time of arrival of the signal from an emitter to several receivers, using one of them as a reference, this time difference is proportional to distance differences and using multilateration algorithm the 3D position of the mobile agent is determined. This method requires four or more receivers. Examples of work jobs that use TDoA and optical signals are Gorostiza et al. (2011) and Salido-Monzú et al. (2014). The biggest handicap of the TDoA method is the multipath (MP) effect, as well as the synchronization in the acquisition of the signals captured from the different sensors. The accuracy obtained using simultaneous analogical-digital converters with the same reference clock is in the range of centimeters.

Concerning other potential functions of light-emitting diode (LED) light in the context of visible light communication (VLC) can be wireless communication and positioning, which make them an attractive research topic. The duplicate use of LEDs for lighting eliminates the cost of installing a positioning system based on VLC. Furthermore, the absence of electromagnetic interferences makes the use of LEDs particularly interesting; positioning based on VLC can be used as an indoor navigation system for location tracking, finding objects, controlling the movement of agents, etc.

Interesting works for the reader such as Huynh and Yoo (2016) propose a design for an IPS using LEDs, an image sensor (IS), and an accelerometer from mobile devices. The proposed scheme consists of four LEDs mounted on the ceiling transmitting their own 3D world coordinates and an IS at an unknown position receiving and demodulating the signals. Based on the 3D world coordinates and the 2D image coordinate of LEDs, the position of the mobile device is determined. To further improve accuracy, we propose a mechanism to reduce the IS noise. With the assumption that the IS coordinate space is parallel to the world reference plane, they evaluate the accuracy obtaining values below 10 cm. They also make a comparison with other state-of-the-art techniques such as Yoshino et al. (2008) and Nakazawa et al. (2013) showing the reaching of similar accuracy results.

Another example is presented in Xu et al. (2016), in which a visible light-based IPS was designed using multiple photodiodes in the environment and the received signal strength to determine the position of the mobile agent, obtaining mean error of 7 cm in a range of 2 m. The main problem encountered was the low signal-noise ratio as distance and angles increase. Work in Lin et al. (2017) proposes an indoor VLC and positioning system using the orthogonal frequency division multiplexing access scheme, in which the signals transmitted by LEDs are encoded with allocated subcarrier, respectively, and the receiver recovers all transmitted signals using a discrete Fourier transformation operation. The feasibility of the scheme is demonstrated in a room of size $20 \times 20 \times 15\,\text{cm}^3$. They show that the proposed scheme offers a mean positioning error of 1.68 cm, overcoming 2 cm of maximum error.

In another similar work (Kumaki et al., 2016), the authors also use four LEDs and IR camera. The paper proposes a fast coding/decoding scheme to be able to use low-cost artificial vision devices (e.g., smartphones). The research work (Sakai et al., 2016) proposes a different system, based on infrared light-emitting diode (IRED) beacons from the mobile agents and deploying an array of photodiodes as detectors. The 3D positioning is obtained by measuring the angle between the emitter and several photodiode receivers. The tests were done in a coverage area of $7 \times 2\,\text{m}^2$ obtaining errors up to 0.7 m.

The main drawback of these technologies concerns the MP by which the signal reaches the receivers. Looking solely at optical signals, the model of near-infrared signal reflection reported in De-La-Llana-Calvo et al. (2017) enabled us to model and analyze how this affects AoA and difference phase of arrival measurement techniques and take this into account in this study.

Concerning position sensitive device (PSD)-based systems, there are a few and also old research works for the indoor positioning of mobile agents, for example, Park et al. (2006) and Salomon et al. (2006). In Park et al. (2006), the authors propose a PSD-based IPS placed on a height of 2.5 m from floor to ceiling, making use of a Kalman filter to track a mobile robot that moves at a constant speed on a ground floor. The location results give a maximum error of 8.97 cm and average error of 1.97 cm, in a coverage area of $3 \times 3\ \text{m}^2$. The maximum error comes from the location of the robot in the periphery of the monitored area.

On the other hand, (Salomon et al., 2006) propose a multiple PSD system located around the surveillance area where a mobile robot with an emitter moves freely. The positioning algorithm is done by means of trilateration, obtaining the distances from each PSD sensor and the robot through stereoscopic measurements. The results achieved in the test indicate that the error in the distance measurements increases linearly from 70 to 440 cm showing a nonlinear behavior outside of that distance boundary.

In the chapter, an IPS is described. The IPS accurately determines the position of mobile agents based on optical signals, using IREDs or LEDs emitters and a PSD sensor. The computational load of the system and measurement methods proposed is very low, making it possible to obtain a very high position update rate. We present in the chapter two different proposals for IPS with their corresponding evaluation. Section 2 describes the use of a PSD sensor and the mathematical model in a IPS. Section 3 is devoted to propose the calibration of the PSD sensor. Section 4 describes two different positioning methods. The last section summarizes the main conclusions of using this type of IPS.

2 Description and Modeling of the Optical Sensor System

We present an IPS for detecting mobile agents based on a single PSD sensor. The main goal is to develop an alternative IPS to other sensing technologies, cheaper, easier to install, and with a low computational load to obtain a high rate of measurements per second. The proposed IPS has the capacity to accurately determine 3D position from the AoA of the signal received at the PSD sensor.

The simplified receiver system, consisting of a PSD sensor and lens coupled, is shown in Fig. 1, where coordinate (x, y) represents the image of the (X, Y, Z) point on PSD sensor surface, (C_x, C_y) is the optical center, f is the focal length, (θ_x, θ_y) are the angular components of AoA, and (X, Y, Z) are the coordinates where the LED is located.

From the diagram given in Fig. 1 one can see that to compute the AoA we need to know the focal length value, center of projection, and the point of impact. Fig. 1 is a

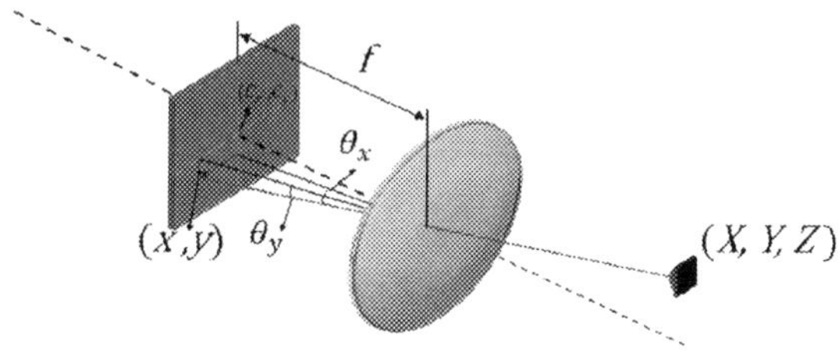

FIG. 1 Receiver based on position sensitive device sensor and lens coupled.

simplified representation, pinhole projection. In fact, the point of impact will be affected by distortions of the lenses, distortions of the PSD sensor, and misalignment of the ideal coupling of PSD and lens. Therefore, the optical center will not be in the center of the PSD sensor. So, in order to measure AoA with good accuracy, it is necessary to perform a previous calibration of the sensor (PSD + lens) system to correct these distortions. In other words, this calibration will obtain the values of the distortion parameters and the parameters involved in the determination of the impact point, such as the optical center and focal length.

2.1 PSD Sensor

Fig. 2 shows the equivalent electrical circuit of a 2D pincushion type for the PSD sensor. As we can see, PSD sensors have a common cathode and four anodes. The output current intensity for each anode will be inversely proportional to the distance between the point of impact of beam light and the position of the anode.

The impact point on the PSD surface is determined from the difference of the output currents in the four anodes, considering the origin of coordinates the center of the PSD sensor. Eqs. (1), (2) are used to calculate the impact point coordinates (x, y):

$$x = \frac{L_X}{2} \frac{(I_{X2} + I_{Y1}) - (I_{X1} + I_{Y2})}{I_{X1} + I_{X2} + I_{Y1} + I_{Y2}} \quad (1)$$

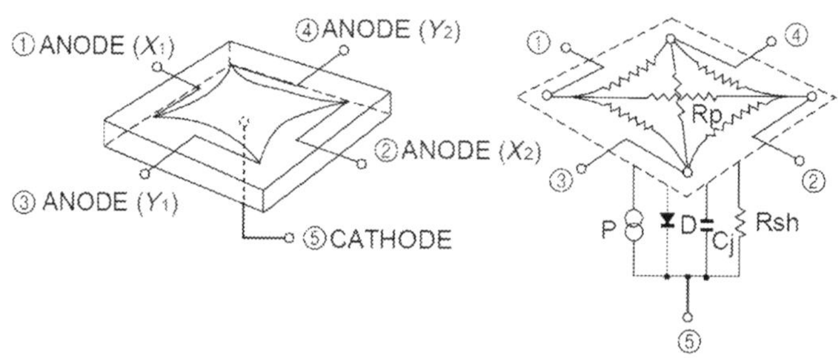

FIG. 2 Equivalent circuit of the PSD sensor pincushion. *(Courtesy of Hamamatsu.)*

$$y = \frac{L_Y}{2} \frac{(I_{X2} + I_{Y2}) - (I_{X1} + I_{Y1})}{I_{X1} + I_{X2} + I_{Y1} + I_{Y2}} \quad (2)$$

where I_{X1}, I_{X2}, I_{Y1}, and I_{Y2} are the output currents from the PSD sensor anode pins and L_X and L_Y are the sensor dimensions.

As will be discussed in the following section, it is not possible to calculate the point of incidence with these expressions because the PSD sensor output currents are very small and must be conditioned for digitization.

2.2 Electrical System Modeling

Due to the small value for the output currents of the PSD sensor, it is convenient to amplify them, first. Therefore, an amplification stage for each channel of the PSD sensor is included for further digitization. These stages will introduce some unbalances, noise, and quantification noise factors to the acquisition system. A typical electrical circuit model used, including the amplification stage, is shown in Fig. 3.

Eqs. (3), (4) will be used for calculating the impact point considering the amplifier's output voltage.

$$x = \frac{L_X}{2} \frac{(V_{X2} + V_{Y1}) - (V_{X1} + V_{Y2})}{V_{X1} + V_{X2} + V_{Y1} + V_{Y2}} \quad (3)$$

$$y = \frac{L_Y}{2} \frac{(V_{X2} + V_{Y2}) - (V_{X1} + V_{Y1})}{V_{X1} + V_{X2} + V_{Y1} + V_{Y2}} \quad (4)$$

In Eqs. (3), (4), $V_{\{X1,X2,Y1,Y2\}}$ are equal to $I_{\{X1,X2,Y1,Y2\}} \cdot K_{\{X1,X2,Y1,Y2\}}$, where $K_{\{X1,X2,Y1,Y2\}}$ are the gain factors of the amplification stages of each output channel of the PSD sensor.

Since these gain factors have a slightly different value for each channel, the calculation of the impact point will be affected. To mitigate this effect, it is proposed to add to the impact point equations a polynomial function, which will be the relationship between the gain factor of a channel (e.g., $G_1 = 1$ if it is used as a reference) and the other three.

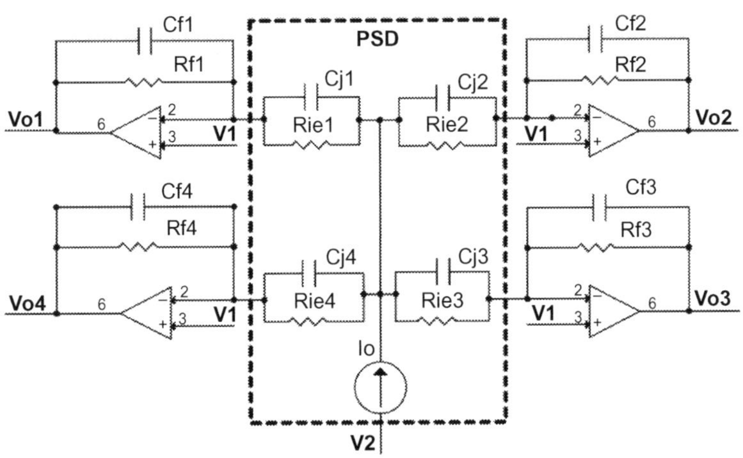

FIG. 3 Amplifier circuit from PSD sensor.

The difference between gain factors will be adjusted by a polynomial grade of order 2 (Eq. 5); however, in several cases it may be sufficient to use a linear approximation depending on the components used.

$$G_j = g_{1j}f^2 + g_{2j}f + g_{3j} \tag{5}$$

where $j = \{1, 2, 3, 4\}$ is related with each amplification stage, g are the coefficients of the polynomial, and f represents the frequency.

The new equations for the impact point considering the before gain model are as follows:

$$x = \frac{L_X}{2} \frac{(V_{X2}G_2 + V_{Y1}G_3) - (V_{X1}G_1 + V_{Y2}G_4)}{V_{X1}G_1 + V_{X2}G_2 + V_{Y1}G_3 + V_{Y2}G_4} \tag{6}$$

$$y = \frac{L_Y}{2} \frac{(V_{X2}G_2 + V_{Y2}G_4) - (V_{X1}G_1 + V_{Y1}G_3)}{V_{X1}G_1 + V_{X2}G_2 + V_{Y1}G_3 + V_{Y2}G_4} \tag{7}$$

Section 3 describes the calibration procedures for obtaining parameters $g_{i,j}$.

2.3 Optical System Modeling

A simple pinhole model has been proposed for the optical system modeling. This is the most common perspective projection for an optical system, which relates the environment coordinates (3D) with the image formed in the PSD sensor plane (2D).

Fig. 4 represents the pinhole model, where (X_W, Y_W, Z_W) are the axes of the world reference system, (X_R, Y_R, Z_R) are the PSD sensor reference system axes, (x, y) is the impact point on the PSD sensor surface, f is the focal length, (C_x, C_y) is the optical center, R is the rotation matrix, and T is the translation vector which relates the different coordinate systems.

The relationship between the world reference system and the receiver reference system is related to the 3×3 rotation matrix (R) and the 3×1 translation vector (T), according to Eq. (8).

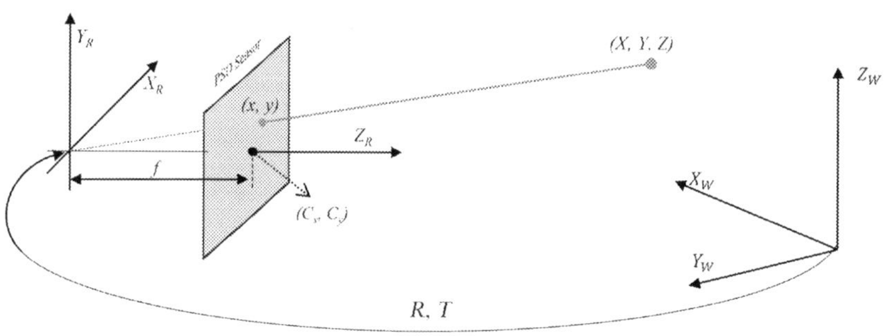

FIG. 4 Pinhole model.

$$\begin{pmatrix} X_R \\ Y_R \\ Z_R \end{pmatrix} = R \begin{pmatrix} X_W \\ Y_W \\ Z_W \end{pmatrix} + T \tag{8}$$

Next we show the relationship between the receiver's reference system and the image plane, Eq. (9) being the one that represents this relationship.

$$\begin{pmatrix} sx \\ sy \\ s \end{pmatrix} = \begin{pmatrix} f & 0 & C_x \\ 0 & f & C_y \\ 0 & 0 & 1 \end{pmatrix} \begin{pmatrix} X_R \\ Y_R \\ Z_R \end{pmatrix} \tag{9}$$

where s represents the scale factor that relates the 3D to 2D projection.

The matrix system (10) presents the mathematical model of the receiver, without taking into account the distortions produced by the lens and the PSD sensor.

$$\begin{pmatrix} sx \\ sy \\ s \end{pmatrix} = \underbrace{\begin{pmatrix} f & 0 & C_x \\ 0 & f & C_y \\ 0 & 0 & 1 \end{pmatrix}}_{A} \underbrace{\begin{pmatrix} r_{11} & r_{12} & r_{13} & T_x \\ r_{21} & r_{22} & r_{23} & T_y \\ r_{31} & r_{32} & r_{33} & T_z \end{pmatrix}}_{RT} \begin{pmatrix} X_W \\ Y_W \\ Z_W \\ 1 \end{pmatrix} \tag{10}$$

The elements of matrix A together with the parameters that model the distortions, which will be shown below, are called intrinsic parameters, since they depend solely on the configuration of the sensor (PSD + lens), regardless of its position and orientation in the world reference system.

Additionally, the RT matrix contains the extrinsic parameters, since they depend on the position and orientation between the two reference systems independently of the sensor configuration used.

As mentioned above, this is a linear model and therefore does not take into account the lens distortion or sensor distortion; therefore, it is necessary to add nonlinear effects caused by the lens and PSD sensor. Due to the low distortion produced by pincushion sensors, it will be the distortion of the lens that predominates, so only this factor will be included in the model.

Typically, the distortion of lenses is modeled taking into account two different effects: radial and tangential distortions. Beginning with radial distortion, Eqs. (11), (12) model this type of distortion.

$$D_x^r = (x^d - C_x)(a_1 r^2 + a_2 r^4 + \cdots + a_n r^{2n}) \tag{11}$$

$$D_y^r = (y^d - C_y)(a_1 r^2 + a_2 r^4 + \cdots + a_n r^{2n}) \tag{12}$$

where r is the Euclidean distance from the distorted point (x^d, y^d) to the optical center (C_x, C_y), and $a_{\{1,2,\ldots,n\}}$ are the parameters that model the radial distortion of the lenses.

In the case of tangential distortion, this is modeled according to Eqs. (13), (14).

$$D_x^t = p_1\left(r^2 + 2\left(x^d - C_x\right)^2\right) + 2p_2\left(x^d - C_x\right)\left(y^d - C_y\right) \tag{13}$$

$$D_y^t = p_2\left(r^2 + 2\left(y^d - C_y\right)^2\right) + 2p_1\left(x^d - C_x\right)\left(y^d - C_y\right) \tag{14}$$

where p_1 and p_2 represent the parameters that model the tangential distortion of the lens.

Therefore, by adding the distortion parameters to the linear model, we obtain Eqs. (15), (16), being (x, y) the impact point considering that there is no distortion.

$$x^d = x + D_x^r + D_x^t \tag{15}$$

$$y^d = y + D_y^r + D_y^t \tag{16}$$

To retrieve the AoA and impact point we require the use of all the parameters considered in the sensor system model. Next section describes the calibration of the sensor system to obtain the parameter values of the given sensor system model.

3 Sensor System Calibration

As mentioned, it is necessary to calibrate the system in order to make precise measurements. The calibration process is based on two stages: first, an electrical calibration and then, a geometric calibration.

3.1 Electrical Calibration Process

In previous section, it has been introduced that electrical circuit for signal conditioning adds errors in the impact point determination; the most common error factors are differences in channels gain, noise, and quantification noise factors. In Rodríguez-Navarro et al. (2016) each one of these factors is analyzed, concluding that the system noise introduces random errors that can be reduced using digital band-pass filters, while the gain differences are systematic errors that can be corrected following the process indicated below.

To correct gain factor misadjustments, an electrical calibration must be carried out. This calibration is based on a uniform illumination of the PSD sensor whole area. The uniform illumination should give the same value for the PSD sensor output currents. So, any difference will be considered an unbalanced amplifiers gain.

To illuminate the PSD sensor uniformly, it is recommended to use a IRED with the widest radiation pattern possible placed in front of the PSD sensor at a distance that makes the solid angle less than 0.01 Sr. Considering a squared sensor with a radius of 10 mm, the emitter should be located at a minimum distance equal to $d = \sqrt{\frac{10 \cdot 10^{-3}}{0 \cdot 01}} = 1\,\text{m}$. This scenario ensures uniform illumination throughout the PSD sensor (Fig. 5).

The root mean square (RMS) values of the signals are compared with the reference signal value, obtaining the relationship between the reference and the other signals. This should be adjusted or modeled for the entire working frequency range of the modulated emitted signal. If a single frequency is used, the ratio is a constant; if the IPS will be used for

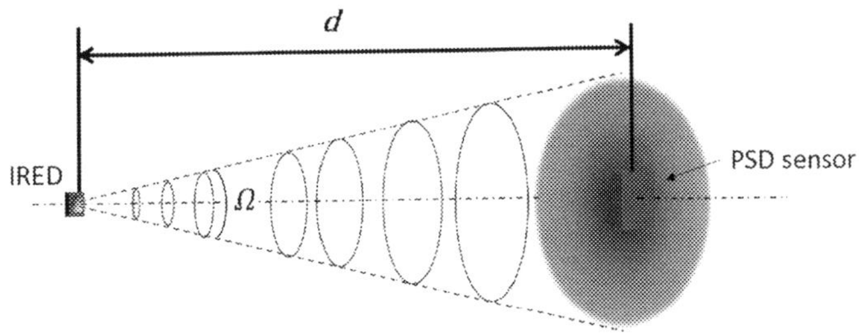

FIG. 5 Uniform illumination of the PSD sensor.

different signal frequencies, an expression of the variation of the gain factor as a function of frequency should be obtained.

Once the RMS values of each signal for a frequency range have been obtained, the coefficients of the polynomial equation (5) for least squares error minimization are calculated, using the equation $Ax = b$. The standard solution that minimizes this equation is $x = (A'A)^{-1}A'b$, where

$$A = \begin{pmatrix} \mathbf{F}^2 & \mathbf{F} & 1 \end{pmatrix}$$

$$x = \begin{pmatrix} a_{11} & a_{12} & a_{13} & a_{14} \\ a_{21} & a_{22} & a_{23} & a_{24} \\ a_{31} & a_{32} & a_{33} & a_{34} \end{pmatrix}$$

$$b = \begin{pmatrix} \frac{\mathbf{V_{ref}}}{\mathbf{V_1}} & \frac{\mathbf{V_{ref}}}{\mathbf{V_2}} & \frac{\mathbf{V_{ref}}}{\mathbf{V_3}} & \frac{\mathbf{V_{ref}}}{\mathbf{V_4}} \end{pmatrix}$$

where $\mathbf{F}$ is a column vector with those frequencies at which the RMS values have been calculated, $\mathbf{V_{ref}}$ the column vector with the RMS values of the reference signal, $\mathbf{V}_{\{1,2,3,4\}}$ the column vectors with the RMS values of the sensor output channels and the matrix x containing the coefficients of the fitting polynomials.

Once the channel imbalances have been compensated, next stage is geometrical calibration. If the gain factors are unbalanced, it is important to perform an electrical calibration beforehand, because errors at the impact point will affect the geometric calibration critically.

3.2 Geometric Calibration

The objective of this stage is to calculate the optical system parameter values, using non-linear system algorithms (Fletcher, 2013) (Gauss-Newton, Levenberg-Marquardt, gradient descent, etc.). One problem associated with these techniques is that they may converge on a solution that is mathematically correct, but that does not conform to reality. In order to avoid this situation, it is necessary to establish a good starting point, based on an approximate knowledge of the real parameter values or obtaining an approximation analytically.

The flow diagram for geometric calibration of the system is shown in Fig. 6. This process has been subdivided into four steps, which are acquisition of images, calculation of the projection matrix, extraction of intrinsic and extrinsic parameters, and iterative method to obtain distortion parameters.

- *Image acquisition.* The acquisition of images refers to the acquisition of impact points from a set of IREDs placed in a 2D pattern. Fig. 7 shows an example of a 2D IRED pattern, in which the coordinates of these points are known. The minimum requirement for the calibration procedure is that the calibration pattern has at least five points and takes three images for the whole pattern in different locations. However, it is advisable to have a pattern with 15–20 points and take between 8 and 12 images, depending on the noise in the system and the lenses distortion.

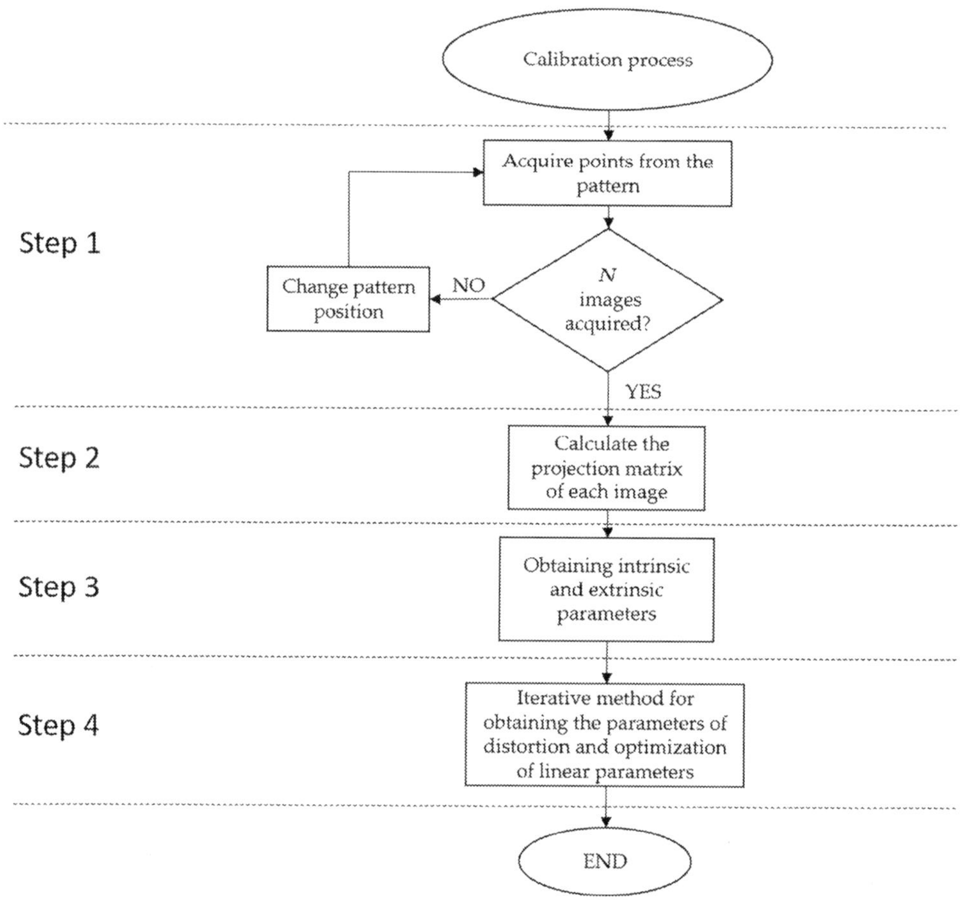

FIG. 6 The flow diagram for geometric calibration of the system.

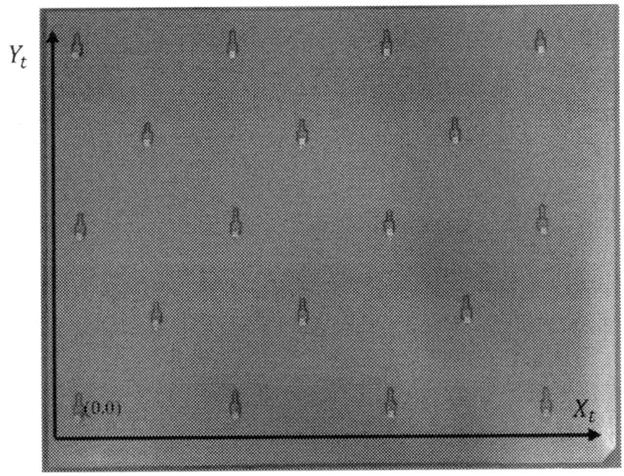

FIG. 7 Example of a calibration pattern including 18 IREDs.

- *Projection matrix for each captured image.* Once all images have been captured, it is necessary to obtain the corresponding projection matrix for each one. Starting from Eq. (17) that relates the pattern points with the impact points captured in the sensor.

$$\begin{pmatrix} sx \\ sy \\ s \end{pmatrix} = M \begin{pmatrix} X_t \\ Y_t \\ 1 \end{pmatrix} = \begin{pmatrix} m_{11} & m_{12} & m_{13} \\ m_{21} & m_{22} & m_{23} \\ m_{31} & m_{32} & m_{33} \end{pmatrix} \begin{pmatrix} X_t \\ Y_t \\ 1 \end{pmatrix} \tag{17}$$

where (x, y) are the coordinates of the impact point in the sensor, M is a 3×3 homography and (X_t, Y_t) the coordinates of corresponding point in the calibration template. By means of Eq. (17), from each image we can obtain 2 equations like those shown in Eqs. (18 and 19).

$$m_{11}X_t' + m_{12}Y_t' + m_{13} - m_{31}X_t'x' - m_{32}Y_t'x' - m_{33}x' = 0 \tag{18}$$

$$m_{21}X_t' + m_{22}Y_t' + m_{23} - m_{31}X_t'y' - m_{32}Y_t'y' - m_{33}y' = 0 \tag{19}$$

where (X_t', Y_t') are the column vectors of the coordinates of the 2D IRED locations in the calibration pattern and (x', y') are the column vectors of the projection on the sensor of the impact points of the corresponding IREDs. In matrix form it can be expressed as:

$$\begin{pmatrix} X_t' & Y_t' & 1 & 0 & 0 & 0 & -x'X_t' & -x'Y_t' - x' \\ 0 & 0 & 0 & X_t' & Y_t' & 1 & -y'X_t & -y'Y_t' - y' \end{pmatrix} \begin{pmatrix} m_{11} \\ m_{12} \\ \vdots \\ m_{33} \end{pmatrix} = 0 \tag{20}$$

This expression is solved by singular-value decomposition (SVD), thus obtaining the projection matrix.

- *Obtaining intrinsic and extrinsic parameters.* The projection matrix is composed of intrinsic parameters (A) multiplied by the extrinsic parameters matrix $\begin{pmatrix} r_1 & r_2 & t \end{pmatrix}$ as shown in Eq. (21).

$$\begin{pmatrix} sx \\ sy \\ s \end{pmatrix} = M \begin{pmatrix} X_t \\ Y_t \\ 1 \end{pmatrix} = A\begin{pmatrix} r_1 & r_2 & t \end{pmatrix}\begin{pmatrix} X_t \\ Y_t \\ 1 \end{pmatrix} \tag{21}$$

Knowing that $[m_1 \quad m_2 \quad m_3] = A[r_1 \quad r_2 \quad t]$, where m_i are the columns of matrix M, A is the matrix containing the intrinsic parameters, r_i are the columns of the rotation matrix, and t is the translation vector, we know that since the rotation matrix is orthonormal, the following relationships can be applied: $r_1^T r_2 = 0$ and $r_1^T r_1 = r_2^T r_2$. Consequently, from these constraints, we obtain Eqs. (22), (23).

$$m_1^T A^{-T} A^{-1} m_2 = 0 \tag{22}$$

$$m_1^T A^{-T} A^{-1} m_2 = 0 \tag{23}$$

where $A^{-T}A^{-1}$ results in the matrix B given in Eq. (24), which is the proposal for our model, where the focal length parameter is considered of equal value for the two coordinate axes.

$$B = A^{-T}A^{-1} = \begin{pmatrix} \frac{1}{f^2} & 0 & \frac{-C_x}{f^2} \\ 0 & \frac{1}{f^2} & \frac{-C_y}{f^2} \\ \frac{-C_x}{f^2} & \frac{-C_y}{f^2} & \frac{-C_x^2}{f^2} + \frac{-C_y^2}{f^2} + 1 \end{pmatrix} = \begin{pmatrix} b_{11} & b_{12} & b_{13} \\ b_{21} & b_{22} & b_{23} \\ b_{31} & b_{32} & b_{33} \end{pmatrix} \tag{24}$$

As the matrix B is symmetrical, it is only necessary to obtain six elements, so the vector is $b = [\, b_{11} \quad b_{12} \quad b_{22} \quad b_{13} \quad b_{23} \quad b_{33} \,]^T$. The elements of matrix B are traditionally obtained using Eq. (25), where
$V_{ij} = [m_{i1}m_{j1} \quad m_{i1}m_{j2} + m_{i2}m_{j1} \quad m_{i2}m_{j2} \quad m_{i3}m_{j1} + m_{i1}m_{j3} \quad m_{i3}m_{j2} + m_{i2}m_{j3} \quad m_{i3}m_{j3}]$, where m_{ij} are the elements of the projection matrix obtained in previous step.

$$\begin{bmatrix} V_{12}^T \\ (V_{11} - V_{22})^T \end{bmatrix} b = 0 \tag{25}$$

Once the matrix system (25) has been solved and according to B matrix, the system's linear intrinsic parameters are obtained by means of the following equations:

$$C_x = -b_{11}/b_{13} \tag{26}$$

$$C_y = -b_{22}/b_{23} \tag{27}$$

$$f = \sqrt{1}/b_{11} \tag{28}$$

Then, the projection matrices for each image are used to calculate the extrinsic parameters as follows:

$$r_1 = \lambda A^{-1} m_1 \tag{29}$$

$$r_2 = \lambda A^{-1} m_2 \tag{30}$$

$$r_3 = r_1 \times r_2 \tag{31}$$

$$t = \lambda A^{-1} m_3 \tag{32}$$

where $\lambda = \left\| \frac{1}{A^{-1} m_1} \right\| = \left\| \frac{1}{A^{-1} m_2} \right\|$. Due to image noise and nonlinearity, these parameters contain error because we are using linear techniques to solve a nonlinear problem.

- *Iterative method.* Having obtained approximate linear parameter values, the extrinsic and intrinsic parameters that model the system are obtained/optimized, in our case using the Levenberg-Marquardt algorithm.
  The function to be minimized is the following:

$$f(x) = \sum_{i=1}^{n} \left\| Q_i - \hat{Q}_i(A, D, R_i, T_i, P) \right\|^2 \tag{33}$$

  where n is the number of images, Q are the points in the image plane, A is the intrinsic matrix containing the parameters, such as the focal length and optical center, vector D contains the parameters that model distortion, R and T are the rotation matrix and the translation vector, respectively, P are the calibration template points, and $\hat{Q}$ are the points in the image plane obtained from the intrinsic and extrinsic parameters and the calibration template points.

Once all parameters have been computed, the distortion caused by the lens can be corrected and the AoA calculated. With the calibration performed, we proceed to describe the 3D positioning methods using AoA.

4 Three-Dimensional Position Determination Using AoA

This section describes two methods for determining the 3D position using only one sensor and AoA measurement, and possible error factors in the 3D measurement.

4.1 Method 1: IPS Located in the Environment and IRED on Board of the Mobile Agent

Fig. 8 shows the IPS consisting of the sensor located on the ceiling and the IREDs on the mobile agents. Assuming the plane in which agents are moving is known, it is possible to determine the 3D position of the *agent* by means of the intersection of the line given by the corresponding AoA and plane of movement.

In Fig. 8, (X, Y, Z) are the environment coordinates with reference to the origin of the receiver position and (A, B, C, D) are the parameters that model the plane.

The proposed method is based on the following three steps:

- compute the parameters of the movement plane (offline);
- obtain the AoAs; and
- determine the 3D position of each mobile agent by means of the intersection of the plane and the line given by each AoA.

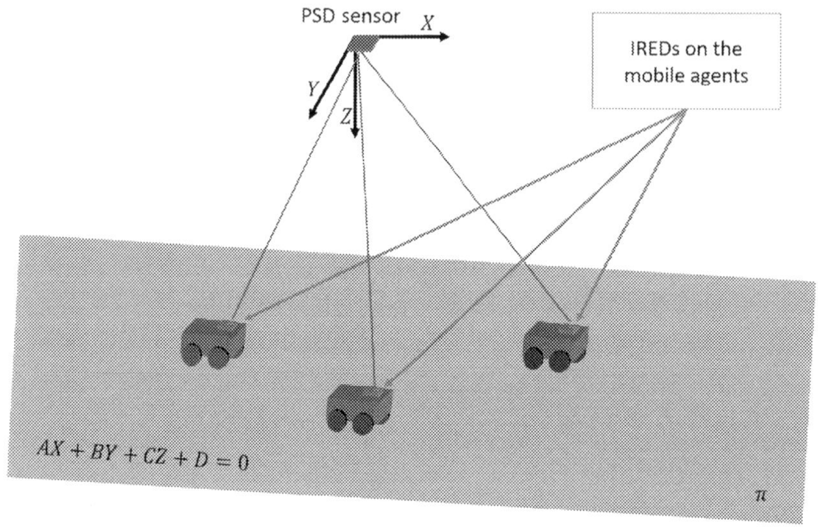

FIG. 8 IPS with receiver located in the environment and IREDs on mobile agents.

To compute the movement plane parameters is necessary to have at least four points belonging to the plane. Then, the plane equation parameters are directly obtained from an SVD procedure.

The 3D coordinates of those plane points are obtained by taking the image from four or more IREDs in different positions on the plane covering as much area as possible, and thus obtaining the rotation matrix and translation vector.

Therefore, the first step is to obtain the projection matrix, as it was previously described in Section 3 for the geometric calibration, using Eq. (20) considering now that (X_t', Y_t') are the coordinates of the points in the plane (taking one as the world reference system), and (x', y') are the corresponding impact point projections on the sensor with the distortion corrected, by means of the projection matrix M.

Since the intrinsic parameters (matrix A) and the elements of matrix M are already known from the geometric calibration, by means of Eqs. (29)–(32) the rotation matrix (R) and the translation vector (T), the extrinsic parameters for the movement plane are determined.

Finally, multiplying the (X_t', Y_t') points by the rotation matrix and adding the translation vector, the 3D coordinates for the points of the plane with reference to the sensor origin can be obtained.

Once the 3D points in the plane have been determined, the matrix equation to obtain the parameters of the plane is as follows:

$$(X' \; Y' \; Z') \begin{pmatrix} A \\ B \\ C \\ D \end{pmatrix} = 0 \tag{34}$$

where (X', Y', Z') are column vectors of the 3D coordinates for the plane points obtained earlier.

Once the plane equation and the intrinsic parameters are known, it is possible to obtain the angular components of each AoA using Eqs. (35), (36).

$$\theta_x = \tan^{-1}\frac{(x - C_x)}{f} \tag{35}$$

$$\theta_y = \tan^{-1}\frac{(y - C_y)}{f} \tag{36}$$

where (x, y) is the impact point with corrected distortion, (C_x, C_y) is the optical center, f is the focal length, and (θ_x, θ_y) are the angular components of AoA.

After computing the AoA values, the equation of straight line that goes from the corresponding emitter to the impact point in the PSD sensor can be obtained as:

$$X_i = \tan\theta_{xi} Z_i \tag{37}$$

$$Y_i = \tan\theta_{yi} Z_i \tag{38}$$

where i represents each mobile agent, and (X, Y, Z) are the 3D point coordinates refereed to the PSD sensor coordinate system.

Combining Eqs. (37), (38) with the plane equation we can obtain the Z coordinate of the line-plane intersection.

$$Z_i = -\frac{D}{A\tan\theta_{xi} + B\tan\theta_{yi} + C} \tag{39}$$

Therefore, replacing Z in Eqs. (37), (38) the 3D coordinate of the mobile agent with reference to the sensor reference system is obtained.

It is possible to change the reference coordinate system. It can be done by means of the matrix system (40), where (X_i, Y_i, Z_i) are the coordinates of each mobile agent, (X_i^p, Y_i^p, Z_i^p) are the 3D coordinates referred to the movement plane (where Z_i^p is null), R is the rotation matrix, and T is the translation vector for the point plane chosen as the origin of the world reference system.

$$\begin{pmatrix} X_i^p \\ Y_i^p \\ Z_i^p \end{pmatrix} = R\begin{pmatrix} X_i \\ Y_i \\ Z_i \end{pmatrix} + T \tag{40}$$

4.2 Method 2: PSD Sensor on Board of Each Mobile Agent and Emitters in the Environment

In this method the location of the PSD sensors and the emitters are exchanged; that is, the emitters are located in the ceiling of the environment and each mobile agent has its own PSD sensor. With this IPS configuration, the lighting infrastructure based on LEDs could also be used for positioning, resulting in a lower cost.

This type of IPS is shown in Fig. 9. In this case, assuming that the plane of movement is parallel to the ceiling; each PSD sensor (mobile agent) must receive a signal from three or more LEDs simultaneously in order to obtain an absolute positioning solution.

Considering the diagram given in Fig. 9, the position of a mobile agent (x, y) with respect to each of the emitters provides two equations given by Eqs. (41), (42).

$$x_i = \cos\beta_i \tan\theta_i h + P_{xi} \tag{41}$$

$$y_i = \sin\beta_i \tan\theta_i h + P_{yi} \tag{42}$$

where θ is the AoA, β is the angle of orientation, and (P_x, P_y, h) is the position of the LED considering the ceiling is at a height h and the LED is located in the point (P_x, P_y) respect the plane $x - y$, and i is the current mobile agent.

It is worth noting that system to be solved contains one variable (h) plus another one for each LED (β). In other words, in the case of three LEDs there would be four variables to be determined and six equations (positions of the emitters are known). However, depending on the locations of the LEDs it could happen that some equations are dependent and therefore there is more than one possible solution. This is the reason why we need at least three emitters.

Due to errors, intersections among beams from three or more emitters do not cross at a single point, so we use an optimization method for determining the 3D position of the mobile agent with the least error. Function to be minimized is given by Eq. (43).

$$\varepsilon(\beta_i, h) = \sum_{i=1}^{n-1}\sum_{j=1}^{n}(\Phi_i - \Phi_{i+j})^2 \tag{43}$$

where Φ represents Eqs. (41), (42).

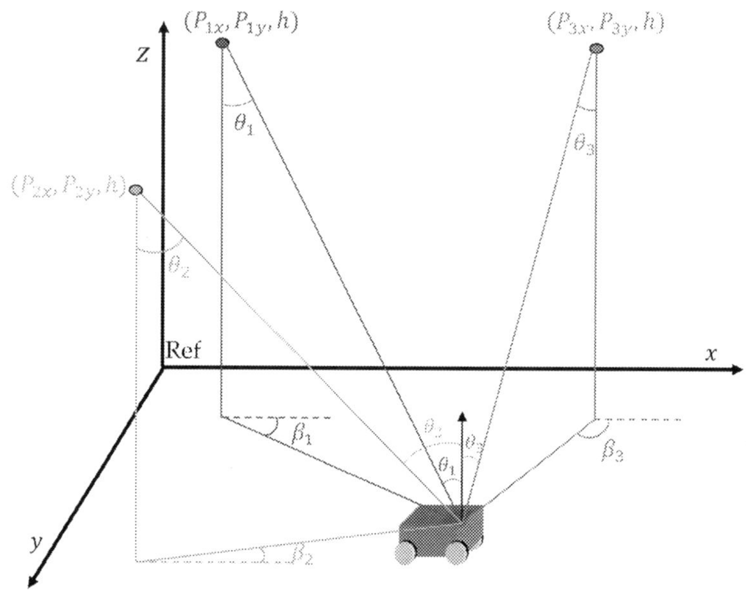

FIG. 9 IPS using LEDs for illuminating and receiver on mobile agent.

5 Discussion

Two proposals have been presented for indoor positioning; the first of them uses a single PSD sensor and a single IRED (or LED), but it is necessary to calibrate the plane of movement in order to be able to determine the intersection of the straight line given by the obtained AoA and the plane; the second method requires three LEDs and a PSD sensor on board each mobile agent, with the advantage that it is not necessary to calibrate the plane and that already installed lighting LEDs can be used.

Both methods are based on AoA measurements so that for a precise positioning it is necessary to calibrate the PSD sensor system. To do so, a procedure divided in two stages has been described: first the electrical calibration, which compensates any deviation in the signal conditioning stages, and second the geometric calibration where the values of the parameters that model the PSD sensor plus lens system are obtained also considering the distortion parameters to correctly determine the AoA.

In conclusion, when choosing an IPS system to locate an agent in a large area (e.g., a room inside a building) any of the existing alternatives can be used. However, when the position must be obtained with accuracy of centimeters at a high rate of measurements, suitable solutions are those using ultrasound and infrared. Infrared-based systems have the advantage that they are immune to electromagnetic noise and much less sensitive to MP effects, making them ideal for narrow spaces such as corridors.

Acknowledgments

This research was supported by the Spanish research program through the Indoor Positioning and Indoor Navigation Spanish Network (REPNIN) (TEC2015-71426-REDT) and through the ALCOR project (DPI2013-47347-C2-1-R).

References

De-La-Llana-Calvo, Á., Lázaro-Galilea, J.L., Gardel-Vicente, A., Rodríguez-Navarro, D., Bravo-Muñoz, I., Tsirigotis, G., Iglesias-Miguel, J., 2017. Modeling infrared signal reflections to characterize indoor multipath propagation. Sensors 17, 4. https://doi.org/10.3390/s17040847. Available from: http://www.mdpi.com/1424-8220/17/4/847.

Fletcher, R., 2013. Practical Methods of Optimization. John Wiley & Sons, New York, NY.

Gorostiza, E.M., Lázaro Galilea, J.L., Meca Meca, F.J., Salido Monzú, D., Espinosa Zapata, F., Pallarés Puerto, L., 2011. Infrared sensor system for mobilerobot positioning in intelligent spaces. Sensors 11 (5), 5416–5438. https://doi.org/10.3390/s110505416. Available from: http://www.mdpi.com/1424-8220/11/5/5416.

Huynh, P., Yoo, M., 2016. VLC-based positioning system for an indoor environment using an image sensor and an accelerometer sensor. Sensors 16 (6). https://doi.org/10.3390/s16060783. Available from: http://www.mdpi.com/1424-8220/16/6/783.

Kashani, A.G., Olsen, M.J., Parrish, C.E., Wilson, N., 2015. A review of LIDAR radiometric processing: from ad hoc intensity correction to rigorous radiometric calibration. Sensors 15 (11), 28099–28128. https://doi.org/10.3390/s151128099. Available from: http://www.mdpi.com/1424-8220/15/11/28099.

Kumaki, H., Akiyama, T., Hashizume, H., Sugimoto, M., 2016. 3D indoor positioning and rapid data transfer using modulated illumination. In: Proceedings of the International Conference on Indoor Positioning and Indoor Navigation (IPIN). Alcalá de Henares, Spain, pp. 4–7.

Lee, C., Chang, Y., Park, G., Ryu, J., Jeong, S.-G., Park, S., Park, J.W., Lee, H.C., Shik Hong, K., Lee, M.H., 2004. Indoor positioning system based on incident angles of infrared emitters. In: 30th Annual Conference of IEEE Industrial Electronics Society, 2004. IECON 2004, vol. 3, pp. 2218–2222.

Lin, B., Tang, X., Ghassemlooy, Z., Lin, C., Li, Y., 2017. Experimental demonstration of an indoor VLC positioning system based on OFDMA. IEEE Photonics J. 9 (2), 1–9. https://doi.org/10.1109/JPHOT.2017.2672038.

Nakazawa, Y., Makino, H., Nishimori, K., Wakatsuki, D., Komagata, H., 2013. Indoor positioning using a high-speed, fish-eye lens-equipped camera in visible light communication. In: International Conference on Indoor Positioning and Indoor Navigation, pp. 1–8.

Park, J.-H., Won, D.-H., Park, K.-Y., Baeg, S.-H., Baeg, M.-H., 2006. Development of a real time locating system using PSD under indoor environments. In: 2006 SICE-ICASE International Joint Conference, pp. 4251–4255.

Rodríguez-Navarro, D., Lázaro-Galilea, J.L., Bravo-Muñoz, I., Gardel-Vicente, A., Tsirigotis, G., 2016. Analysis and calibration of sources of electronic 460 error in PSD sensor response. Sensors 16 (5). https://doi.org/10.3390/s16050619. Available from: http://www.mdpi.com/1424-8220/16/5/619.

Sakai, N., Zempo, K., Mizutani, K., Wakatsuki, N., 2016. Linear positioning system based on IR beacon and angular detection photodiode array. In: Proceedings of the International Conference on Indoor Positioning and Indoor Navigation (IPIN). Alcalá de Henares, Spain, pp. 4–7.

Salido-Monzú, D., Martín-Gorostiza, E., Lázaro-Galilea, J.L., Martos-Naya, E., Wieser, A., 2014. Delay tracking of spread-spectrum signals for indoor optical ranging. Sensors 14 (12), 23176–23204. https://doi.org/10.3390/s141223176. Available from: http://www.mdpi.com/1424-8220/14/12/23176.

Salomon, R., Schneider, M., Wehden, D., 2006. Low-cost optical indoor localization system for mobile objects without image processing. In: 2006 IEEE Conference on Emerging Technologies and Factory Automation, pp. 629–632.

Want, R., Hopper, A., Falcão, V., Gibbons, J., 1992. The active badge location system. ACM Trans. Inf. Syst. 10 (1), 91–102. https://doi.org/10.1145/128756.128759.

Xu, Y., Zhao, J., Shi, J., Chi, N., 2016. Reversed three-dimensional visible light indoor positioning utilizing annular receivers with multi-photodiodes. Sensors 16 (8), 1254. https://doi.org/10.3390/s16081254. Available from: http://www.mdpi.com/1424-8220/16/8/1254.

Yoshino, M., Haruyama, S., Nakagawa, M., 2008. High-accuracy positioning system using visible LED lights and image sensor. In: 2008 IEEE Radio and Wireless Symposium, pp. 439–442.

Index

Note: Page number followed by *f* indicate figure and *t* indicate tables.

Y

Z